Thermodynamics of One-Dimensional Solvable Models

Exactly solvable models are very important in physics. They are important not just from a theoretical point of view but also from the experimentalist's perspective, because in such cases theoretical results and experimental results can be compared without ambiguity. This is a book about an important class of exactly solvable models in physics.

The subject area is the Bethe-ansatz approach for a number of one-dimensional models, and the setting up of equations within this approach to determine the thermodynamics of these systems. It is a topic that crosses the boundaries between condensed matter physics, mathematics and field theory. The derivation and application of thermodynamic Bethe-ansatz equations for one-dimensional models are explained in detail. This technique is indispensable for physicists studying the low-temperature properties of one-dimensional substances.

This book, written by one of the top physicists in this field, and the originator of much of the work in the field, will be of great interest to theoretical condensed matter physicists.

T0192869

CAMBRIDGE UNIVERSITY PRESS
Cambridge, New York, Melbourne, Madrid, Cape Town, Singapore, São Paulo

Cambridge University Press
The Edinburgh Building, Cambridge CB2 2RU, UK

Published in the United States of America by Cambridge University Press, New York

www.cambridge.org
Information on this title: www.cambridge.org/9780521551434

First published 1999
This digitally printed first paperback version 2005

A catalogue record for this publication is available from the British Library

Library of Congress Cataloguing in Publication data
Takahashi, Minoru, 1944–
Thermodynamics of one-dimensional solvable models /
Minoru Takahashi.
p.   cm.
Includes bibliographical references and index.
ISBN 0 521 55143 9 (hb)
1. Bethe-ansatz technique.   2. Statistical thermodynamics.
3. Mathematical physics.   I. Title.
QC20.7.b47T35   1999
530.15–dc21   98-34997   CIP

ISBN-13 978-0-521-55143-4 hardback
ISBN-10 0-521-55143-9 hardback

ISBN-13 978-0-521-01979-8 paperback
ISBN-10 0-521-01979-6 paperback

THERMODYNAMICS OF ONE-DIMENSIONAL SOLVABLE MODELS

Minoru Takahashi

*Institute for Solid State Physics,
University of Tokyo,
Tokyo, Japan*

Contents

Contents

Preface

The subject of this book is on the borders of condensed matter physics, mathematics and field theory. It is believed that the N body problem is soluble only for $N = 2$, but is not soluble for systems with $N \geq 3$. But, considering one dimensional systems, there are many solvable models, like the XXZ model. In condensed matter theory, the essential problem is that of solving many-body interacting systems and it is rare that we encounter solvable cases. But in these rare cases we can compare the theoretical results and experimental fact in detail. This kind of work is very valuable. Actually, orbits of planets and the energy spectrum of a hydrogen atom are treated in the regime of the exactly solvable case of two-body systems. Thus, the knowledge of exactly solvable systems is important not only for theorists but also for experimentalists. This book is planned for readers who have taken an elementary course of statistical mechanics and quantum mechanics.

There are also solvable two-dimensional classical systems like the six-vertex model. The Hamiltonians of 1D quantum systems and the transfer matrices of 2D classical systems sometimes have common eigenstates. In many cases we can write down many-body eigenfunctions of these matrices by the method of the Bethe-ansatz. The N-body wave function is represented as a linear combination of $N!$ plane waves with N quasi-momenta. The energy eigenvalue of the lowest energy state in the thermodynamic limit is reduced to a distribution function of the quasi-momenta. The distribution function must satisfy a linear integral equation. The energy per unit length is obtained by solving this integral equation. Elementary excitations from the ground state are expressed by the deviation of the distribution of quasi-momenta from the equilibrium.

In many cases the quasi-momenta are all real numbers for the ground state. But for general eigenstates, quasi-momenta are complex numbers. Fortunately for the delta-function Boson case, quasi-momenta are always real. Yang and Yang introduced the thermodynamic theory for this case. The total energy is represented by the distribution function of quasi-momenta. Entropy is represented by the distribution functions of quasi-momenta and holes. By the condition of minimizing the free energy, we can derive equations for these two distribution functions. These are coupled non-linear equations. We then obtain the thermodynamic potential as a function of the chemical potential and temperature. We can calculate all thermodynamic quantities such as energy, entropy, pressure, specific heat and so on.

This method can be extended to other Bethe-ansatz soluble models. Then next simplest case is the spin-1/2 Heisenberg model. In this case one cannot obtain the correct free energy assuming that the quasi-momenta are all real. But complex quasi-momenta group together to form a string in the rapidity plane in the limit of a big system. The length of the string is an arbitrary natural number 1, 2, 3, ... We must then consider the distribution functions for strings with length 1, 2, ... This is a set of coupled integral equations which has an infinite number of unknown functions. It looks very complicated, but we can solve this set of coupled equations analytically in some special limit. The asymptotic behaviour of the distribution function can be calculated for a very long string. We can thus approximate this set of equations by one with a finite number of unknown functions and get thermodynamic quantities numerically for a given temperature. In almost all cases, the thermodynamic Bethe-ansatz equations have an infinite number of unknown functions, as in delta-function fermions, the Kondo problem and the Hubbard model. The string assumption is indispensable for determining the thermodynamics of Bethe-ansatz soluble models.

In this book we mainly treat the delta-function quantum gas, the 1D Heisenberg model and the Hubbard model. We stress the thermodynamic properties of these models. Except in the case of a boson gas system, strings appear in the complex plane of rapidity. For general eigenstates, these strings should be considered. The length of the string is dependent on the parameters of the system. But once we accept the string assumption we can construct the equation for the free energy of these models at a given temperature. In many cases the equation contains an infinite number of unknown functions. But, surprisingly, this set of equations can be solved in some special limits and we obtain the exact free energy of these complex models in the limit of an infinite system. We can treat the problem using numerical methods for the finite systems. Bethe-ansatz equations give very interesting results which cannot be predicted using direct numerical analysis. For example the susceptibility exponent of a 1D Heisenberg ferromagnet was expected to be less than two for a long time. But it was established that γ is exactly two by the thermodynamic Bethe-ansatz method. This was later confirmed by the numerical Monte Carlo method. Very recently the anomalies of the susceptibility of a Heisenberg antiferromagnet was also found by this method. For these soluble models many analytical and important results are obtained. The ground state energy properties are also derived as a special case of the zero temperature limit. We can calculate low temperature properties of these models. To get analytical results we need various mathematical tools. In the process of calculation of Bethe-ansatz equations, many mathematical tools have been invented. Physically new results are also obtained by these methods.

The author is grateful to Dr. E. Williams for his critical reading of the manuscript.

Part one

Thermodynamics of non-interacting systems and ground states of interacting systems

1

Free energy and correlation functions of the XY model

1.1 The isotropic XY model

1.1.1 Introduction and historical overview

As the simplest case of a solvable model we consider the following spin 1/2 Hamiltonian,

$$\mathscr{H} = J \sum_{i=1}^{N} S_i^x S_{i+1}^x + S_i^y S_{i+1}^y - 2h \sum_{i=1}^{N} S_i^z, \quad S_{N+1}^\alpha \equiv S_1^\alpha, \tag{1.1}$$

$$[S_k^\alpha, S_l^\beta] = i\delta_{kl}\epsilon_{\alpha\beta\gamma}S_k^\gamma. \tag{1.2}$$

Each site has two states: an up-spin state and a down-spin state. S_i^x and S_i^y are represented by the Pauli matrices,

$$S_j^x = \frac{1}{2}\sigma_j^x, \quad S_j^y = \frac{1}{2}\sigma_j^y, \quad S_j^z = \frac{1}{2}\sigma_j^z,$$

$$\sigma_j^x = \begin{pmatrix} 0 & 1 \\ 1 & 0 \end{pmatrix}, \quad \sigma_j^y = \begin{pmatrix} 0 & -i \\ i & 0 \end{pmatrix}, \quad \sigma_j^z = \begin{pmatrix} 1 & 0 \\ 0 & -1 \end{pmatrix}. \tag{1.3}$$

For this Hamiltonian there are 2^N states, and it can be transformed as follows:

$$\mathscr{H} = \frac{J}{2} \sum_{i=1}^{N} S_i^+ S_{i+1}^- + S_i^- S_{i+1}^+ - 2h \sum_{i=1}^{N} S_i^z, S_k^\pm \equiv S_k^x \pm i S_k^y. \tag{1.4}$$

Lieb, Schultz and Mattis[61] and Katsura[47] investigated this model in detail. The operators $S_k^\pm$ satisfy the relations

$$\{S_i^+, S_i^-\} = 1, \quad [S_i^+, S_j^-] = 0 \quad \text{for} \quad i \neq j. \tag{1.5}$$

1

The above commutation relations are neither fermionic nor bosonic. But if we introduce the operators

$$
c_k = \prod_{i=1}^{k-1}(2S_i^+ S_i^- - 1)S_k^+, \quad c_k^\dagger = S_k^- \prod_{i=1}^{k-1}(2S_i^+ S_i^- - 1),
$$

these satisfy the fermionic commutation relations

$$
\{c_k^\dagger, c_l\} = \delta_{kl}, \quad \{c_k, c_l\} = \{c_k^\dagger, c_l^\dagger\} = 0. \tag{1.6}
$$

Fortunately one can find the expression for spin operators using these fermionic operators

$$
S_k^- = \prod_{i=1}^{k-1}(1 - 2c_i^\dagger c_i)c_k^\dagger, \quad S_k^+ = \prod_{i=1}^{k-1}(1 - 2c_i^\dagger c_i)c_k. \tag{1.7}
$$

The Hamiltonian (1.1) is transformed as

$$
\mathcal{H} = -\frac{J}{2}\sum_{i=1}^{N-1} c_{i+1}^\dagger c_i + c_i^\dagger c_{i+1} + \frac{J}{2}\alpha(c_1^\dagger c_N + c_N^\dagger c_1) - hN + 2h\sum_{i=1}^{N} c_i^\dagger c_i, \tag{1.8}
$$

where $\alpha \equiv \prod_{k=1}^{N}(1 - 2c_k^\dagger c_k)$. The total number of down-spins M is a constant of motion. The value of α is $(-1)^M$. We introduce the Fourier transformation of these fermionic operators:

$$
c_q = \frac{1}{\sqrt{N}}\sum_{k=1}^{N} \exp(-iqk)c_k, \quad q = 2\pi n/N, \tag{1.9}
$$

where n is integer (half-odd integer) for odd (even) M. These operators satisfy

$$
\{c_q, c_p\} = \{c_q^\dagger, c_p^\dagger\} = 0, \quad \{c_q, c_p^\dagger\} = \delta_{pq}. \tag{1.10}
$$

The Hamiltonian is the same as that of one-dimensional spinless fermions

$$
\mathcal{H} = -hN + \sum_q (2h - J\cos q)c_q^\dagger c_q. \tag{1.11}
$$

The lowest energy state at fixed M is

$$
\prod_{l=1}^{M} c_{\pi(M+1-2l)/N}^\dagger |0\rangle, \tag{1.12}
$$

and the total energy is

$$
-h(N - 2M) - J\sum_{l=1}^{M}\cos(\pi(M+1-2l)/N). \tag{1.13}
$$

In this way the XY model is treated by the transformation of spin operators to fermion operators. (1.7) is called the Jordan–Wigner transformation.

1.1.2 Energy eigenvalues of the Hamiltonian and the partition function

General eigenstates for fixed M are

$$\prod_{j=1}^{M} c^\dagger(2\pi I_j/N)|0\rangle. \tag{1.14}$$

Here $\{I_1, I_2, ..., I_M\}$ is a set of different integers or half-odd integers. If two sets of integers are different, the two corresponding states are orthogonal. The total number of states with M down-spins and $N - M$ up-spins is given by the binomial coefficient

$$C_N^M = \frac{N!}{M!(N-M)!}. \tag{1.15}$$

The total number of states represented by (1.14) is

$$C_N^1 + C_N^2 + ... + C_N^N = 2^N. \tag{1.16}$$

Thus the states represented by (1.14) give a complete orthonormal set of eigenstates of the Hamiltonian (1.1).

The partition function of this system is

$$Z = z^{-N/2}\left(\frac{1}{2}\left[\prod_{l=1}^{N}(1 + ze^{\frac{J\cos(2\pi l/N)}{T}}) - \prod_{l=1}^{N}(1 - ze^{\frac{J\cos(2\pi l/N)}{T}})\right]\right.$$
$$\left. + \frac{1}{2}\left[\prod_{l=1}^{N}(1 + ze^{\frac{J\cos(\pi(2l-1)/N)}{T}}) + \prod_{l=1}^{N}(1 - ze^{\frac{J\cos(\pi(2l-1)/N)}{T}})\right]\right), \tag{1.17}$$

where $z \equiv \exp(-2h/T)$. The first bracket gives the odd M states and the second gives the even M states. The second term in each bracket is much smaller than the first term, and so we can neglect them in the thermodynamic limit.

In the case of the lowest energy state, the energy per site in the large N, M limit is

$$e = -h + \frac{1}{2\pi}\int_{-\pi M/N}^{\pi M/N}(2h - J\cos q)dq$$
$$= -h(1 - 2M/N) + \left(\frac{J\sin(\pi M/N)}{\pi}\right). \tag{1.18}$$

The free energy per site in the thermodynamic limit is the same as for non-interacting fermions

$$f = -\frac{T}{N} \ln Z = -h - T \int_{-\pi}^{\pi} \ln[1 + \exp(-(2h - J\cos q)/T)] \frac{dq}{2\pi}. \quad (1.19)$$

The entropy per site of this system is as follows,

$$s = -\frac{\partial f}{\partial T} = \int \frac{dq}{2\pi} u\left(\frac{2h - J\cos q}{T}\right),$$

$$u(x) \equiv \ln(2\cosh x/2) - \frac{x}{2}\tanh x/2. \quad (1.20)$$

The function $u(x)$ is a symmetric and rapidly decreasing function of x

$$u(x) = u(-x), \qquad \int_{-\infty}^{\infty} u(x)dx = \frac{\pi^2}{3}. \quad (1.21)$$

Thus the low-temperature entropy is

$$s \simeq \frac{1}{\pi} \frac{T}{J\cos^{-1}(2h/J)} \int_{-\infty}^{\infty} u(x)dx = \frac{\pi T}{3J\cos^{-1}(2h/J)}. \quad (1.22)$$

The specific heat per site is

$$C = T\frac{\partial s}{\partial T} = \frac{\pi T}{3J\cos^{-1}(2h/J)}. \quad (1.23)$$

On the other hand the velocity of a low energy excitation is

$$v_s = J\cos^{-1}(2h/J).$$

Then the specific heat is written as

$$C = \frac{\pi T}{3v_s}. \quad (1.24)$$

1.1.3 Correlation functions

The static correlation function $\langle S_l^z S_m^z \rangle$ is called the longitudinal correlation function and $\langle S_l^- S_m^+ \rangle$ is the transverse correlation function. These can be calculated analytically[65]. In the fermion representation, the correlation functions are written as follows:

$$\langle S_l^z S_m^z \rangle = \langle (1 - 2c_l^\dagger c_l)(1 - 2c_m^\dagger c_m) \rangle /4 = \frac{1}{4} - \frac{M}{N} + \langle c_l^\dagger c_l c_m^\dagger c_m \rangle, \quad (1.25)$$

$$\langle S_l^- S_m^+ \rangle = \langle c_l^\dagger \prod_{k=l+1}^{m-1} (1 - 2c_k^\dagger c_k)c_m \rangle. \quad (1.26)$$

It should be noted that the longitudinal correlation function is represented by averages of four fermion operators, but the transverse correlation function is represented by averages of many fermion operators. The value of $\langle c_l^\dagger c_l c_m^\dagger c_m \rangle$ is given by $\langle c_l^\dagger c_l \rangle \langle c_m^\dagger c_m \rangle - \langle c_l^\dagger c_m \rangle \langle c_m^\dagger c_l \rangle$. Here we have used Wick's theorem. Thus we have

$$\langle S_l^z S_m^z \rangle = \left(\frac{1}{2} - \frac{M}{N} \right)^2 - u_{lm} u_{ml}, \tag{1.27}$$

$$u_{lm} \equiv \langle c_l^\dagger c_m \rangle. \tag{1.28}$$

In principle we can decompose the thermal average of complicated operators of non-interacting fermions into products of averages of two fermion operators using Wick's theorem. Thus we can calculate (1.26) analytically. The highest order term is

$$(-2)^{m-l-1} \langle c_l^\dagger \prod_{k=l+1}^{m-1} (c_k^\dagger c_k) c_m \rangle.$$

This is decomposed as

$$2^{m-l-1} \det \begin{bmatrix} u_{l,l+1} & u_{l+1,l+1} & u_{l+2,l+1} & \cdots & u_{m-1,l+1} \\ u_{l,l+2} & u_{l+1,l+2} & u_{l+2,l+2} & \cdots & u_{m-1,l+2} \\ u_{l,l+3} & u_{l+1,l+3} & u_{l+2,l+3} & \cdots & u_{m-1,l+3} \\ \cdot & \cdot & \cdot & \cdots & \cdot \\ \cdot & \cdot & \cdot & \cdots & \cdot \\ u_{l,m} & u_{l+1,m} & u_{l+2,m} & \cdots & u_{m-1,m} \end{bmatrix}.$$

The other terms are also written as determinants of this kind. Summing up all terms we have the transverse correlation function,

$$\langle S_l^- S_m^+ \rangle = 2^{m-l-1} \times \det$$

$$\begin{bmatrix} u_{l,l+1} & u_{l+1,l+1} - \frac{1}{2} & u_{l+2,l+1} & \cdot & u_{m-1,l+1} \\ u_{l,l+2} & u_{l+1,l+2} & u_{l+2,l+2} - \frac{1}{2} & \cdot & u_{m-1,l+2} \\ u_{l,l+3} & u_{l+1,l+3} & u_{l+2,l+3} & \cdot & u_{m-1,l+3} \\ \cdot & \cdot & \cdot & \cdot & \cdot \\ u_{l,m-1} & u_{l+1,m-1} & u_{l+2,m-1} & \cdot & u_{m-1,m-1} - \frac{1}{2} \\ u_{l,m} & u_{l+1,m} & u_{l+2,m} & \cdot & u_{m-1,m} \end{bmatrix}.$$

In the limit $N \to \infty$ we have u_{lm} for the ground state at $N = 2M$

$$u_{lm} = \frac{1}{2\pi} \int_{-\pi/2}^{\pi/2} e^{ik(l-m)} \mathrm{d}k = \begin{cases} \frac{1}{2} & \text{for } l = m, \\ \frac{\sin \pi(l-m)/2}{\pi(l-m)} & \text{for } l \neq m. \end{cases}$$

Then we have the longitudinal correlation function

$$\langle S_l^z S_m^z \rangle = \begin{cases} \frac{1}{4} & \text{for } l = m, \\ -(1 - (-1)^{l-m})\frac{1}{\pi^2(l-m)^2} & \text{for } l \neq m. \end{cases} \tag{1.29}$$

and a determinant expression of the transverse correlation function,

$$S_{xy}(m - l) = 2\langle S_l^- S_m^+ \rangle =$$

$$2^{m-l}\pi^{-(m-l)}\det \begin{bmatrix} 1 & 0 & 1 & 0 & -\frac{1}{3} & \cdots \\ 0 & 1 & 0 & 1 & 0 & \cdots \\ -\frac{1}{3} & 0 & 1 & 0 & 1 & \cdots \\ \cdot & \cdot & & \cdots & & \cdot \\ \cdot & \cdot & & \cdots & & \cdot \\ \cdot & \cdot & -\frac{1}{3} & 0 & 1 & 0 \\ \cdot & \cdot & 0 & -\frac{1}{3} & 0 & 1 \end{bmatrix}. \tag{1.30}$$

The determinant is represented by A_n^2 for $l - m = 2n$ and $A_n A_{n-1}$ for $l - m = 2n - 1$, where A_n is the determinant of an $n \times n$ matrix with elements $x_{i,j} = (-1)^{i-j}/(2(i - j) + 1)$,

$$A_n \equiv \det \begin{bmatrix} 1 & 1 & -\frac{1}{3} & \cdots & \frac{(-1)^n}{2n-3} \\ -\frac{1}{3} & 1 & 1 & \cdots & \\ \cdot & \cdot & \cdot & \cdots & \cdot \\ \cdot & \cdot & \cdot & \cdots & \cdot \\ \cdot & \cdot & -\frac{1}{3} & 1 & 1 \\ \frac{(-1)^{n-1}}{2n-1} & \cdot & \cdot & -\frac{1}{3} & 1 \end{bmatrix}. \tag{1.31}$$

One can eliminate $x_{1,j}, j \geq 2$ by putting $x'_{i,j} = x_{i,j} - x_{i,1}x_{1,j}$ without changing the value of the determinant. The new elements are

$$\frac{4(i - 1)(j - 1)(-1)^{i-j}}{(2i - 1)(2j - 3)(2(i - j) + 1)}, \quad j \geq 2.$$

By this operation one gets the following recursion relation:

$$A_1 = 1, \quad A_n = \frac{((2n - 2)!!)^2}{(2n - 1)!!(2n - 3)!!}A_{n-1}. \tag{1.32}$$

Then we have

$$A_n = \prod_{j=1}^{n-1} \frac{(2n - 2j)^{2j}}{(2n + 1 - 2j)^{2j-1}}. \tag{1.33}$$

If we define $B_n = (2/\pi)^n A_n$, the two-point function is given as

$$S_{xy}(2n) = B_n^2, \quad S_{xy}(2n - 1) = B_n B_{n-1}.$$

B_n satisfies:

$$\frac{B_{n+1}}{B_n} = \frac{2}{\pi} \prod_{j=1}^{n} \left(1 - \left(\frac{1}{2j}\right)^2\right)^{-1}.$$

Using $2/\pi = \prod_{j=1}^{\infty}(1 - (2j)^{-2})$, we have:

$$\ln \frac{B_{n+1}}{B_n} = \sum_{j=n+1}^{\infty} \ln(1 - (2j)^{-2}) \simeq -\frac{1}{4n}. \tag{1.34}$$

Then we can expect that B_n behaves as $n^{-1/4}$. The two-point function $S_{xy}(n)$ decays as $n^{-1/2}$. On the other hand, $S_{zz}(n)$ decays as n^{-2}. Thus the correlation exponents are different for S_{zz} and S_{xy}. We find that two-point functions decay algebraically and not exponentially at zero temperature. At finite temperature these decay exponentially.

1.2 The anisotropic XY model

We consider the anisotropic case of the XY model:

$$\mathscr{H} = \sum_{j=1}^{N} J_x S_j^x S_{j+1}^x + J_y S_j^y S_{j+1}^y - 2h \sum_{j=1}^{N} S_j^z. \tag{1.35}$$

This is written in terms of $S^{\pm}$ operators as follows:

$$\mathscr{H} = \frac{1}{2} \sum_{j=1}^{N} J(S_j^+ S_{j+1}^- + S_j^- S_{j+1}^+) + J'(S_j^+ S_{j+1}^+ + S_j^- S_{j+1}^-)$$

$$-2h \sum_{j=1}^{N} S_j^z, \quad J \equiv \frac{J_x + J_y}{2}, \quad J' \equiv \frac{J_x - J_y}{2}. \tag{1.36}$$

This Hamiltonian changes the number of down-spins by two. Thus space is divided by the parity of number of down-spins. By the Jordan–Wigner transformation (1.7) we have

$$\mathscr{H} = -\frac{J}{2} \left[-\alpha(c_1^\dagger c_N + c_N^\dagger c_1) + \sum_{j=1}^{N-1} c_{j+1}^\dagger c_j + c_j^\dagger c_{j+1}\right]$$

$$-\frac{J'}{2} \left[-\alpha(c_1^\dagger c_N^\dagger + c_N c_1) + \sum_{j=1}^{N-1} c_{j+1}^\dagger c_j^\dagger + c_j c_{j+1}\right]$$

$$-hN + 2h \sum_{j=1}^{N} c_j^\dagger c_j, \quad \alpha = \prod(1 - 2c_j^\dagger c_j). \tag{1.37}$$

We can show that $\alpha^2 = 1$ and $[\mathscr{H}, \alpha] = 0$. The Hamiltonian and α are simultaneously diagonalized and the eigenvalue of α is ± 1. The Fourier transformation of these fermionic operators is

$$c_q = \frac{1}{\sqrt{N}} \frac{1-i}{\sqrt{2}} \sum_{k=1}^{N} \exp(-iqk)c_k, \quad q = 2\pi n/N, \tag{1.38}$$

where n is integer (half-odd integer) for $\alpha = -1(+1)$.

$$\mathscr{H} = -hN + \sum_{q}{}' (2h - J\cos q)(c_q^\dagger c_q + c_{-q}^\dagger c_{-q}) + J'\sin q(c_q^\dagger c_{-q}^\dagger + c_{-q}c_q). \tag{1.39}$$

Here $\sum_q'$ means the sum over $0 < q < \pi$.

1.2.1 The subspace $\alpha = 1$

In this Hamiltonian, particles with momentum q and $-q$ are coupled. We apply the following transformation for fermion operators c_q and c_{-q},

$$c_q = \cos\theta_q \eta_q + \sin\theta_q \eta_{-q}^\dagger, \quad c_{-q} = -\sin\theta_q \eta_q^\dagger + \cos\theta_q \eta_{-q},$$

$$\eta_q = \cos\theta_q c_q - \sin\theta_q c_{-q}^\dagger, \quad \eta_{-q} = +\sin\theta_q c_q^\dagger + \cos\theta_q c_{-q}. \tag{1.40}$$

$Nq/2\pi$ is half-odd integer. The Hamiltonian (1.36) is transformed as

$$\mathscr{H} = -hN + \sum_{q}{}' 2\sin^2\theta_q(2h - J\cos q) + 2\sin\theta_q\cos\theta_q J'\sin q$$

$$+[(2h - J\cos q)\cos 2\theta_q - J'\sin q\sin 2\theta_q](\eta_q^\dagger \eta_q + \eta_{-q}^\dagger \eta_{-q})$$

$$+[(2h - J\cos q)\sin 2\theta_q + J'\sin q\cos 2\theta_q](\eta_q^\dagger \eta_{-q}^\dagger + \eta_{-q}\eta_q). \tag{1.41}$$

The last term is removed if we put

$$\tan 2\theta_q = \frac{J'\sin q}{J\cos q - 2h}. \tag{1.42}$$

Thus the Hamiltonian becomes

$$\mathscr{H} = \sum_{q} \epsilon(q)(\eta_q^\dagger \eta_q - \frac{1}{2}),$$

$$\epsilon(q) = \sqrt{(J\cos q - 2h)^2 + (J'\sin q)^2}. \tag{1.43}$$

The lowest energy state $|\Psi\rangle$ must satisfy $\eta_q|\Psi\rangle = 0$. The following state satisfies this condition,

$$|\Psi\rangle = \prod_{q}{}' (\cos\theta_q + \sin\theta_q c_q^\dagger c_{-q}^\dagger)|0\rangle. \tag{1.44}$$

As we are considering the case $\alpha = 1$, the total number of quasi particles must be even. So the number of states which belong to this subspace is not 2^N but $\sum_j C_{2j}^N = 2^{N-1}$.

1.2.2 The subspace $\alpha = -1$

The number of particles must be odd in this subspace. Then the Hamiltonian is the same form as (1.43) but $qN/2\pi$ must be an integer. The lowest energy state in this subspace is

$$|\Psi\rangle = c_0^\dagger {\prod_q}' (\cos\theta_q + \sin\theta_q c_q^\dagger c_{-q}^\dagger)|0\rangle. \tag{1.45}$$

The general states are given by an even number of excitations from this state. The number of states is also 2^{N-1}. These states are orthogonal to each other and therefore all these states together form a complete set of wave vectors.

1.2.3 The free energy

Using the results of 1.2.1 and 1.2.2, one obtains the partition function of the system,

$$Z = \exp\left(\frac{\sum_q \epsilon(q)}{2T}\right) \frac{1}{2}\left\{\prod_q(1 + e^{-\epsilon(q)/T}) + \prod_q(1 - e^{-\epsilon(q)/T})\right\}$$

$$+ \exp\left(\frac{\sum_{q'} \epsilon(q')}{2T}\right) \frac{1}{2}\left\{\prod_{q'}(1 + e^{-\epsilon(q')/T}) - \prod_{q'}(1 - e^{-\epsilon(q')/T})\right\}, \tag{1.46}$$

where $qN/2\pi$ is a half-odd integer and $q'N/2\pi$ is an integer. In the thermodynamic limit the second term in $\{...\}$ is much smaller than the first term and we obtain the free energy per site. The free energy is given by the logarithm of the partition function

$$f = -\frac{T}{N}\ln Z$$

$$= e_0 - \frac{T}{2\pi}\int_{-\pi}^{\pi} \ln(1 + \exp(-\sqrt{(J\cos q - 2h)^2 + (J'\sin q)^2}/T))dq. \tag{1.47}$$

Here e_0 is the ground state energy per site

$$e_0 = -\frac{1}{4\pi}\int_{-\pi}^{\pi} \sqrt{(J\cos q - 2h)^2 + (J'\sin q)^2}dq.$$

2

Systems with a delta-function potential

2.1 The boson problem

2.1.1 The $c = 0$ case

Here we consider the system

$$\mathcal{H} = -\sum_{i=1}^{N} \frac{\partial^2}{\partial x_i^2} + 2c \sum_{i<j} \delta(x_i - x_j). \tag{2.1}$$

At first we consider the cases $c = 0$ and $c = \infty$. We assume periodic boundary conditions.

In second quantized form, the Hamiltonian (2.1) at $c = 0$ is written as

$$\mathcal{H} = \sum_k k^2 a_k^\dagger a_k, \quad k = 2\pi n/L,$$

$$[a_k, a_q^\dagger] = \delta_{k,q}, \quad [a_k, a_q] = [a_k^\dagger, a_q^\dagger] = 0. \tag{2.2}$$

The eigenstates and eigenvalues are given by

$$\prod_k (n_k!)^{-1/2} (a_k^\dagger)^{n_k} |0\rangle, \quad \sum_k k^2 n_k. \tag{2.3}$$

A set of integers $\{n_k\}$ gives an eigenstate. The ground state of this system is $n_0 = N$, $n_{k \neq 0} = 0$. The partition function of the grand canonical ensemble at chemical potential A is

$$\Xi = \prod_k \sum_{n_k = 0,1,2,\dots} \exp(-(k^2 - A)n_k/T) = \prod_k (1 - \exp[-(k^2 - A)/T])^{-1}. \tag{2.4}$$

The Gibbs free energy is given by

$$G = -T \ln \Xi = T \sum_k \ln(1 - \exp[-(k^2 - A)/T]). \tag{2.5}$$

10

The number of particles is given by

$$N = -\frac{\partial G}{\partial A} = \sum_k \frac{1}{\exp[(k^2 - A)/T] - 1}. \tag{2.6}$$

The pressure is

$$p = -\frac{\partial G}{\partial L} = -T\frac{\partial}{\partial L}\sum_{n=-\infty}^{\infty} \ln[1 - \exp((-(2\pi n/L)^2 + A)/T)]$$

$$= \frac{1}{L}\sum_k \frac{2k^2}{\exp((k^2 - A)/T) - 1}. \tag{2.7}$$

In the thermodynamic limit we have

$$p = \frac{1}{2\pi}\int \frac{2k^2 dk}{\exp(k^2 - A)/T - 1}$$

$$= -\frac{T}{2\pi}\int \ln(1 - \exp[(A - k^2)/T])dk = \lim_{L\to\infty} -G/L, \tag{2.8}$$

by partial differentiation with respect to k.

2.1.2 *The $c = \infty$ case*

It is difficult to treat this case via the second quantization representation. So we assume that the wave function is represented by the Bethe-ansatz wave function. We assume N quasi-momenta $k_1, k_2, k_3, ..., k_N$,

$$\psi(x_1, x_2, ..., x_N) = \sum_P A(P)\exp(i\sum_l k_{Pl}x_l), \tag{2.9}$$

in the region $x_1 < x_2 < x_3 < ... < x_N$. In the other $N! - 1$ regions the wave function is determined by the full symmetry of the wave function. The energy eigenvalue must be given by $E = \sum_l k_l^2$. The wave function must be zero at $x_n = x_{n+1}$. This condition is satisfied if $A(P) = \epsilon(P)$. This means that

$$A(12) = 1, \quad A(21) = -1, \quad \text{for} \quad N = 2,$$
$$A(123) = A(231) = A(312) = 1, A(213) = A(132) = A(321) = -1,$$
$$\text{for} \quad N = 3. \tag{2.10}$$

The periodic boundary condition is

$$\psi(x_1, x_2, ..., x_N) = \psi(x_2, x_3, ..., x_N, x_1 + L). \tag{2.11}$$

This condition is satisfied by $\exp(ik_l) = (-1)^{N-1}$. This means that the quasi-momenta are $2\pi/L\times$integer for odd N and $2\pi/L\times$ half-odd integer for even N. If two quasi-momenta are the same, the wave function vanishes, thus they must be always different. The ground state is given by the set of quasi-momenta

$$\{k\} = \{(N-1)\pi/L, (N-3)\pi/L, ..., -(N-1)\pi/L\}.$$

The partition function of this system for the ground canonical ensemble is

$$\Xi = \frac{1}{2}\left[\prod_n(1 + e^{(A-((2n-1)\pi/L)^2)/T}) + \prod_n(1 - e^{(A-((2n-1)\pi/L)^2)/T})\right]$$

$$+ \frac{1}{2}\left[\prod_n(1 + e^{(A-(2n\pi/L)^2)/T}) - \prod_n(1 - e^{(A-(2n\pi/L)^2)/T})\right]. \qquad (2.12)$$

In the first bracket, even order terms with respect to fugacity $z \equiv \exp(A/T)$ survive. This is the contribution of states with even number of particles. The second bracket is the contribution of states with odd number of particles. This formula resembles that of the free fermion model but is slightly complicated because the parity of the quasi-momenta change according to the number of particles. In the thermodynamic limit the second terms in each bracket are much less than the first terms. The Gibbs free energy per unit length is

$$g = \lim_{L \to \infty} -T \ln \Xi/L = -\frac{T}{2\pi} \int \ln(1 + \exp[-(k^2 - A)/T])dk. \qquad (2.13)$$

Thus the boson problem at $c = \infty$ resembles that of noninteracting fermions. The momentum distribution problem of this case was investigated by Girardeau[31] and Lenard[58].

2.1.3 Scattering states of Bose particles with finite interaction

The applicability of the Bethe-ansatz method was first shown by Lieb and Liniger[60,59]. At first we consider the scattering problem of two particles,

$$\mathcal{H} = -\frac{\partial^2}{\partial x_1^2} - \frac{\partial^2}{\partial x_2^2} + 2c\delta(x_1 - x_2). \qquad (2.14)$$

We assume that the wave function is represented by a superposition of plane waves

$$f(x_1, x_2) = A(12)\exp(i(k_1 x_1 + k_2 x_2)) + A(21)\exp(i(k_2 x_1 + k_1 x_2)), \qquad (2.15)$$

for $x_1 \leq x_2$. By the symmetry condition of the Boson wave function we have

$$f(x_1, x_2) = A(12)\exp(i(k_1 x_2 + k_2 x_1)) + A(21)\exp(i(k_2 x_2 + k_1 x_1)), \qquad (2.16)$$

at $x_1 \geq x_2$. This wave function is continuous at $x_1 = x_2$ but its derivative is discontinuous,

$$\left[-\frac{\partial^2}{\partial x_1^2} - \frac{\partial^2}{\partial x_2^2}\right] f(x_1, x_2) = (k_1^2 + k_2^2)f + 2i\delta(x_1 - x_2)$$
$$\times (A(12) - A(21))(k_1 - k_2)\exp(i(k_1 + k_2)x_1). \tag{2.17}$$

Thus if $i(k_1 - k_2)(A(12) - A(21)) + c(A(12) + A(21)) = 0$ is satisfied, the function f satisfies the eigenvalue condition $\mathscr{H}f = (k_1^2 + k_2^2)f$. This means that the amplitudes $A(12)$ and $A(21)$ must satisfy the following condition,

$$\frac{A(12)}{A(21)} = -\frac{k_1 - k_2 + ic}{k_1 - k_2 - ic}. \tag{2.18}$$

Next we treat the three-body problem. We assume the wave function is of the form

$$f(x_1, x_2, x_3) =$$
$$A(123)e^{i(k_1 x_1 + k_2 x_2 + k_3 x_3)} + A(213)e^{i(k_2 x_1 + k_1 x_2 + k_3 x_3)}$$
$$+ A(132)e^{i(k_1 x_1 + k_3 x_2 + k_2 x_3)} + A(321)e^{i(k_3 x_1 + k_2 x_2 + k_1 x_3)}$$
$$+ A(231)e^{i(k_2 x_1 + k_3 x_2 + k_1 x_3)} + A(312)e^{i(k_3 x_1 + k_1 x_2 + k_2 x_3)},$$

$$\tag{2.19}$$

for $x_1 < x_2 < x_3$. We have six conditions for six amplitudes $A(ijk)$:

$$A(123)/A(213) = -(k_1 - k_2 + ic)/(k_1 - k_2 - ic),$$
$$A(123)/A(132) = -(k_2 - k_3 + ic)/(k_2 - k_3 - ic),$$
$$A(213)/A(231) = -(k_1 - k_3 + ic)/(k_1 - k_3 - ic),$$
$$A(132)/A(312) = -(k_1 - k_3 + ic)/(k_1 - k_3 - ic),$$
$$A(321)/A(231) = -(k_3 - k_2 + ic)/(k_3 - k_2 - ic),$$
$$A(321)/A(312) = -(k_2 - k_1 + ic)/(k_2 - k_1 - ic). \tag{2.20}$$

These relations must be dependent, because only five independent relations are possible. Rewriting (2.20), we get

$$A(123) = Y_{12}A(213), A(123) = Y_{23}A(132), A(213) = Y_{13}A(231),$$

$$A(132) = Y_{13}A(312), A(321) = Y_{32}A(231), A(321) = Y_{21}A(312),$$

$$Y_{ab} \equiv -(k_a - k_b + ic)/(k_a - k_b - ic). \tag{2.21}$$

To get $A(cba)$ from $A(abc)$ we have two paths

$$A(cba) = Y_{ab}Y_{ac}Y_{bc}A(abc), \quad A(cba) = Y_{bc}Y_{ac}Y_{ab}A(abc).$$

The final results become equivalent if

$$Y_{ab} = Y_{ba}, \quad Y_{ab}Y_{ac}Y_{bc} = Y_{bc}Y_{ac}Y_{ab}. \tag{2.22}$$

One can prove these relations from the definition. The boson problem is the simplest example of a Yang–Baxter relation. The answer is

$$
\begin{aligned}
A(123) &= C(k_1 - k_2 + ic)(k_1 - k_3 + ic)(k_2 - k_3 + ic), \\
A(213) &= -C(k_2 - k_1 + ic)(k_1 - k_3 + ic)(k_2 - k_3 + ic), \\
A(132) &= -C(k_1 - k_2 + ic)(k_1 - k_3 + ic)(k_3 - k_2 + ic), \\
A(321) &= -C(k_2 - k_1 + ic)(k_3 - k_1 + ic)(k_3 - k_2 + ic), \\
A(231) &= C(k_2 - k_1 + ic)(k_3 - k_1 + ic)(k_2 - k_3 + ic), \\
A(312) &= C(k_1 - k_2 + ic)(k_3 - k_1 + ic)(k_3 - k_2 + ic).
\end{aligned} \tag{2.23}
$$

For the case of an N-body problem the relations (2.22) are sufficient to construct a consistent solution. For the N-body problem one must determine $N!$ coefficients $A(P)$ in the wave function which holds for $x_1 < x_2 < ... < x_N$. The answer is

$$f = \sum_P A(P) \exp[i(k_{P1}x_1 + k_{P2}x_2 + ... + k_{PN}x_N)].$$

$$A(P) = C\epsilon(P) \prod_{j<k}(k_{Pj} - k_{Pk} + ic). \tag{2.24}$$

The energy eigenvalue is

$$E = \sum_{j=1}^{N} k_j^2. \tag{2.25}$$

The total momentum operator is defined by

$$\mathcal{K} = \sum_{j=1}^{N} -i\frac{\partial}{\partial x_j}.$$

The wave function (2.24) is also an eigenstate of this operator,

$$\mathcal{K}f = Kf, \quad K = \sum_{j=1}^{N} k_j. \tag{2.26}$$

2.1.4 Periodic boundary conditions

In the previous section we considered the N-body problem in a one-dimensional space of infinite length. Here we consider the N-body problem on a ring with length L. The wave function must satisfy

$$f(x_1, x_2, ..., x_j, ..., x_N) = f(x_1, x_2, ..., x_j + L, x_{j+1}, ..., x_N),$$
$$j = 1, ..., N. \tag{2.27}$$

This condition is satisfied if

$$\exp(ik_{PN}L) = \frac{A(PN, P1, P2, ..., P(N-1))}{A(P1, P2, P3, ..., PN)}, \tag{2.28}$$

for any permutation P. This condition gives

$$\exp(ik_j L) = \prod_{l \neq j} \left(\frac{k_j - k_l + ic}{k_j - k_l - ic} \right), \quad j = 1, 2, ..., N. \tag{2.29}$$

These are N conditions for N numbers. Taking the logarithm of these equations, we have

$$k_j L = 2\pi I_j - \sum_{l=1}^{N} 2 \tan^{-1} \left(\frac{k_j - k_l}{c} \right). \tag{2.30}$$

In the limit $c \to \infty$ we have $k_j = 2\pi I_j / L$. For the ground state (lowest energy state with fixed N)

$$\{I_j\} = \{(N-1)/2, (N-3)/2, ..., -(N-1)/2\}, \quad \text{or} \quad I_j = (N+1)/2 - j. \tag{2.31}$$

This is a set of integers or half-odd integers which determines an eigenstate. One can show that the set of integers gives a unique real solution for the k_j s. The ground state is always unique and is never degenerate with other states. Thus the integers (2.31) give the ground state.

2.1.5 Linear integral equation for the distribution of quasi-momenta

We take the limit of infinite N and L, setting the density as finite. The distribution function of ks is denoted as $\rho(k)$. (2.30) becomes

$$k = 2\pi \int_0^k \rho(q)dq - \int_{-B}^B 2 \tan^{-1} \frac{(k-q)}{c} \rho(q)dq.$$

Differentiating with respect to k, we have

$$\rho(k) = \frac{1}{2\pi} + \int_{-B}^B \frac{c/\pi}{c^2 + (k-q)^2} \rho(q)dq. \tag{2.32}$$

This is called the Lieb–Liniger equation. The energy and number of particles per unit length are

$$e = \int_{-B}^B k^2 \rho(k)dk, \quad n = \int_{-B}^B \rho(k)dk. \tag{2.33}$$

Equation (2.32) is a Fredholm type linear integral equation. If one puts $p = Bx, \lambda = c/B$, we have

$$\rho(x) = \frac{1}{2\pi} + \frac{1}{\pi} \int_{-1}^{1} \frac{\lambda}{\lambda^2 + (x - y)^2} \rho(y)\mathrm{d}y,$$

$$e = B^3 F(\lambda), \quad F(\lambda) \equiv \int_{-1}^{1} x^2 \rho(x)\mathrm{d}x,$$

$$n = BG(\lambda), \quad G(\lambda) \equiv \int_{-1}^{1} \rho(x)\mathrm{d}x. \tag{2.34}$$

Then we have

$$e/n^3 = F(\lambda)/G^3(\lambda), \quad c/n = \lambda/G(\lambda). \tag{2.35}$$

Thus we find that energy per unit length e is expressed as

$$e = n^3 u(c/n), \tag{2.36}$$

by a certain function $u(x)$. The perturbation expansion for a 1D boson system with delta-function potential gives

$$u(x) = x - \frac{4\pi}{3} x^{3/2} + \left(\frac{1}{6} - \frac{1}{\pi^2}\right)x^2 + O(x^{5/2}). \tag{2.37}$$

The Bethe-ansatz solution of delta-function bosons is used to check approximate theories for Bose systems[99, 100]. These results are compared in Fig. (2.1).

2.1.6 Bound states in the case $c < 0$

In the case of an attractive interaction, $c < 0$, we have solutions with complex quasi momenta. We consider the two-body wave function (2.15) in the infinite one-dimensional space,

$$f(x_1, x_2) = \exp(i(k_1 x_1 + k_2 x_2))[A(12)\theta(x_1 - x_2) + A(21)\theta(x_2 - x_1)]$$

$$+ \exp(i(k_2 x_1 + k_1 x_2))[A(21)\theta(x_1 - x_2) + A(12)\theta(x_2 - x_1)].$$

The term $\exp(i(k_1 x_1 + k_2 x_2))$ is rewritten as $\exp(i(k_1 + k_2)(x_1 + x_2)/2 + i(k_1 - k_2)(x_1 - x_2)/2)$. This term diverges if $\Im(k_1 + k_2) \neq 0$. Here $\Im(x)$ means the imaginary part of x. Moreover if $\Im(k_1 - k_2) > 0$, we should have $A(12) \neq 0$ and $A(21) = 0$. Then from equation (2.18) we have $k_1 = k_2 + i|c|$ and therefore

$$k_1 = \alpha + i|c|/2, \quad k_2 = \alpha - i|c|/2,$$

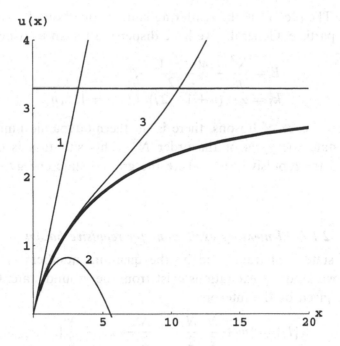

Fig. 2.1. The function $u(x)$ determines the zero temperature properties of a delta-function Bose gas, the thick line is the result of numerical calculation of (2.34). Line 1 is the result of primitive perturbation. Line 2 is the result of Bogoliubov theory. Line 3 is the perturbation result (2.37) by the author[99].

where α is a real number. The energy, momentum and wave function of this state is

$$E = 2\alpha^2 - c^2/2, \quad K = 2\alpha,$$
$$f(x_1, x_2) = \exp(iXK)\exp(-c|x_1 - x_2|/2). \quad (2.38)$$

Then we have $E = K^2/2 - c^2/2$. This means that the mass is doubled and the binding energy is $-c^2/2$. This can be regarded as a bound state of two particles.

Next we consider the three-body problem in the infinite one-dimensional space, and we assume $\Im k_1 > \Im k_2 > \Im k_3$. In this case equation (2.19) is normalizable if $A(312) = A(231) = A(321) = 0$. This is satisfied by the following two cases,

$$k_1 = k_2 - ic, \quad k_3 = k_2 + ic,$$
$$k_1 = \alpha - ic/2, \quad k_3 = \alpha + ic/2, \quad k_2 = \beta.$$

The former is the bound state of three particles. We call this the string of length three. The second is the scattering state of one bound state and one independent particle. Generally we have dispersion for an *n*-string

$$E = \frac{K^2}{n} - \frac{n(n^2-1)c^2}{12},$$
$$k_j = \alpha - (n+1-2j)c/2, \quad j = 1,...,n. \tag{2.39}$$

In the attractive case of bosons, there is no thermodynamic limit, because the ground state energy is of the order N^3. This situation is completely different from the repulsive case, where the ground state energy is of order N.

2.1.7 Elementary excitations for repulsive bosons

The ground state is characterized by the quantum numbers I_j which are successive. Two kinds of excitations exist from the ground state. One is the particle state given by the integers

$$\{I_j\} = \{m + \frac{N}{2}, \frac{N}{2} - 1, \frac{N}{2} - 2, ..., -\frac{N}{2}\}. \tag{2.40}$$

Here the number of particles changes from N to $N+1$. The total momentum of this state is $2\pi m/L$. We put the solution of (2.30) for the ground state as $\{k_1, k_2, ..., k_N\}$ and for this excited state as $\{q, k'_1, k_2, ..., k'_N\}$. Putting $\Delta k_j = k'_j - k_j$ we can write down the equation for Δk_j,

$$\Delta k_j L = -\pi + 2\tan^{-1}\frac{q-k_j}{c} - \sum_{j=1}^{N} 2\left(\tan^{-1}\frac{k'_j - k'_l}{c} - \tan^{-1}\frac{k_j - k_l}{c}\right). \tag{2.41}$$

As Δk_j is of the order of N^{-1}, we can write this equation as follows:

$$\Delta k_j = -\frac{1}{L}\left(\pi - 2\tan^{-1}\frac{q-k_j}{c} - 2\sum_{j=1}^{N}\frac{c}{c^2 + (k_j - k_l)^2}(\Delta k_j - \Delta k_l)\right). \tag{2.42}$$

This equation becomes

$$\Delta k_j\left(1 + \frac{1}{L}\sum_{l=1}^{N}\frac{2c}{c^2 + (k_j - k_l)^2}\right) = \frac{1}{L}\left(\pi - 2\tan^{-1}\frac{q-k_j}{c}\right)$$
$$-\frac{1}{L}\sum\frac{2c}{c^2 + (k_j - k_l)^2}\Delta k_l. \tag{2.43}$$

Using (2.32) this is written as

$$2\pi\Delta k_j\rho(k_j) = -\frac{1}{L}\left(\pi - 2\tan^{-1}\frac{q-k_j}{c}\right)$$

$$+ \int_{-B}^{B} \frac{2c}{c^2 + (k_j - k_l)^2} \Delta k_l \rho(k_l) dk_l. \tag{2.44}$$

Then we can define the back flow $J(k, q) = L\Delta(k)\rho(k)$,

$$J(k, q) - \int_{-B}^{B} \frac{1}{\pi} \frac{c}{c^2 + (k - k')^2} J(k', q) dk' = \frac{1}{2\pi} \left(\pi - 2 \tan^{-1} \frac{q - k}{c} \right). \tag{2.45}$$

The energy change $\Delta E(q)$ and momentum change $\Delta K(q)$ are

$$\Delta E(q) = q^2 + \int_{-B}^{B} 2k J(k, q) dk,$$

$$\Delta K(q) = q + \int_{-B}^{B} J(k, q) dk, \quad q > B. \tag{2.46}$$

We define the Green function for this type of integral equation:

$$U(k, k') - \int_{-B}^{B} \frac{1}{\pi} \frac{c dk''}{c^2 + (k - k'')^2} U(k', k'') = \delta(k - k'). \tag{2.47}$$

Using this Green function we can write $\rho(k)$ and $J(k, q)$,

$$\rho(k) = \frac{1}{2\pi} \int_{-B}^{B} U(k, k') dk',$$

$$J(k, q) = - \int_{-B}^{B} U(k, k') \frac{1}{2\pi} \left(\pi - 2 \tan^{-1} \frac{q - k'}{c} \right) dk'. \tag{2.48}$$

Considering that $U(k, k') = U(k', k)$ we can calculate $\Delta K(q)$,

$$\Delta K(q) = q - \int_{-B}^{B} \rho(k) \left(\pi - 2 \tan^{-1} \frac{q - k}{c} \right) dk$$

$$= -\pi n + q + \int_{-B}^{B} 2 \tan^{-1} \frac{q - k}{c} \rho(k) dk$$

$$= -\pi n + \int_{0}^{q} dk' \left\{ 1 + \int_{-B}^{B} \frac{2c}{c^2 + (k' - k)^2} \rho(k) dk \right\}$$

$$= -\pi n + 2\pi \int_{0}^{q} \rho(k) dk. \tag{2.49}$$

At $q = B$ we have $\Delta K = 0$. The energy change ΔE is the chemical potential at zero temperature. This is called a type I excitation, and corresponds to the Bogoliubov excitation.

Next we consider the hole state of elementary excitation, given by integers

$$\{I_j\} = \left\{ \frac{N}{2}, \frac{N}{2} - 1, ..., \frac{N}{2} - m + 1, \frac{N}{2} - m - 1, ..., -\frac{N}{2} + 1 \right\}. \tag{2.50}$$

The number of particles is $N - 1$. The energy and momentum change are given by

$$\Delta K(q) = -q + \int_{-B}^{B} J(k,q)dk,$$

$$\Delta E(q) = -q^2 + \int_{-B}^{B} 2kJ(k,q)dk,$$

$$J(k,q) - \int_{-B}^{B} \frac{1}{\pi} \frac{c}{c^2 + (k-k')^2} J(k',q)dk'$$

$$= \frac{1}{2\pi} \left(\pi + 2\tan^{-1} \frac{k-q}{c} \right), \quad q \le B. \tag{2.51}$$

This excitation is a hole excitation in the fermionic distribution. Ishikawa and Takayama[40] identified that this excitation is solitonic in nature.

It is noteworthy that the boson system resembles a spinless fermion system. The excitations near the ground state are described by the sum of these two excitations.

2.2 The fermion problem

2.2.1 The $c = 0$ case

In second quantization form the Hamiltonian is written as

$$\mathcal{H} = \sum_{k} k^2 (a_{k\uparrow}^{\dagger} a_{k\uparrow} + a_{k\downarrow}^{\dagger} a_{k\downarrow}), \quad k = 2\pi n/L,$$

$$\{a_{k\sigma}^{\dagger}, a_{q\sigma'}\} = \delta_{k,q}\delta_{\sigma,\sigma'}, \quad \{a_{k\sigma}, a_{q\sigma'}\} = \{a_{k\sigma}^{\dagger}, a_{q\sigma'}^{\dagger}\} = 0. \tag{2.52}$$

The state is given by $\prod_{k\sigma}(a^{\dagger})^{n_{k\sigma}}|0\rangle$. $n_{k\sigma}$ is the number of particles in state k,σ and takes only the values 0 or 1. The total energy is given by $E = \sum k^2 n_{k\sigma}$. The total number of particles is $\sum n_{k\sigma}$. The grand partition function at chemical potential A is

$$\Xi = \sum_{n_{k,\sigma}=0,1} \exp[(E - AN)/T] = \prod_{k\sigma}(1 + \exp(-(k^2 - A)/T)). \tag{2.53}$$

The free energy is given by

$$G(T, A, L) = -2T \sum_{n=-\infty}^{\infty} \ln(1 + \exp[-((2\pi n/L)^2 - A)/T]). \tag{2.54}$$

In the limit of $L \to \infty$ this becomes

$$G = -2TL \int_{-\infty}^{\infty} (2\pi)^{-1} \ln(1 + \exp[-(k^2 - A)/T])dk. \tag{2.55}$$

The pressure is given by $-\partial G/\partial L$,

$$p = 2T \int_{-\infty}^{\infty} (2\pi)^{-1} \ln(1 + \exp[-(k^2 - A)/T]) dk. \qquad (2.56)$$

2.2.2 The two-body problem

The wave function of particles with spin has not only coordinates for position but also spin coordinates,

$$f(x_1 s_1, x_2 s_2, x_3 s_3, ..., x_N s_N). \qquad (2.57)$$

So we have four functions to represent an eigenstate of the two $S = 1/2$ particles system

$$f_{\uparrow\uparrow}(x_1, x_2), f_{\uparrow\downarrow}(x_1, x_2), f_{\downarrow\uparrow}(x_1, x_2), f_{\downarrow\downarrow}(x_1, x_2). \qquad (2.58)$$

The fermion's wave function must be antisymmetric and we have

$$f_{\uparrow\uparrow}(x_1, x_2) = -f_{\uparrow\uparrow}(x_2, x_1),$$
$$f_{\uparrow\downarrow}(x_1, x_2) = -f_{\downarrow\uparrow}(x_2, x_1),$$
$$f_{\downarrow\downarrow}(x_1, x_2) = -f_{\downarrow\downarrow}(x_2, x_1). \qquad (2.59)$$

Thus we obtain $f_{\uparrow\uparrow}(x_1, x_1) = 0$ and $f_{\downarrow\downarrow}(x_1, x_1) = 0$. This means that in the fully polarized case the delta-function potential has no physical effect. The wave function is represented by a determinant

$$f_{\uparrow\uparrow}(x_1, x_2) = C\det \begin{pmatrix} \exp(ik_1 x_1) & \exp(ik_1 x_2) \\ \exp(ik_2 x_1) & \exp(ik_2 x_2) \end{pmatrix}. \qquad (2.60)$$

Next we consider the case where one fermion has spin up and the other has spin down,

$$f_{\uparrow\downarrow}(x_1, x_2) =$$
$$\theta(x_2 - x_1)\left([12, 12]e^{i(k_1 x_1 + k_2 x_2)} + [12, 21]e^{i(k_2 x_1 + k_1 x_2)}\right)$$
$$+\theta(x_1 - x_2)\left([21, 12]e^{i(k_2 x_1 + k_1 x_2)} + [21, 21]e^{i(k_2 x_2 + k_1 x_1)}\right), \qquad (2.61)$$

$$\left[-\frac{\partial^2}{\partial x_1^2} - \frac{\partial^2}{\partial x_2^2}\right] f_{\uparrow\downarrow}(x_1, x_2) = (k_1^2 + k_2^2) f_{\uparrow\downarrow}(x_1, x_2) + e^{i(k_1 + k_2)x_1}$$
$$\times \left[2i\delta(x_1 - x_2)(k_1 - k_2)([12, 12] - [12, 21] + [21, 12] - [21, 21])\right.$$
$$\left. -\delta'(x_1 - x_2)([12, 12] + [12, 21] - [21, 12] - [21, 21])\right]. \qquad (2.62)$$

To satisfy $\mathcal{H} f_{\uparrow\downarrow}(x_1, x_2) = E f_{\uparrow\downarrow}(x_1, x_2)$, we put

$$E = k_1^2 + k_2^2, \quad [12, 12] + [12, 21] - [21, 12] - [21, 21] = 0,$$
$$2i(k_1 - k_2)\big([12, 12] - [21, 12] + [12, 21] - [21, 21]\big)$$
$$+ 2c\big([12, 12] + [12, 21]\big) = 0. \tag{2.63}$$

From these equations we have

$$\begin{pmatrix} [12, 12] \\ [21, 12] \end{pmatrix} = \begin{pmatrix} u_{21} - 1 & u_{21} \\ u_{21} & u_{21} - 1 \end{pmatrix} \begin{pmatrix} [12, 21] \\ [21, 21] \end{pmatrix}, \quad u_{jl} \equiv \frac{k_j - k_l}{k_j - k_l + ic}. \tag{2.64}$$

If we write $\xi(12) = \begin{pmatrix} [12, 12] \\ [21, 12] \end{pmatrix}$ we have

$$\xi(12) = [(u_{21} - 1)I + u_{21}P_{12}]\xi(21), \tag{2.65}$$

where I and P_{12} are identity operator and pair-exchange operator, respectively. One solution is

$$\xi(12) = \begin{bmatrix} k_1 - \Lambda - ic' \\ -(k_2 - \Lambda + ic') \end{bmatrix}, \quad \xi(21) = \begin{bmatrix} -(k_2 - \Lambda - ic') \\ k_1 - \Lambda + ic' \end{bmatrix}, \quad c' \equiv c/2. \tag{2.66}$$

2.2.3 The three-body problem

In this case there are $3! \times 3!$ coefficients $[Q, P]$. The wave function is given by

$$f(x_1, x_2, x_3) = \sum_P [Q, P] \exp[i(k_{P1}x_{Q1} + k_{P2}x_{Q2} + k_{P3}x_{Q3})], \tag{2.67}$$

at $x_{Q1} \leq x_{Q2} \leq x_{Q3}$. We write ξ_P as a $3!(= 6)$ dimensional vector. The relation between $\xi(123)$ and $\xi(213)$ is $\xi(123) = Y_{21}^{12} \xi(213)$

$$\xi(123) = \begin{bmatrix} [123, 123] \\ [132, 123] \\ [321, 123] \\ [213, 123] \\ [231, 123] \\ [312, 123] \end{bmatrix}, \quad \xi(213) = \begin{bmatrix} [123, 213] \\ [132, 213] \\ [321, 213] \\ [213, 213] \\ [231, 213] \\ [312, 213] \end{bmatrix},$$

$$Y_{21}^{12} = \begin{bmatrix} u_{21} - 1 & 0 & 0 & u_{21} & 0 & 0 \\ 0 & u_{21} - 1 & 0 & 0 & 0 & u_{21} \\ 0 & 0 & u_{21} - 1 & 0 & u_{21} & 0 \\ u_{21} & 0 & 0 & u_{21} - 1 & 0 & 0 \\ 0 & 0 & u_{21} & 0 & u_{21} - 1 & 0 \\ 0 & u_{21} & 0 & 0 & 0 & u_{21} - 1 \end{bmatrix}.$$

The other relations are

$$\xi(123) = Y_{32}^{23}\xi(132), \quad \xi(132) = Y_{31}^{12}\xi(312), \quad \xi(321) = Y_{23}^{12}\xi(231),$$
$$\xi(321) = Y_{12}^{23}\xi(312), \quad \xi(213) = Y_{31}^{23}\xi(231). \tag{2.68}$$

The inverses of these relations are

$$\xi(213) = Y_{12}^{12}\xi(123), \quad \xi(132) = Y_{23}^{23}\xi(123), \quad \xi(312) = Y_{13}^{12}\xi(132),$$
$$\xi(231) = Y_{32}^{12}\xi(321), \quad \xi(312) = Y_{21}^{23}\xi(321), \quad \xi(231) = Y_{13}^{23}\xi(213). \tag{2.69}$$

Here

$$Y_{jl}^{ab} = (u_{jl} - 1)I + u_{jl}P_{ab}, u_{jl} = \frac{k_j - k_l}{k_j - k_l + ic} \tag{2.70}$$

and P_{ab} is the operator which exchanges elements a and b. Sometimes we write P_{ab} as (ab).

In general N the coefficients $[Q, P]$ should satisfy

$$[Q, P] = (u_{P(l+1)Pl} - 1)[Q, P(l, l+1)] + u_{P(l+1)Pl}[Q(l, l+1), P(l, l+1)]. \tag{2.71}$$

As $(l, l+1)$ is the exchange of l and $l+1$, $P(l, l+1)$ is a permutations as follows:

$$\begin{pmatrix} 1 & 2 & \dots & l & l+1 & \dots & N \\ P1 & P2 & \dots & P(l+1) & Pl & \dots & PN \end{pmatrix}.$$

In the same way $Q(l, l+1)$ is

$$\begin{pmatrix} 1 & 2 & \dots & l & l+1 & \dots & N \\ Q1 & Q2 & \dots & Q(l+1) & Ql & \dots & QN \end{pmatrix}.$$

To get $\xi(321)$ from $\xi(123)$ there are two paths,

$$\xi(123) \rightarrow \xi(213) \rightarrow \xi(231) \rightarrow \xi(321),$$

$$\xi(123) \rightarrow \xi(132) \rightarrow \xi(312) \rightarrow \xi(321).$$

For the two paths to be equivalent, the Ys must satisfy

$$Y_{23}^{12}Y_{13}^{23}Y_{12}^{12} = Y_{12}^{23}Y_{13}^{12}Y_{23}^{23}. \tag{2.72}$$

The left-hand side is

$$[(u_{23} - 1)(u_{13} - 1)(u_{12} - 1) + (u_{13} - 1)u_{23}u_{12}]I$$
$$+[u_{23}(u_{13} - 1)(u_{12} - 1) + u_{12}(u_{23} - 1)(u_{13} - 1)]P_{12}$$
$$+(u_{23} - 1)u_{13}(u_{12} - 1)P_{23} + u_{23}u_{13}u_{12}P_{13}$$
$$+(u_{23} - 1)u_{13}u_{12}P_{23}P_{12} + u_{23}u_{13}(u_{12} - 1)P_{12}P_{23}. \tag{2.73}$$

The right-hand side is

$$[(u_{12} - 1)(u_{13} - 1)(u_{23} - 1) + (u_{13} - 1)u_{12}u_{23}]I$$
$$+[(u_{12} - 1)(u_{13} - 1)u_{23} + u_{12}(u_{13} - 1)(u_{23} - 1)]P_{23}$$
$$+(u_{12} - 1)(u_{23} - 1)u_{13}P_{12} + u_{12}u_{13}u_{23}P_{13}$$
$$+u_{12}u_{13}(u_{23} - 1)P_{23}P_{12} + (u_{12} - 1)u_{13}u_{23}P_{12}P_{23}. \qquad (2.74)$$

Here we have used $P_{12}P_{23}P_{12} = P_{23}P_{12}P_{23} = P_{13}$ and $P_{12}^2 = I$. It is apparent that the coefficients of $I, P_{13}, P_{23}P_{12}, P_{12}P_{23}$ are the same. We can show that $u_{23}(u_{13}-1)(u_{12}-1)+u_{12}(u_{23}-1)(u_{13}-1) = u_{12}u_{13}(u_{23}-1)$ from the definition of u_{jl}. The coefficients of $P(12)$ and $P(23)$ coincide in the l.h.s. and r.h.s. Thus we have proved the identity (2.72). Moreover $\xi(123) \to \xi(213) \to \xi(123)$ requires that

$$Y_{jl}^{a,a+1} Y_{lj}^{a,a+1} = I. \qquad (2.75)$$

The l.h.s. is

$$[(u_{jl} - 1)(u_{lj} - 1) + u_{jl}u_{lj}]I + [(u_{jl} - 1)u_{lj} + u_{jl}(u_{lj} - 1)]P_{jl}.$$

One can show that $(u_{jl}-1)(u_{lj}-1)+u_{jl}u_{lj} = 1$ and $(u_{jl}-1)u_{lj}+u_{jl}(u_{lj}-1) = 0$. Thus if $\xi(123)$ is given, we can construct all the ξ_P consistently.

We assume that particles 1 and 2 have spin up and particle 3 has spin down. Then $f(x_1, x_2, x_3) = -f(x_2, x_1, x_3)$ and

$$[123, P] = -[213, P], [321, P] = -[312, P], [132, P] = -[231, P]. \qquad (2.76)$$

Therefore we should determine 6×3 coefficients

$$\Phi(1; P) = \epsilon(P)[312, P], \quad \Phi(2; P) = \epsilon(P)[321, P], \quad \Phi(3; P) = \epsilon(P)[123, P]. \qquad (2.77)$$

The relations among the coefficients are

$$\Phi(y; P) = \Phi(y; P(l, l + 1)), \quad \text{for} \quad y \neq l, l + 1,$$

$$\Phi(l; P) = (1 - u_{Pl,P(l+1)})\Phi(l; P(l, l + 1)) + u_{Pl,P(l+1)}\Phi(l + 1; P(l, l + 1)),$$

$$\Phi(l+1; P) = (1 - u_{Pl,P(l+1)})\Phi(l+1; P(l, l+1)) + u_{Pl,P(l+1)}\Phi(l; P(l, l+1)). \qquad (2.78)$$

A solution of these equations is

$$\Phi(1; P) = D(k_{P2} - \Lambda - ic')(k_{P3} - \Lambda - ic'),$$
$$\Phi(2; P) = D(k_{P1} - \Lambda + ic')(k_{P3} - \Lambda - ic'),$$
$$\Phi(3; P) = D(k_{P1} - \Lambda + ic')(k_{P2} - \Lambda + ic'). \qquad (2.79)$$

2.2.4 *The M = 1 and arbitrary N case*

The solution of $N - 1$ up-spin electrons and 1 down-spin electron was solved by McGuire[66,67],

$$\Phi(y; P) = DF_P(y, \Lambda), \tag{2.80}$$

$$F_P(y, \Lambda) \equiv \prod_{j=1}^{y-1}(k_{Pj} - \Lambda + ic') \prod_{l=y+1}^{N}(k_{Pl} - \Lambda - ic'). \tag{2.81}$$

The case $M = 2$ was solved by Flicker and Lieb[26].

2.2.5 *The arbitrary M and N case*

Later Gaudin[29] and Yang[117] derived the solution for a general number of down-spins. Assume that there are M down-spins in an N-particle system. The relations of $\Phi(y_1, y_2, ..., y_M; P)$ are as follows:

$$\Phi(y_1, y_2, ..., y_M; P) = \Phi(y_1, y_2, ..., y_M; P(l, l + 1)),$$
$$\text{for} \quad y_\alpha \neq l, l + 1,$$
$$\Phi(y_1, y_2, ..., y_M; P) = (1 - u_{Pl,P(l+1)})\Phi(y_1, ..., y_\alpha, ..., y_M; P(l, l + 1))$$
$$+ u_{Pl,P(l+1)}\Phi(y_1, ..., y_\alpha + 1, ..., y_M; P(l, l + 1)),$$
$$\text{for} \quad y_\alpha = l \quad \text{and} \quad y_{\alpha+1} \neq l + 1$$
$$\Phi(y_1, y_2, ..., y_M; P) = (1 - u_{Pl,P(l+1)})\Phi(y_1, ..., y_\alpha, ..., y_M; P(l, l + 1))$$
$$+ u_{Pl,P(l+1)}\Phi(y_1, ..., y_\alpha - 1, ..., y_M; P(l, l + 1)),$$
$$\text{for} \quad y_\alpha = l + 1 \quad \text{and} \quad y_{\alpha-1} \neq l,$$
$$\Phi(y_1, y_2, ..., y_M; P) = \Phi(y_1, y_2, ..., y_M; P(l, l + 1)),$$
$$\text{for} \quad y_\alpha = l \quad \text{and} \quad y_{\alpha+1} = l + 1.$$

Regarding $y_1, y_2, ..., y_M$ as particle coordinates on the lattice with N sites, we assume the generalized Bethe-ansatz,

$$\Phi(y_1, ..., y_M; P) = \sum_R A(R)F_P(\Lambda_{R1}, y_1)F_P(\Lambda_{R2}, y_2)...F_P(\Lambda_{RM}, y_M). \tag{2.82}$$

Here the R s are permutations of $1, 2, ..., M$,

$$A(R)F_P(\Lambda_{R\alpha}, y_\alpha)F_P(\Lambda_{R(\alpha+1)}, y_\alpha + 1)$$
$$+ A(R')F_P(\Lambda_{R(\alpha+1)}, y_\alpha)F_P(\Lambda_{R\alpha}, y_\alpha + 1)$$
$$= A(R)F_{P'}(\Lambda_{R\alpha}, y_\alpha)F_{P'}(\Lambda_{R(\alpha+1)}, y_\alpha + 1)$$
$$+ A(R')F_{P'}(\Lambda_{R(\alpha+1)}, y_\alpha)F_{P'}(\Lambda_{R\alpha}, y_\alpha + 1), \tag{2.83}$$

and we put $P' \equiv P(y_\alpha, y_\alpha + 1)$, $R' \equiv R(\alpha, \alpha + 1)$. Thus we have

$$A(R)(\Lambda_{R(\alpha+1)} - \Lambda_{R\alpha} - ic) + A(R(\alpha, \alpha + 1))(\Lambda_{R\alpha} - \Lambda_{R(\alpha+1)} - ic) = 0. \quad (2.84)$$

This is satisfied if

$$A(R) = \epsilon(R) \prod_{j<l} (\Lambda_{Rj} - \Lambda_{Rl} - ic). \quad (2.85)$$

Thus we can construct a wave function which satisfies $\mathcal{H}\Psi = E\Psi$ for arbitrary set of numbers $k_1, ..., k_N$ and $\Lambda_1, ..., \Lambda_M$.

2.2.6 Periodic boundary conditions

The periodic boundary condition requires

$$\xi(123) = P_{23}P_{12}\xi(231)\exp(ik_1 L), \quad \xi(132) = P_{23}P_{12}\xi(321)\exp(ik_1 L),$$

$$\xi(213) = P_{23}P_{12}\xi(132)\exp(ik_2 L), \quad \xi(231) = P_{23}P_{12}\xi(312)\exp(ik_2 L),$$

$$\xi(312) = P_{23}P_{12}\xi(123)\exp(ik_3 L), \quad \xi(321) = P_{23}P_{12}\xi(213)\exp(ik_3 L). \quad (2.86)$$

These six equations yield three equations for $\xi(123)$:

$$\exp(ik_1 L)\xi(123) = X_{21}X_{31}\xi(123), \quad \exp(ik_2 L)\xi(123) = X_{32}X_{12}\xi(123),$$

$$\exp(ik_3 L)\xi(123) = X_{13}X_{23}\xi(123), \quad X_{jl} \equiv P_{jl}Y_{jl}^{jl}. \quad (2.87)$$

Here we have used $P_{jl}^2 = I, P_{jk}P_{jl} = P_{jl}P_{lk}$. One can show that the three operators $X_{21}X_{31}, X_{32}X_{12}, X_{13}X_{23}$ commute with each other using

$$X_{ij}X_{ji} = I, \quad X_{jk}X_{ik}X_{ij}X_{kj}X_{ki}X_{ji} = I, \quad X_{ij}X_{kl} = X_{kl}X_{ij}, \quad (2.88)$$

where i, j, k, l are all unequal. Thus the three operators can be diagonalized simultaneously.

The periodic boundary condition for the wave function is

$$\psi(x_1, x_2, ..., x_N) = \psi(x_N - L, x_1, x_2, ..., x_{N-1}). \quad (2.89)$$

This is satisfied by

$$[Q1, Q2, ..., QN; P1, P2, ..., PN]\exp(ik_{PN}L)$$
$$= [QN, Q1, ..., Q(N-1); PN, P1, ..., P(N-1)]. \quad (2.90)$$

This is rewritten as

$$\Phi(y_1, y_2, ..., y_M; P)\exp(ik_{PN}L)$$
$$= \Phi(y_1 + 1, y_2 + 1, ..., y_M + 1; PN, P1, ..., P(N-1)), \quad (2.91)$$
$$\Phi(y_1, ..., y_{M-1}, N + 1; P) \equiv \Phi(1, y_1, y_2, ..., y_{M-1}; P). \quad (2.92)$$

Assume that $y_M \neq N$. Using (2.81) and (2.82) we have

$$\exp(ik_{PN}L) = \prod_{\alpha=1}^{M} \frac{k_{PN} - \Lambda_\alpha + ic'}{k_{PN} - \Lambda_\alpha - ic'}. \tag{2.93}$$

In the case $y_M = N$ we have

$$\Phi(y_1, ..., y_{M-1}, N; P) \prod_{\alpha=1}^{M} \frac{k_{PN} - \Lambda_\alpha + ic'}{k_{PN} - \Lambda_\alpha - ic'}$$
$$= \Phi(1, y_1 + 1, y_2 + 1, ..., y_{M-1} + 1; PN, P1, ..., P(N-1)), \tag{2.94}$$

by substituting (2.93) into (2.91). This is satisfied if

$$A(R) = A(RM, R1, ..., R(M-1)) \prod_{j=1}^{N} \left(\frac{k_{Pj} - \Lambda_{RM} - ic'}{k_{Pj} - \Lambda_{RM} + ic'} \right). \tag{2.95}$$

Substituting (2.85) we get

$$\prod_{j=1}^{N} \left(\frac{k_j - \Lambda_{RM} + ic'}{k_j - \Lambda_{RM} - ic'} \right) = -\prod_{j=1}^{M} \left(\frac{\Lambda_{RM} - \Lambda_j - ic}{\Lambda_{RM} - \Lambda_j + ic} \right). \tag{2.96}$$

Thus the periodic boundary condition is satisfied by

$$\exp(ik_j L) = \prod_{\alpha=1}^{M} \left(\frac{k_j - \Lambda_\alpha + ic'}{k_j - \Lambda_\alpha - ic'} \right), \quad j = 1, ..., N, \tag{2.97}$$

$$\prod_{j=1}^{N} \left(\frac{\Lambda_\alpha - k_j + ic'}{\Lambda_\alpha - k_j - ic'} \right) = -\prod_{\beta=1}^{M} \left(\frac{\Lambda_\alpha - \Lambda_\beta + ic}{\Lambda_\alpha - \Lambda_\beta - ic} \right), \quad \alpha = 1, ..., M. \tag{2.98}$$

These are $N + M$ coupled equations for $N + M$ unknowns $\{k_j\}$ and $\{\Lambda_\alpha\}$. Taking the logarithm of (2.97) and (2.98) we have

$$k_j L = 2\pi I_j - 2 \sum_{\alpha=1}^{M} \tan^{-1} \frac{k_j - \Lambda_\alpha}{c'}, \tag{2.99}$$

$$\sum_{j=1}^{N} 2 \tan^{-1} \frac{\Lambda_\alpha - k_j}{c'} = 2\pi J_\alpha + 2 \sum_{\beta=1}^{M} \tan^{-1} \frac{\Lambda_\alpha - \Lambda_\beta}{c}, \tag{2.100}$$

where I_j is an integer (half-odd integer) for even (odd) M, and J_α is an integer (half-odd integer) for odd (even) $N - M$. The total momentum is

$$K = \sum_{j=1}^{N} k_j = \frac{2\pi}{L} \left(\sum_j I_j + \sum_\alpha J_\alpha \right). \tag{2.101}$$

2.2.7 The ground state for $c > 0$

We consider the ground state in the case $c \to \infty$. In this limit the Λs are proportional to c. We denote $x_\alpha = \Lambda_\alpha / c'$. The ks remain finite. Then equations (2.99), (2.100) become

$$k_j L = 2\pi I_j + y_1 - \frac{k_j}{c} y_2,$$

$$y_1 = 2 \sum_\alpha \tan^{-1} x_\alpha, \quad y_2 = 2 \sum_\alpha \frac{1}{1 + x_\alpha^2},$$

$$2N \tan^{-1} x_\alpha = 2\pi J_\alpha + \sum_{\beta=1}^{M} 2 \tan^{-1} \frac{x_\alpha - x_\beta}{2}. \tag{2.102}$$

The total energy is

$$E = \sum_j k_j^2 = \sum_j \left(\frac{2\pi I_j}{L} + y_1 \right)^2 \left[1 - \frac{2 y_2}{c} + O(c^{-2}) \right]. \tag{2.103}$$

To minimize this in the limit of very large c for even N and odd M, I_j and J_α should be taken as follows,

$$I_j = (N+1)/2 - j, \quad J_\alpha = (M+1)/2 - \alpha. \tag{2.104}$$

In the case of even N and odd M the lowest energy state is unique. The determination of x_α is the same problem as the antiferromagnetic Heisenberg chain with N sites and M down-spins in section 3.1.

We put the distribution function of ks as $\rho(k)$ and that of Λs as $\sigma(\Lambda)$. From equations (2.102) and (2.104) we have

$$k = 2\pi \int^k \rho(t) dt - \int_{-B}^{B} \tan^{-1} \frac{2(k - \Lambda)}{c} \sigma(\Lambda) d\Lambda, \tag{2.105}$$

$$\int_{-Q}^{Q} 2 \tan^{-1} \frac{2(\Lambda - k)}{c} \rho(k) dk$$

$$= 2\pi \int^\Lambda \sigma(x) dx + \int_{-B}^{B} 2 \tan^{-1} \frac{\Lambda - \Lambda'}{c} \sigma(\Lambda') d\Lambda'. \tag{2.106}$$

Differentiating with respect to k and Λ we have

$$\rho(k) = \frac{1}{2\pi} + \int_{-B}^{B} \frac{2c\sigma(\Lambda) d\Lambda}{\pi(c^2 + 4(k - \Lambda)^2)}, \tag{2.107}$$

$$\sigma(\Lambda) + \int_{-B}^{B} \frac{c\sigma(\Lambda') d\Lambda'}{\pi(c^2 + (\Lambda - \Lambda')^2)} = \int_{-Q}^{Q} \frac{2c\rho(k) dk}{\pi(c^2 + 4(k - \Lambda)^2)}. \tag{2.108}$$

We put the energy, number of particles and number of down-spin particles per unit length as $e, n, n_\downarrow$,

$$e = \int_{-Q}^{Q} k^2 \rho(k) dk, \quad n = \int_{-Q}^{Q} \rho(k) dk, \quad n_\downarrow = \int_{-B}^{B} \sigma(\Lambda) d\Lambda. \tag{2.109}$$

In the case of $B = \infty$ equation (2.108) can be treated by the Fourier transformation,

$$\sigma(\Lambda) = \frac{1}{2c} \int_{-Q}^{Q} \text{sech}\left(\frac{\pi(k - \Lambda)}{c}\right) \rho(k) dk. \tag{2.110}$$

This is a special point and we have

$$n_\downarrow = \int_{-\infty}^{\infty} \sigma(\Lambda) d\Lambda = \frac{1}{2} \int_{-Q}^{Q} \rho(k) dk = n/2. \tag{2.111}$$

In the limit $c = 0+$ we have $\sigma(\Lambda) = \rho(\Lambda)/2 = (2\pi)^{-1}$ for $|\Lambda| < Q$. Then

$$e = \frac{2Q^3}{3\pi}, \quad n = \frac{2Q}{\pi}. \tag{2.112}$$

2.2.8 The ground state for $c < 0$

In the case that N is even and M is odd, the ground state at $c = 0+$, the solution of (2.93), becomes

$$\Lambda_\alpha = \frac{\pi}{L}(M + 1 - 2\alpha), \quad k_\alpha = k_{\alpha+M} = \Lambda_\alpha,$$

$$k_{2M+n} = -k_{N+1-n} = \frac{\pi}{L}(M - 1 + 2n). \tag{2.113}$$

Then at $c = 0$, two real ks and one Λ coincide. In the case $c = 0-$ these two ks become complex conjugates of each other. Λ remains on the real axis. In the limit of $L \to \infty$ these two ks and Λ form the following bound state,

$$k_\alpha = \Lambda_\alpha + \frac{c}{2}i + O(\exp(-\delta L)), \quad k_{M+\alpha} = \Lambda_\alpha - \frac{c}{2}i + O(\exp(-\delta L)). \tag{2.114}$$

Thus in the ground state $2M$ ks become complex and $N - 2M$ ks are real. From equation (2.93) we have

$$k_j = 2\pi I_j + 2 \sum_{\alpha=1}^{M} \tan^{-1}[2(k_j - \Lambda_\alpha)/|c|]; j = 2M + 1, ..., N. \tag{2.115}$$

For k_α, equation (2.93) is

$$\exp(ik_\alpha L) = \frac{k_\alpha - \Lambda_\alpha + ic'}{k_\alpha - \Lambda_\alpha - ic'} \prod_{\beta \neq \alpha} \left(\frac{\Lambda_\alpha - \Lambda_\beta + ic}{\Lambda_\alpha - \Lambda_\beta}\right).$$

The first term of the r.h.s. should be treated carefully. In the same way we have

$$\exp(ik_{\alpha+M}L) = \frac{k_\alpha - \Lambda_\alpha + ic'}{k_\alpha - \Lambda_\alpha - ic'} \prod_{\beta \neq \alpha} \left(\frac{\Lambda_\alpha - \Lambda_\beta}{\Lambda_\alpha - \Lambda_\beta - ic}\right).$$

The product of these two equations is

$$\exp(2i\Lambda_\alpha l) = \frac{k_\alpha - \Lambda_\alpha + ic'}{k_\alpha - \Lambda_\alpha - ic'} \frac{k_\alpha - \Lambda_\alpha + ic'}{k_\alpha - \Lambda_\alpha - ic'} \prod_{\beta \neq \alpha} \left(\frac{\Lambda_\alpha - \Lambda_\beta + ic}{\Lambda_\alpha - \Lambda_\beta - ic}\right).$$

From (2.98) we have

$$\frac{k_\alpha - \Lambda_\alpha + ic'}{k_\alpha - \Lambda_\alpha - ic'} \frac{k_{\alpha+M} - \Lambda_\alpha + ic'}{k_{\alpha+M} - \Lambda_\alpha - ic'} = \prod_{j=2M+1}^{N} \frac{k_j - \Lambda_\alpha - ic'}{k_j - \lambda_\alpha + ic'}.$$

Thus

$$\exp(2i\Lambda_\alpha l) = \prod_{j=2M+1}^{N} \frac{k_j - \Lambda_\alpha - ic'}{k_j - \lambda_\alpha + ic'} \prod_{\beta \neq \alpha} \left(\frac{\Lambda_\alpha - \Lambda_\beta + ic}{\Lambda_\alpha - \Lambda_\beta - ic}\right).$$

Taking the logarithm we have

$$2\Lambda_\alpha L = 2\pi J_\alpha + 2 \sum_{\beta=1}^{M} \tan^{-1} \frac{\Lambda_\alpha - \Lambda_\beta}{|c|} + 2 \sum_{j=1}^{N-2M} \tan^{-1} \frac{2(\Lambda_\alpha - k_j)}{|c|};$$

$$\alpha = 1, ..., M. \tag{2.116}$$

The total energy is given by

$$E = \sum_{\alpha=1}^{M} \left[\left(\Lambda_\alpha + \frac{ic}{2}\right)^2 + \left(\Lambda_\alpha - \frac{ic}{2}\right)^2\right] + \sum_{j=1}^{n-2M} k_j^2. \tag{2.117}$$

For the ground state of even N and odd M, the I_j s and the J_α s are

$$I_j = \frac{N - 2M - 1}{2}, \frac{N - 2M - 3}{2}, ..., -\frac{N - 2M - 1}{2}, \tag{2.118}$$

$$J_\alpha = \frac{M - 1}{2}, \frac{M - 3}{2}, ..., -\frac{M - 1}{2}. \tag{2.119}$$

We set the distribution function of the real ks as $\rho(k)$ and that of the Λs as $\sigma(\Lambda)$. We have equations for these functions:

$$\rho(k) = \frac{1}{2\pi} - \frac{1}{\pi} \int_{-B}^{B} \frac{2|c|\sigma(\Lambda)d\Lambda}{c^2 + 4(k - \Lambda)^2}, \tag{2.120}$$

$$\sigma(\Lambda) + \frac{1}{\pi} \int_{-B}^{B} \frac{|c|\sigma(\Lambda')d\Lambda'}{c^2 + (\Lambda - \Lambda')^2} = \frac{1}{\pi} - \frac{1}{\pi} \int_{-Q}^{Q} \frac{2|c|\rho(k)dk}{c^2 + 4(k - \Lambda)^2}. \tag{2.121}$$

The particle numbers and energy per unit length are

$$n = 2 \int_{-B}^{B} \sigma(\Lambda)d\Lambda + \int_{-Q}^{Q} \rho(k)dk,$$

$$n_\uparrow - n_\downarrow = \int_{-Q}^{Q} \rho(k)dk,$$

$$e = \int_{-B}^{B} (2\Lambda^2 - \frac{c^2}{2})\sigma(\Lambda)d\Lambda + \int_{-Q}^{Q} \rho(k)dk. \tag{2.122}$$

It should be noted that one Λ and two ks form a bound state of two fermions with binding energy $c^2/2$. In contrast with an attractive boson system, there are no bound states of more than two fermions. The system has a thermodynamic limit.

2.2.9 *Expansion from small* $n_\downarrow$

We consider the small B limit of equations (2.107) and (2.108). We define the function $f(n_\uparrow, n_\downarrow; c')$ as the internal energy per unit length as a function of $c' = c/2$. Using the small B expansion we obtain at $c > 0$

$$f(n - x, x; c') = \frac{\pi^2 n^3}{3} - x\left\{2\left(\frac{c'^2}{\pi} + \pi n^2\right) \tan^{-1} \frac{\pi n}{c'} - 2c'n\right\}$$

$$+ 4x^2 n\left\{\tan^{-1} \frac{\pi n}{c'}\right\}^2 + O(x^3). \tag{2.123}$$

The first order term is obtained by McGuire[66, 67] and Yang[117]. the quadratic term is obtained by Suzuki[87]. For $c < 0$ we have from (2.120), (2.121) and (2.122)

$$f(n - x, x; c') = \frac{\pi^2 n^3}{3} - x\left\{2\left(\frac{c'^2}{\pi} + \pi n^2\right)(\pi + \tan^{-1} \frac{\pi n}{c'}) - 2c'n\right\}$$

$$+ 4x^2 n\left\{\pi + \tan^{-1} \frac{\pi n}{c'}\right\}^2 + O(x^3). \tag{2.124}$$

These expansions can be written in a unified form as follows:

$$f(n - x, x; c') = \frac{\pi^2 n^3}{3} - x\left\{2\left(\frac{c'^2}{\pi} + \pi n^2\right)\left(\frac{\pi}{2} - \tan^{-1} \frac{c'}{\pi n}\right) - 2c'n\right\}$$

$$+ 4x^2 n\left\{\frac{\pi}{2} - \tan^{-1} \frac{c'}{\pi n}\right\}^2 + O(x^3). \tag{2.125}$$

The saturation magnetic field is given by

$$h_s = \left(\frac{c'^2}{\pi} + \pi n^2\right)\left(\frac{\pi}{2} - \tan^{-1} \frac{c'}{\pi n}\right) - cn'. \tag{2.126}$$

As a function of c' this is analytic even at $c' = 0$, notwithstanding that the two sets of equations seem to be quite different. We can construct a unified integral equation which analytically continues from $c > 0$ to $c < 0$. Then we can show that the internal energy $f(n_\uparrow, n_\downarrow; c)$ is analytic on the real axis of c when $n_\uparrow \neq n_\downarrow$.

2.2.10 A unified form of the integral equations

We can eliminate $\rho(k)$ from the integral equations for (2.107)–(2.108) or (2.120)–(2.121),

$$\sigma(\Lambda) + \int_{-B}^{B} K(\Lambda, \Lambda')\sigma(\Lambda')d\Lambda' = \frac{1}{2\pi}g(\Lambda), \tag{2.127}$$

with

$$K(\Lambda, \Lambda') = \frac{c^2}{\pi^2} \int_{|x|>Q} \frac{dx}{(c^2 + (\Lambda - x)^2)(c^2 + (x - \Lambda')^2)},$$

$$g(\Lambda) = 1 - \frac{1}{\pi}\left(\tan^{-1}\frac{c'}{Q+\Lambda} + \tan^{-1}\frac{c'}{Q-\Lambda}\right). \tag{2.128}$$

The internal energy density f, particle density n and $n_\downarrow$ are

$$f = \frac{Q^3}{3\pi} + \int_{-B}^{B} \frac{d\Lambda}{\pi}\left\{2Qc' - \Lambda c \log\frac{c'^2 + (Q+\lambda)^2}{c'^2 + (Q-\Lambda)^2}\right.$$
$$\left. + \pi(\Lambda^2 - c'^2)g(\Lambda)\right\}\sigma(\Lambda), \tag{2.129}$$

$$n = \frac{Q}{\pi} + \int_{-B}^{B} d\Lambda g(\Lambda)\sigma(\lambda), \tag{2.130}$$

$$n_\downarrow = \int_{-B}^{B} \sigma(\Lambda)d\Lambda. \tag{2.131}$$

As far as $B < Q$, the integration kernel K and inhomogeneous term analytically continue from $c' > 0$ to $c' < 0$. Then if $n_\uparrow \neq n_\downarrow$, f also continues from $c > 0$ to $c < 0$. This means that $f(n, n - x, c')$ is expanded by a Taylor series of c', and that series has a finite convergence radius for $x \neq n/2$.

2.3 Bosons and fermions with arbitrary spin

Sutherland showed that the spin $1/2$ fermion theory can be generalized to fermions with larger spins[84] for the repulsive case ($c > 0$). He obtained integral equations for the distribution. The author investigated the case of attractive fermions with arbitrary spin[92]. Lai and Yang investigated the mixture of spin $1/2$ fermions and spin zero bosons with the same mass and repulsive interaction[57]. In the attractive case of bosons, there is no thermodynamic limit.

3

The isotropic Heisenberg model

3.1 The ferromagnetic case

The Heisenberg model was the first model to be treated by the method of the Bethe-ansatz[13]. In the beginning of the 1930s only the ferromagnetic model was considered:

$$\mathscr{H} = -J \sum_{l=1}^{N} S_l^x S_{l+1}^x + S_l^y S_{l+1}^y + S_l^z S_{l+1}^z - 2h \sum_{l=1}^{} S_l^z,$$

$$h \geq 0, \quad J > 0, \quad \mathbf{S}_{N+1} \equiv \mathbf{S}_1. \tag{3.1}$$

This Hamiltonian is defined on a 2^N dimensional space, but the space is classified by the total $S^z = \sum S_l^z$. The ground state is the state where all spins are up and the total S^z is $N/2$,

$$\mathscr{H}|0\rangle = E_0|0\rangle, \quad E_0 = -JN/4 - Nh. \tag{3.2}$$

We write eigenfunctions at $S^z = N/2 - M$ as follows:

$$\Psi = \sum f(n_1, n_2, ..., n_M) S_{n_1}^- S_{n_2}^- ... S_{n_M}^- |0\rangle, \tag{3.3}$$

where $1 \leq n_1 < n_2 < ... < n_M \leq N$ and $2M \leq N$. In terms of $f(n_1, n_2, ..., n_M)$ the eigenvalue equation is

$$-\frac{J}{2} \sum_j (1 - \delta_{n_j+1, n_{j+1}}) \Big\{ f(n_1, ..., n_j + 1, n_{j+1}, ..., n_M)$$

$$+ f(n_1, ..., n_j, n_{j+1} - 1, ..., n_m) \Big\}$$

$$+ \Big\{ E_0 - E + (J + 2h)M - J \sum_j \delta_{n_j+1, n_{j+1}} \Big\} f(n_1, n_2, ..., n_M) = 0. \tag{3.4}$$

Next we assume that the wave function is written as follows:

$$f(n_1, n_2, ..., n_M) = \sum_P^{M!} A(P) \exp(i \sum_{j=1}^{M} k_{Pj} n_j),$$ (3.5)

where P is a permutation of $1, 2, ..., M$,

$$P = \begin{pmatrix} 1, & 2, & ..., & M \\ P1, & P2, & ..., & PM \end{pmatrix}.$$ (3.6)

$k_1, k_2, ..., k_M$ are called quasi-momenta. As $2M \leq N$ there is at least one set of $(n_1, n_2, ..., n_M)$ which satisfies $n_j + 1 < n_{j+1}$. Then we have

$$E = E_0 + \sum_{j=1}^{M} [J(1 - \cos k_j) + h].$$ (3.7)

The situation is almost the same as for δ function bosons, the difference is that here the space is discrete. The shift operator T is defined as follows,

$$Tf(n_1, n_2, ..., n_M) = f(n_1 + 1, n_2 + 1, ..., n_M + 1).$$ (3.8)

This wave function is also an eigenstate of the shift operator $T\Psi = e^{iK}\Psi$. We call K is the total momentum, which is given by

$$K = \sum_{j=1}^{M} k_j.$$ (3.9)

Next we consider the case $n_j + 1 = n_{j+1}$ and $n_l + 1 < n_{l+1}$. f should satisfy

$$f(n_1, ..., n_j + 1, n_j + 1, n_{j+2}, ..., n_M)$$
$$+ f(n_1, ..., n_j, n_j, n_{j+2}, ..., n_M)$$
$$- 2f(n_1, ..., n_j, n_{j+1}, n_{j+2}, ..., n_M) = 0.$$ (3.10)

This is satisfied if

$$0 = A(P)(e^{ik_{Pj}} + e^{-ik_{P(j+1)}} - 2)e^{ik_{P(j+1)}}$$
$$+ A(P(j, j+1))(e^{ik_{P(j+1)}} + e^{-ik_{Pj}} - 2)e^{ik_{Pj}}.$$ (3.11)

There are $N!(N-1)/2$ equations. These equation are satisfied if we put

$$A(P) = \epsilon(P) \prod_{l<j} (e^{i(k_{Pl}+k_{Pj})} + 1 - 2e^{ik_{Pl}}).$$ (3.12)

The periodic boundary condition requires:

$$f(n_1, n_2, ..., n_{M-1}, N+1) = f(1, n_1, n_2, ..., n_{M-1}).$$ (3.13)

This is satisfied if the following condition stands,

$$\exp(ik_{PM}N)A(P1, P2, ..., PM) = A(PM, P1, ..., P(M-1)). \tag{3.14}$$

This is equivalent to

$$\exp(ik_j N) = (-1)^{M-1} \prod_{l \neq j} \frac{\exp[i(k_j + k_l)] + 1 - 2\exp(ik_j)}{\exp[i(k_j + k_l)] + 1 - 2\exp(ik_l)},$$

$$j = 1, 2, ..., M. \tag{3.15}$$

This is a set of complicated coupled equations for M unknowns. If we have a solution of this set, we have one eigenstate and its energy eigenvalue and total momentum. We introduce rapidity parameters $x_j = \cot(k_j/2)$. In terms of the new parameters

$$\exp(ik_j) = \left(\frac{x_j + i}{x_j - i}\right) \tag{3.16}$$

and equation (3.15) becomes

$$\left(\frac{x_j + i}{x_j - i}\right)^N = \prod_{l \neq j}\left(\frac{x_j - x_l + 2i}{x_j - x_l - 2i}\right). \tag{3.17}$$

The r.h.s. is a function of the difference of the x_js. The wave function is

$$f(n_1, n_2, ..., n_M) = \sum_P A(P) \prod_{j=1}^{M} \left(\frac{x_{Pj} + i}{x_{Pj} - i}\right)^{n_j},$$

$$A(P) = D\epsilon(P) \prod_{j<l}(x_{Pj} - x_{Pl} - 2i). \tag{3.18}$$

The energy and momentum are

$$E = E_0 + \sum_{j=1}^{M}\left[\frac{2J}{x_j^2 + 1} + 2h\right], \quad e^{iK} = \prod_{j=1}^{M}\left(\frac{x_j + i}{x_j - i}\right), \tag{3.19}$$

$$K \equiv \pi M - 2\sum_{k=1}^{M} \tan^{-1} x_j \quad \mathrm{mod}(2\pi). \tag{3.20}$$

In the lattice problem the momenta K and $K + 2\pi \times$ integer are equivalent. This point is different from the problems of chapter 2 in the continuum space.

3.2 The string solution of an infinite system

In the $M = 1$ case, the elementary excitation is

$$E - E_0 = J(1 - \cos K) + 2h. \tag{3.21}$$

This is the spin wave excitation of the ferromagnetic Heisenberg model. Bethe[13] found that the bound state of spin waves exists in $M = n$ subspace,

$$x_j = \alpha + i(n + 1 - 2j), j = 1, 2, ..., n. \tag{3.22}$$

The energy and momentum of this excitation are

$$E = E_0 + \frac{2nJ}{\alpha^2 + n^2} + 2nh, \quad K = \frac{1}{i} \ln\left(\frac{\alpha + in}{\alpha - in}\right). \tag{3.23}$$

Thus the dispersion of these excitation is

$$E = E_0 + \frac{J}{n}(1 - \cos K) + 2nh. \tag{3.24}$$

These string solutions play an essential role in the thermodynamics of soluble models.

3.3 The Hulthen solution for an antiferromagnet

The antiferromagnetic case is the $J < 0$ case of equation (3.1). In actual materials, the ferromagnetic case is rare. Typical one-dimensional magnetic substances are antiferromagnetic. From equation (3.17) we have

$$2N \tan^{-1} x_j = 2\pi I_j + 2 \sum_{l=1}^{M} \tan^{-1} \frac{x_j - x_l}{2}, \tag{3.25}$$

where I_j is an integer (half-odd integer) for odd (even) $N - M$. The total momentum is given by:

$$K = \pi(1 - (-1)^M)/2 - \frac{2\pi}{N} \sum_j I_j.$$

One can show that for even N the lowest energy state in the subspace of fixed total S_z is given by

$$I_j = (M + 1 - 2j)/2, \quad j = 1, 2, ..., M. \tag{3.26}$$

In the thermodynamic limit the x_js distribute from $-B$ to B. From equation (3.25) we have:

$$\tan^{-1} x = \pi \int_{-B}^{x} \rho(t)dt + \int_{-B}^{B} \tan^{-1} \frac{x - y}{2} \rho(y)dy. \tag{3.27}$$

Differentiating with respect to x we have

$$\rho(x) = \frac{1}{\pi}\frac{1}{x^2+1} - \int_{-B}^{B}\frac{1}{\pi}\frac{2}{(x-y)^2+4}\rho(y)\mathrm{d}y. \tag{3.28}$$

The energy and magnetization per site are:

$$\frac{E}{N} = \frac{|J|}{4} - h + \int_{-B}^{B}\Big[2h - \frac{2|J|}{x^2+1}\Big]\rho(x)\mathrm{d}x, \tag{3.29}$$

$$\frac{S_z}{N} = \frac{1}{2} - \int_{-B}^{B}\rho(x)\mathrm{d}x. \tag{3.30}$$

This integral equation can be solved in the case of infinite B. We define the Fourier transform of $\rho(x)$ as follows:

$$\tilde{\rho}(\omega) = \int_{-\infty}^{\infty}e^{-i\omega x}\rho(x)\mathrm{d}x. \tag{3.31}$$

Using the formula $\int \pi^{-1}n/(x^2+n^2)\exp(-ix\omega)\mathrm{d}x = \exp(-n|\omega|)$ we can rewrite (3.28) as follows:

$$\rho(\omega)(1 + e^{-2|\omega|}) = e^{-|\omega|}.$$

Then we have $\tilde{\rho}(\omega) = 1/(2\cosh\omega)$ and

$$\rho(x) = \frac{1}{2\pi}\int_{-\infty}^{\infty}e^{i\omega x}\tilde{\rho}(\omega)\mathrm{d}\omega = \frac{1}{4}\mathrm{sech}\Big(\frac{\pi x}{2}\Big). \tag{3.32}$$

Substituting this into (3.30) we have:

$$e = -|J|\Big(\ln 2 - \frac{1}{4}\Big) = -0.443147|J|, \quad s_z = 0. \tag{3.33}$$

Then we find that the case $B = \infty$ is the true ground state at $h = 0$ and that the magnetization is zero. Thus we have an analytical result for a one-dimensional antiferromagnet (Hulthen[39]). The first neighbour correlation is derived from this result

$$\langle S_i^z S_{i+1}^z \rangle = \frac{1}{12}(1 - 4\ln 2) = -0.14771573. \tag{3.34}$$

An analytic expression of the second neighbour correlation function is also known for this model (Takahashi[101]),

$$\langle S_i^z S_{i+2}^z \rangle = \frac{1}{12}(1 - 16\ln 2 + 9\zeta(3)) = 0.06067977. \tag{3.35}$$

3.4 The des Cloizeaux–Pearson mode of an antiferromagnet

We consider excitations from Hulthen's ground state at $M = N/2$, where the Is are

$$-\frac{N'-1}{2}, -\frac{N'-3}{2}, ..., \frac{N'-1}{2}, \quad N' = N/2. \tag{3.36}$$

It is difficult to find a real solution of ks for $S_{\text{total}} = 0$, except for the ground state. One possibility is the following state in $S_{\text{total}} = 1$ subspace

$$\frac{N'}{2}, \frac{N'}{2} - 1, ..., \frac{N'}{2} - r + 1, \frac{N'}{2} - r - 1,$$
$$..., \frac{N'}{2} - s + 1, \frac{N'}{2} - s - 1, -\frac{N'}{2}. \tag{3.37}$$

We put $0 \leq r < s \leq N'$. In this case all $M - 1$ ks can be real numbers. The total momentum is $\pi(1 - M)/2 + 2\pi(N' - r - s)/N$. Then the change of total momentum is $\Delta K = 2\pi(r + s)/N$. This state has two holes at positions x_r and x_s. In the thermodynamic limit we have

$$\Delta K = 2\pi \left(\int_{x_r}^{\infty} \rho(x) dx + \int_{x_s}^{\infty} \rho(x) dx \right)$$
$$= 2 \tan^{-1}(e^{-\pi x_r/2}) + 2 \tan^{-1}(e^{-\pi x_s/2}). \tag{3.38}$$

The change in energy is not $2|J|/(1 + x_r^2) + 2|J|/(1 + x_s^2)$ because the other x_js move of the order of N^{-1}. To get the change of total energy one must calculate the sum of the small shifts of the other ks. This is called the back-flow effect. From equation (3.25) we have

$$2N(\tan^{-1}(x_j + \Delta x_j) - \tan^{-1} x_j)$$
$$-2 \sum_{m=1}^{N'-1} \tan^{-1}\left(\frac{x_j + \Delta x_j - x_m - \Delta x_m}{2}\right) - \tan^{-1}\left(\frac{x_j - x_m}{2}\right)$$
$$= \pi[\text{sign}(x_j - x_r) + \text{sign}(x_j - x_s) - 1] - 2 \tan^{-1} \frac{x_j - x_{N'}}{2}. \tag{3.39}$$

The last term of the r.h.s. is replaced by $-\pi$ because $x_{N'} = -\infty$. From this we have

$$\frac{2\Delta x_j}{x_j^2 + 1} - \frac{1}{N} \sum_{m=1}^{N'-1} \frac{4(\Delta x_j - \Delta x_m)}{(x_j - x_m)^2 + 4} = \frac{\pi}{N}[\text{sign}(x_j - x_r) + \text{sign}(x_j - x_s) - 2]. \tag{3.40}$$

If we put $N\Delta x \rho(x) = J(x)$ this equation becomes

$$\frac{2}{x^2+1}\frac{J(x)}{\rho(x)} - \frac{J(x)}{\rho(x)}\int_{-\infty}^{\infty}\frac{4\rho(y)dy}{(x-y)^2+4} + \int_{-\infty}^{\infty}\frac{4J(y)dy}{(x-y)^2+4}$$
$$= \pi(\text{sign}(x-x_r) + \text{sign}(x-x_s) - 2). \qquad (3.41)$$

Using (3.28) we have

$$J(x) + \frac{1}{\pi}\int_{-\infty}^{\infty}\frac{2}{(x-y)^2+4}J(y)dy = \int_x^{\infty}[-\delta(u-x_r) - \delta(u-x_s)]du. \quad (3.42)$$

This is the integral equation for the back flow $J(x)$. The analytic solution of this equation is

$$J(x) = \int_x^{\infty}[-\delta(u-x_r) + R(u-x_r) - \delta(u-x_s) + R(u-x_s)]du,$$

$$R(x) \equiv \frac{1}{\pi}\sum_{n=1}^{\infty}(-1)^{n-1}\frac{2n}{x^2+4n^2}. \qquad (3.43)$$

Thus the change of energy ΔE is

$$\Delta E = 2|J|\left[\frac{1}{x_{N'}^2+1} - \int_{-\infty}^{\infty}\frac{d}{dx}\frac{1}{x^2+1}J(x)dx\right].$$

By partial differentiation,

$$\Delta E = 2|J|\int_{-\infty}^{\infty}\frac{\delta(x-x_r) + \delta(x-x_s) - R(x-x_r) - R(x-x_s)}{x^2+1}dx$$

$$= \frac{\pi|J|}{2}\left(\text{sech}\frac{\pi x_r}{2} + \text{sech}\frac{\pi x_s}{2}\right) = 2|J|\pi(\rho(x_r) + \rho(x_s)). \qquad (3.44)$$

Putting $q_1 = 2\tan^{-1}e^{-\pi x_r/2}, q_2 = 2\tan^{-1}e^{-\pi x_s/2}$, we have

$$\Delta E = \frac{\pi|J|}{2}(\sin q_1 + \sin q_2), \quad \Delta K = q_1 + q_2. \qquad (3.45)$$

This spectrum is shown in Fig. (3.1). Thus $S_{\text{total}} = 1$ states with real ks are composites of two excitations. One spinon with momentum q has energy $(\pi|J|/2)\sin q$ and spin $1/2$. Actually the spinon number must be even. Des Cloizeaux and Pearson[19] obtained analytic expressions for these excitations by putting one spinon at zero energy $q_1 = 0$ or π,

$$\Delta E = \frac{\pi|J|}{2}|\sin K|.$$

This is the lowest energy state with momentum K. This is called the des Cloizeaux–Pearson mode. This means that the ground state has excitations with an infinitesimal energy gap. We find the velocity v_s of this elementary excitation at low momentum,

$$v_s = \lim_{K\to 0} d\Delta E/dK = \pi|J|/2. \qquad (3.46)$$

(E-E0)/J

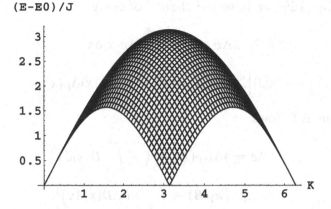

Fig. 3.1. Spinon excitation of XXX model at $T = 0$ and $h = 0$.

According to the linear spin-wave theory (Anderson[2], Kubo[54]) the excitation spectrum is

$$\Delta E = |J||\sin K|. \tag{3.47}$$

Then spin-wave theory gives a slower velocity than the exact Bethe-ansatz solution. But one should note that the coincidence of spinon lower edge and spin wave spectrum is very accidental. In a one-dimensional XXX antiferromagnet there is no Neel order even at zero temperature.

3.5 The magnetic susceptibility and magnetization curve for $J < 0$

In a sufficiently strong magnetic field the ground state is completely magnetized. The energy of the ground state is $E_0 = N(\frac{|J|}{4} - h)$, and the energy of an excited state is

$$E = E_0 + [2h - |J|(1 - \cos k)]. \tag{3.48}$$

The minimum of this energy is $2h - 2|J|$ at $k = \pi$. Then at $h < |J|$ the completely magnetized state cannot be the lowest energy state. To minimize the energy we should have $\partial e/\partial B = 0$ for equation (3.29). We must consider the change $B \to B + \Delta B$ and the change $\rho(x) \to \rho(x) + \Delta\rho(x)$. Then the equation for $\Delta\rho(x)$ is

$$\Delta\rho(x) + \int_{-B}^{B} a_2(x - y)\Delta\rho(y)\mathrm{d}y$$
$$= -(a_2(x - B) + a_2(x + B))\rho(B)\Delta B,$$
$$a_n(x) \equiv \frac{1}{\pi}\frac{n}{x^2 + n^2}. \tag{3.49}$$

From equation (3.29) we have the change of energy

$$\Delta e = 2h\left[2\Delta B\rho(B) + \int_{-B}^{B}\Delta\rho(x)dx\right]$$

$$-2\pi|J|\left[2\Delta B\rho(B)a_1(B) + \int_{-B}^{B}a_1(x)\Delta\rho(x)dx\right]. \tag{3.50}$$

This is written as follows,

$$\Delta e = 4\Delta B\rho(B)\left[h\left(1 - \int_{-B}^{B}D(x)dx\right)\right.$$

$$\left. -\pi|J|\left(a_1(B) - \int_{-B}^{B}a_1(x)D(x)dx\right)\right], \tag{3.51}$$

where $D(x)$ is defined as the solution of the equation

$$D(x) + \int_{-B}^{B}a_2(x-y)D(y)dy = a_2(x-B). \tag{3.52}$$

The expression inside the second parenthesis of the r.h.s. of (3.51) is $\rho(B)$. The expression inside the first parenthesis is $F(B)$, where $F(x)$ is defined as the solution of the equation

$$F(x) + \int_{-B}^{B}a_2(x-y)F(y)dy = 1. \tag{3.53}$$

The magnetic field h for a given B is

$$h = |J|\frac{\pi\rho(B)}{F(B)}. \tag{3.54}$$

The magnetization is given by (3.30). Thus one can plot the magnetization curve as a function of magnetic field by changing B from 0 to ∞ (Griffiths[32]). See Fig. (3.2a).

In a small magnetic field the energy becomes minimum at large B. The equation (3.28) can be transformed as follows

$$\rho(x) + \int_{-\infty}^{\infty}a_2(x-y)\rho(y)dy = a_1(x) + \int_{|y|>B}a_2(x-y)\rho(y)dy. \tag{3.55}$$

$$\rho(x) = \rho_0(x) + \int_{|y|>B}R(x-y)\rho(y)dy, \quad \rho_0(x) \equiv \frac{1}{4}\mathrm{sech}(\frac{\pi x}{2}). \tag{3.56}$$

$\rho_0(x)$ is the solution at $B = \infty$ or $h = 0$. Substituting this into (3.30) we have

$$s_z = \frac{1}{2} - \int_{-\infty}^{\infty}\rho(y)dy + \int_{|y|>B}\rho(y)dy = \frac{1}{2}\int_{|y|>B}\rho(y)dy. \tag{3.57}$$

Here we have used $\int \rho_0(x)dx = \int R(x)dx = 1/2$. The energy change is

$$
\begin{aligned}
e - e_0 &= -2hs_z + 2\pi|J|\left[\int_{-\infty}^{\infty} a_1(x)\rho_0(x)dx - \int_{-B}^{B} a_1(x)\rho(x)dx\right] \\
&= -2hs_z + 2\pi|J|\int_{|x|>B} \rho_0(x)\rho(x)dx.
\end{aligned}
\tag{3.58}
$$

Putting $p(x) = 2\exp(\pi B/2)\rho(x + B)$, equation (3.56) becomes

$$
p(x) = e^{-\pi x/2} + \int_0^{\infty} (R(x - y) + R(x + y + 2B))p(y)dy,
\tag{3.59}
$$

and

$$
s_z = e^{-\pi B/2}a(B), \quad e - e_0 = -2hs_z + |J|e^{-\pi B}b(B),
$$
$$
a(B) = (1/2)\int_0^{\infty} p(x)dx, \quad b(B) = \pi\int_0^{\infty} e^{-\pi x/2}p(x)dx.
\tag{3.60}
$$

The functions $a(B)$ and $b(B)$ behave for $B \gg 1$ as follows:

$$
a(B) = a_0\left[1 + \frac{1}{2\pi B} + O\left(\frac{\ln B}{B^2}\right)\right], \quad b(B) = b_0\left[1 + O(B^{-2})\right].
\tag{3.61}
$$

In the thermodynamic equilibrium $\partial e/\partial B = 0$. Thus we have

$$
h = (b_0/a_0)|J|e^{-\pi B/2}[1 + (2\pi B)^{-1}]^{-1},
$$
$$
\chi \equiv \frac{2s_z}{h} = \frac{2a_0^2}{b_0|J|}(1 + (2\pi B)^{-1})^2 + O(\ln B/B^2)
$$
$$
= \frac{2a_0^2}{b_0|J|}\left[1 + \frac{1}{2\ln(b_0|J|/(a_0 h))} + O\left(\frac{\ln|\ln h|}{(\ln h)^2}\right)\right].
\tag{3.62}
$$

There is a logarithmic singularity in the magnetic field as a function of the magnetic field. χ has an infinite slope. See Fig. (3.2b–c). Griffiths[32] estimated numerically that $a_0 = 0.48394$, $b_0 = 1.15573$ and predicted that

$$
\frac{2a_0^2}{b_0} = \frac{4}{\pi^2}.
\tag{3.63}
$$

Yang and Yang[120] proved this analytically using the Wiener–Hopf factorization as shown in the next section. Thus χ is related to the excitation velocity obtained in the previous section:

$$
\chi = \frac{2}{\pi v_s}.
\tag{3.64}
$$

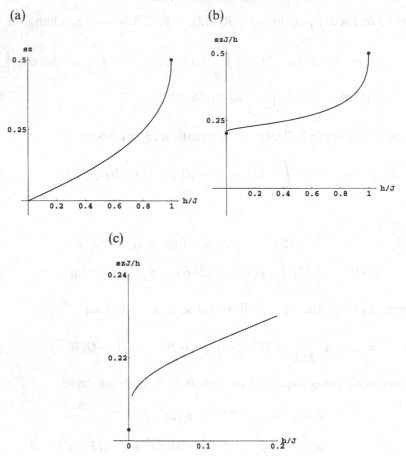

Fig. 3.2. Magnetization curve (a) and $s_z/h - h$ curve (b), (c) for the XXX chain. Near zero field Griffiths found a logarithmic anomaly in the $s_z/h - h$ curve. It has infinite slope near $h = 0$.

3.6 Wiener–Hopf type integral equation

We should consider the solution of the following integral equation,

$$p(x) - \int_0^\infty R(x - y)p(y)\mathrm{d}y = e^{-\pi x/2}. \tag{3.65}$$

We define

$$\tilde{p}(\omega) = \int_0^\infty p(x)e^{i\omega x}\mathrm{d}x. \tag{3.66}$$

The l.h.s. of (3.65) is

$$\int_{-\infty}^{\infty} \frac{\mathrm{d}\omega}{2\pi} e^{-i\omega x} \frac{1}{\exp(-2|\omega|) + 1} \tilde{p}(\omega). \tag{3.67}$$

Here we define $|\omega| \equiv \text{sign}(\Re(\omega))\omega$. The denominator has zeros at $\omega = (n + 1/2)\pi i$ and a branch cut on the imaginary axis. We factorize the denominator

$$\exp(-2|\omega|) + 1 = G_+(\omega)G_-(\omega), \tag{3.68}$$

where $G_+(\omega)$ and $G_-(\omega)$ are analytic and non-zero in the upper and lower half-plane respectively, plus the real axis,

$$G_+(\omega) = G_-(-\omega), \quad G_+(\infty) = 1. \tag{3.69}$$

The analytic expression of $G_+(\omega)$ has a discontinuity or branch cut on the negative imaginary axis only. Using (E.43) and (E.46) (see Appendix E) we find the following function satisfies this condition,

$$G_+(\omega) = \frac{\sqrt{2\pi}}{\Gamma(z + \frac{1}{2})} z^z e^{-z}, z = -i\omega/\pi. \tag{3.70}$$

The solution of equation (3.65) is

$$\tilde{p}(\omega) = \frac{G_+(\omega)G_-(-\pi i/2)}{\pi/2 - i\omega} = \frac{\sqrt{2/e} z^z e^{-z}}{(z + \frac{1}{2})\Gamma(z + \frac{1}{2})}. \tag{3.71}$$

By the definition we have

$$a_0 = \tilde{p}(0)/2 = \sqrt{2/(\pi e)}, \quad b_0 = \pi \tilde{p}(\pi i/2) = \pi/e. \tag{3.72}$$

Using these values we have (3.63). In the detailed calculation we have

$$\chi = \frac{4}{J\pi^2}\left[1 + \frac{1}{2\ln(h_0/h)} - \frac{\ln\ln(h_0/h)}{4(\ln(h_0/h))^2} + O\left(\frac{1}{(\ln(h_0/h))^2}\right)\right],$$

$$h_0 = J\sqrt{\pi^3/(2e)} = 2.388155328J. \tag{3.73}$$

The same kind of anomalies are also found in the Hubbard model and other itinerant electron models[50].

4

The XXZ model

4.1 Symmetry of the Hamiltonian

In this section we consider the following Hamiltonian:

$$\mathcal{H}(J, \Delta, h) = -J \sum_{l=1}^{N} S_l^x S_{l+1}^x + S_l^y S_{l+1}^y + \Delta S_l^z S_{l+1}^z - 2h \sum_{l=1}^{\infty} S_l^z,$$

$$h \geq 0, \quad \mathbf{S}_{N+1} \equiv \mathbf{S}_1. \tag{4.1}$$

This Hamiltonian contains a new parameter Δ. The case $\Delta = 1$ is the XXX model treated in the previous section. The case $\Delta = 0$ is the XY model treated in chapter 1. The limit of very large Δ is the Ising model. The generalization of Bethe's method to $\Delta \neq 1$ was done by Orbach[77] and Walker[114]. Yang and Yang investigated the ground state of this model in detail[119–122]. Bonner and Fisher[16] investigated this model using the exact diagonalization method up to $N = 12$. In this Hamiltonian the magnetic field is applied in the z-direction. For a magnetic field in another direction, the exact solution is not known.

Let us consider the following unitary transformation,

$$\mathcal{H}(J, \Delta, h) = U_1 \mathcal{H}(J, \Delta, -h) U_1^{-1}, \quad U_1 \equiv \prod_{l=1}^{\infty} 2S_l^x = U_1^{-1}. \tag{4.2}$$

By this unitary transformation S_{total}^z changes its sign and thus we can treat the $N \geq M > N/2$ case. In the case of even N we can show that

$$\mathcal{H}(-J, -\Delta, h) = U_2 \mathcal{H}(J, \Delta, h) U_2^{-1}, \quad U_2 \equiv \prod_{l=\text{even}} 2S_l^z = U_2^{-1}. \tag{4.3}$$

By this unitary transformation S_l^x, S_l^y, S_l^z change to $-S_l^x, -S_l^y, S_l^z$ at $l = $ even.

4.2 The Bethe-ansatz wave function

We consider the state where all spins are up and the total S^z is $N/2$,

$$\mathscr{H}|0\rangle = E_0|0\rangle, \quad E_0 = -J\Delta N/4 - Nh. \tag{4.4}$$

We assume the wave function of equation (3.3). Corresponding to (3.4) we have

$$0 = -\frac{J}{2}\sum_j(1 - \delta_{n_j+1,n_{j+1}})\big\{f(n_1, ..., n_j + 1, n_{j+1}, ..., n_M)$$

$$+f(n_1, ..., n_j, n_{j+1} - 1, ..., n_m)\big\}$$

$$+\Big\{E_0 - E + (J\Delta + 2h)M - J\Delta\sum_j\delta_{n_j+1,n_{j+1}}\Big\}f(n_1, n_2, ..., n_M). \tag{4.5}$$

The wave function of the type of equation (3.5) can be an eigenstate with

$$E = E_0 + \sum_{j=1}^{M}[J(\Delta - \cos k_j) + h], \tag{4.6}$$

$$0 = A(P)(e^{ik_{Pj}} + e^{-ik_{P(j+1)}} - 2\Delta)e^{ik_{P(j+1)}}$$

$$+A(P(j, j+1))(e^{ik_{P(j+1)}} + e^{-ik_{Pj}} - 2\Delta)e^{ik_{Pj}}, \tag{4.7}$$

$$A(P) = \epsilon(P)\prod_{l<j}(e^{i(k_{Pl}+k_{Pj})} + 1 - 2\Delta e^{ik_{Pl}}). \tag{4.8}$$

The periodic boundary condition is as follows,

$$\exp(ik_jN) = (-1)^{M-1}\prod_{l\neq j}\frac{\exp[i(k_j + k_l)] + 1 - 2\Delta\exp(ik_j)}{\exp[i(k_j + k_l)] + 1 - 2\Delta\exp(ik_l)},$$

$$j = 1, 2, ..., M. \tag{4.9}$$

The transformation (3.16) is not convenient because the phase factor

$$\frac{\exp[i(k_j + k_l)] + 1 - 2\Delta\exp(ik_j)}{\exp[i(k_j + k_l)] + 1 - 2\Delta\exp(ik_l)},$$

cannot be written as a function of $x_j - x_l$ except the case $\Delta = 1$. If we put

$$\exp(ik_j) = \frac{\sin\frac{\phi}{2}(x_j + i)}{\sin\frac{\phi}{2}(x_j - i)}, \tag{4.10}$$

in place of (3.16), the phase factor becomes

$$\frac{\cos\frac{\phi}{2}(x_j + x_l)(\cosh\phi - \Delta) + (\Delta\cos\frac{\phi}{2}(x_j - x_l + 2i) - \cos\frac{\phi}{2}(x_j - x_l))}{\cos\frac{\phi}{2}(x_j + x_l)(\cosh\phi - \Delta) + (\Delta\cos\frac{\phi}{2}(x_l - x_j + 2i) - \cos\frac{\phi}{2}(x_l - x_j))}.$$

Then if we put $\cosh\phi - \Delta = 0$, this phase factor becomes a function of $x_j - x_l$ and independent of $x_j + x_l$,

$$\sin\frac{\phi}{2}(x_j - x_l + 2i)/\sin\frac{\phi}{2}(x_j - x_l - 2i).$$

Then for $\Delta > 1$, (4.9) becomes

$$\left(\frac{\sin\frac{\phi}{2}(x_j + i)}{\sin\frac{\phi}{2}(x_j - i)}\right)^N = \prod_{l\neq j}\frac{\sin\frac{\phi}{2}(x_j - x_l + 2i)}{\sin\frac{\phi}{2}(x_j - x_l - 2i)},$$

$$\phi = \cosh^{-1}\Delta, \quad \phi > 0. \tag{4.11}$$

For $\Delta < -1$ we put

$$\exp(ik_j) = -\frac{\sin\frac{\phi}{2}(x_j + i)}{\sin\frac{\phi}{2}(x_j - i)}. \tag{4.12}$$

Equation (4.9) becomes

$$\left(\frac{\sin\frac{\phi}{2}(x_j + i)}{\sin\frac{\phi}{2}(x_j - i)}\right)^N = \prod_{l\neq j}\frac{\sin\frac{\phi}{2}(x_j - x_l + 2i)}{\sin\frac{\phi}{2}(x_j - x_l - 2i)},$$

$$\phi = \cosh^{-1}(-\Delta), \quad \phi > 0. \tag{4.13}$$

In the case $-1 < \Delta < 1$ we put

$$\exp(ik_j) = -\frac{\sinh\frac{\gamma}{2}(x_j + i)}{\sinh\frac{\gamma}{2}(x_j - i)}. \tag{4.14}$$

Equation (4.9) becomes

$$\left(\frac{\sinh\frac{\gamma}{2}(x_j + i)}{\sinh\frac{\gamma}{2}(x_j - i)}\right)^N = \prod_{l\neq j}\frac{\sinh\frac{\gamma}{2}(x_j - x_l + 2i)}{\sinh\frac{\gamma}{2}(x_j - x_l - 2i)},$$

$$\gamma = \cos^{-1}(-\Delta), \quad \pi > \gamma > 0. \tag{4.15}$$

4.3 The string solution for $\Delta > 1$

By the transformation (4.10), the wave function and eigenvalue are

$$f(n_1, n_2, ..., n_M)$$
$$= \sum_P \epsilon(P) \prod_{j<l} \sin\frac{\phi}{2}(x_{Pj} - x_{Pl} + 2i) \prod_{j=1}^M \left(\frac{\sin\frac{\phi}{2}(x_{Pj} + i)}{\sin\frac{\phi}{2}(x_{Pj} - i)}\right)^{n_j},$$

$$\tag{4.16}$$

$$E = E_0 + \sum_{j=1}^M \left(2h + \frac{J\sinh^2\phi}{\cosh\phi - \cos\phi x_j}\right), \quad K = \sum_j 2\cot^{-1}\frac{\tan(\phi x_j/2)}{\tanh(\phi/2)}. \tag{4.17}$$

From the normalizability of the wave function, the following string solutions are possible for complex x_js for the $N = \infty$ case,

$$x_j = \alpha + (M + 1 - 2j)i, \tag{4.18}$$

where α is a real number, $-\pi < \alpha \leq \pi$. The total momentum and energy are given by

$$K = 2\cot^{-1}\frac{\tan(\phi\alpha/2)}{\tanh(M\phi/2)}, \quad E = E_0 + \frac{J\sinh\phi\sinh M\phi}{\cosh M\phi - \cos\phi\alpha} + 2Mh. \tag{4.19}$$

From these we have:

$$E = E_0 + 2Mh + J\sinh\phi\left[\frac{\cosh M\phi - \cos K}{\sinh M\phi}\right]. \tag{4.20}$$

This excitation energy gives (3.24) in the limit $\Delta \to 1$. In the limit of large Δ, the energy is $J\Delta + 2Mh$. This state is M successive down spins in the sea of up spins. The lowest energy state for $\Delta > 1$ with M down spins is the M string state given by (4.20) with zero total momentum. Thus,

$$E = -\frac{JN\Delta}{4} - (N - 2M)h + J\sinh\phi\tanh\frac{M\phi}{2}. \tag{4.21}$$

4.4 The lowest energy state for $\Delta < -1$

For $\Delta < -1$ equation (4.13) yields

$$N\theta_1(x_j, \phi) = 2\pi I_j + \sum_{l=1}^{M}\theta_2(x_j - x_l, \phi),$$

$$\theta_n(x, \phi) \equiv 2\tan^{-1}\left(\frac{\tan(\phi x/2)}{\tanh n\phi/2}\right) + 2\pi\left[\frac{\phi x + \pi}{2\pi}\right]. \tag{4.22}$$

Example of this function are shown in Fig. (4.1). Here we take the branch of $\tan^{-1} x$ as $-\pi/2 < \tan^{-1} x \leq \pi/2$. x_j moves in the region $-Q < x_j \leq Q$, $Q \equiv \pi/\phi$. The distribution function of the x_js at the lowest energy state should satisfy

$$\theta_1(x, \phi) = 2\pi\int_{-\infty}^{x}\rho(t)dt + \int_{-B}^{B}\theta_2(x - y, \phi)\rho(y)dy.$$

Differentiating with respect to x we have

$$\mathbf{a}_1(x, \phi) = \rho(x) + \int_{-B}^{B}\mathbf{a}_2(x - y, \phi)\rho(y)dy,$$

$$\mathbf{a}_n(x, \phi) \equiv \frac{1}{2\pi}\frac{\partial}{\partial x}\theta_n(x, \phi) = \frac{1}{2\pi}\frac{\phi\sinh n\phi}{\cosh n\phi - \cos\phi x}. \tag{4.23}$$

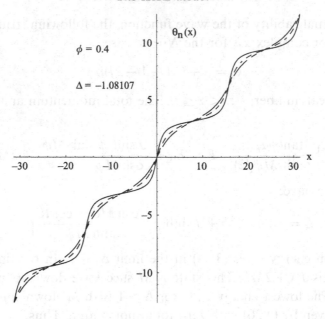

Fig. 4.1. The function $\theta_n(x, \phi)$ at $\phi = 0.4$. The solid line is for $n = 1$, the dashed line is for $n = 2$ and the dotted chain line is for $n = 3$.

Here B varies from 0 to Q. The magnetization and energy are

$$s_z = \frac{1}{2} - \int_{-B}^{B} \rho(x)dx,$$

$$e = -\frac{J\Delta}{4} - h + \int_{-B}^{B}\left(2h - \frac{2\pi J \sinh \phi}{\phi}\mathbf{a}_1(x)\right)\rho(x)dx. \qquad (4.24)$$

At $B = Q$ equation (4.23) can be solved by Fourier series,

$$\tilde{\rho}(n) = \int_{-Q}^{Q} \rho(x)\exp(-in\phi x)dx, \quad n = ..., -2, -1, 0, 1, 2, ...,$$

$$\exp(-|n|\phi) = \tilde{\rho}(n)(1 + \exp(-2|n|\phi)). \qquad (4.25)$$

Thus we have the distribution function for the ground state $\rho_0(x)$,

$$\tilde{\rho}_0(n) = \frac{1}{2\cosh n\phi},$$

$$\rho_0(x) = \frac{1}{2Q}\sum_{n=-\infty}^{\infty} \exp(in\phi x)\tilde{\rho}(n) = \frac{K}{2\pi Q}\mathrm{dn}(\frac{Kx}{Q}, u), \qquad (4.26)$$

where $dn(x, u)$ is the Jacobian elliptic function and the modulus u is determined by

$$\frac{K(\sqrt{1 - u^2})}{K(u)} = \frac{1}{Q} = \frac{\phi}{\pi}, \quad K(u) \equiv \int_0^{\pi/2} \frac{dp}{\sqrt{1 - u^2 \sin^2 p}}. \tag{4.27}$$

The magnetization is zero and the energy density is

$$e = -\frac{J\Delta}{4} - J \sinh \phi \left[\frac{1}{2} + 2 \sum_{n=1}^{\infty} \frac{1}{e^{2n\phi} + 1}\right]. \tag{4.28}$$

4.5 The magnetization curve for a field in the z-direction at $\Delta < -1$

The energy per site, which is given by (4.41) and (4.24) at a given magnetic field, must be minimized. To minimize the energy we should have $\partial e/\partial B = 0$ for equation (4.41). We must consider the change of the region $B \rightarrow B + \Delta B$ and the change $\rho(x) \rightarrow \rho(x) + \Delta\rho(x)$. The equation for $\Delta\rho$ is

$$\Delta\rho(x) + \int_{-B}^{B} a_2(x - y)\Delta\rho(y)dy = -(a_2(x - B) + a_2(x + B))\rho(B)\Delta B. \tag{4.29}$$

Thus the change of energy becomes

$$\Delta e = 2h \left[2\Delta B \rho(B) + \int_{-B}^{B} \Delta\rho(x)dx\right]$$
$$- \frac{2\pi \sinh \phi J}{\phi} \left[2\Delta B a_1(B)\rho(B) + \int_{-B}^{B} a_1(x)\Delta\rho(x)dx\right]. \tag{4.30}$$

This is written as follows

$$\frac{\Delta e}{4\Delta B \rho(B)} = h\left(1 - \int_{-B}^{B} D(x)dx\right)$$
$$- \frac{\pi \sinh \phi J}{\phi}\left(a_1(B) - \int_{-B}^{B} a_1(x)D(x)dx\right), \tag{4.31}$$

where $D(x)$ is defined as the solution of the equation

$$D(x) + \int_{-B}^{B} a_2(x - y)D(y)dy = a_2(x - B). \tag{4.32}$$

The second parentheses of the r.h.s. in (4.31) is equal to $\pi\rho(B)$. The magnetic field h for given B is

$$h = \frac{J\pi \sinh \phi}{\phi} \frac{\rho(B)}{F(B)}, \tag{4.33}$$

where $F(x)$ is defined as the solution of the equation

$$F(x) + \int_{-B}^{B} a_2(x - y)F(y)dy = 1. \tag{4.34}$$

The magnetization is given by (4.24). At $B = 0$ we have $F(B) = 1$, $J(\sinh \phi/\phi)\rho(B) = J(1 - \Delta)/(2\pi)$. Then the corresponding magnetic field is $h = J(1 - \Delta)/2$. Above this magnetic field all spins are up and $s_z = 1/2$.

At $B = Q$ we have $F(x) = 1/2$ and (4.26). Then the corresponding magnetic field is

$$h_c = \frac{J \sinh \phi K(u')u'}{\phi}. \tag{4.35}$$

At zero temperature and $h < h_c$, the state is not magnetized. The phase diagram and magnetization curve are given in Figs. (4.2) and (4.3).

4.6 The lowest energy state for fixed M and $-1 < \Delta < 1$

Assume that $J > 0$. In the case of M down spins, the number of states is C_M^N. The off-diagonal matrix elements of the Hamiltonian (4.1) are $-J$ or 0. Then the lowest energy state wave function is always nodeless and non-degenerate. We take the logarithm of (4.9),

$$k_j N = 2\pi I_j - \sum_{l=1}^{M} \Theta(k_j, k_l),$$

$$\Theta(x, y) \equiv 2 \tan^{-1} \frac{\Delta \sin \frac{1}{2}(x - y)}{\cos \frac{1}{2}(x + y) - \Delta \cos \frac{1}{2}(x - y)}, \tag{4.36}$$

where I_j are different integers (half-odd integers) for odd (even) M. Set

$$I_j = \frac{M + 1}{2} - j, \quad j = 1, ..., M. \tag{4.37}$$

At $\Delta = 0$ this gives $k_j = \pi(M + 1 - 2j)/N$ and the corresponding energy coincides with (1.13). At $-\infty < \Delta < 1$ equations (4.36) with (4.37) have unique real solutions of $\{k_j\}$. Thus the lowest energy state at fixed M at $0 \leq M \leq N/2$ is certainly expressed by Bethe-ansatz. Using the transformation U_2 in (4.3) we find that the lowest energy state at $J < 0, N =$ even, $\Delta > -1$ is expressed by the Bethe ansatz. The analysis for XXX antiferromagnets with $J < 0$, $\Delta = 1$ in §3.3 is justified.

At $-1 < \Delta < 1$ equation (4.37) is written as follows

$$2N \tan^{-1} \left[\frac{\tanh(\gamma x_j/2)}{\tan(\gamma/2)} \right] = 2\pi I_j + 2 \tan^{-1} \left[\frac{\tanh \gamma(x_j - x_l)/2}{\tan \gamma} \right]. \tag{4.38}$$

Writing the distribution function of the x_js as $\rho(x)$, we have

$$2 \tan^{-1}\left[\frac{\tanh(\gamma x/2)}{\tan(\gamma/2)}\right] = 2\pi \int^x \rho(t)dt +$$

$$\int_{-B}^{B} 2 \tan^{-1}\left[\frac{\tanh \gamma(x-y)/2}{\tan \gamma}\right]\rho(y)dy. \tag{4.39}$$

By differentiation

$$a(x,1) = \rho(x) + \int_{-B}^{B} a(x-y,2)\rho(y)dy,$$

$$a(x,n) \equiv \frac{1}{2\pi} \frac{\gamma \sin n\gamma}{\cosh \gamma x - \cos n\gamma}. \tag{4.40}$$

The energy density and magnetization density are

$$s_z = \frac{1}{2} - \int_{-B}^{B} \rho(x)dx,$$

$$e = -\frac{J\Delta}{4} - h + \int_{-B}^{B}\left[2h - \frac{2\pi J \sin^2 \gamma}{\gamma}a(x,1)\right]\rho(x)dx. \tag{4.41}$$

At $B = \infty$ the integral equation can be solved analytically with the use of a Fourier transformation,

$$\frac{\sinh(p_0 - 1)\omega}{\sinh p_0\omega} = \tilde{\rho}(\omega)\left(1 + \frac{\sinh(p_0 - 2)\omega}{\sinh p_0\omega}\right), \quad p_0 \equiv \pi/\gamma. \tag{4.42}$$

Thus we have

$$\tilde{\rho}(\omega) = \frac{1}{2\cosh \omega}, \quad \rho(x) = \frac{1}{4}\operatorname{sech}\left(\frac{\pi x}{2}\right). \tag{4.43}$$

Using (4.41) we have the ground state energy per site at $s_z = 0$,

$$e = -\frac{J\Delta}{4} - J\frac{\sin \gamma}{\gamma} \int_{-\infty}^{\infty} \frac{\sinh(p_0 - 1)\omega}{\cosh \omega \sinh p_0\omega}d\omega. \tag{4.44}$$

4.7 The magnetization curve for a field in the z-direction at $-1 \le \Delta < 1$

The energy per site, which is given by (4.41) at a given magnetic field, must be minimized. To minimize the energy we should have $\partial e/\partial B = 0$ for equation (4.41). We must consider the change of the region $B \to B + \Delta B$ and the change of $\rho(x) \to \rho(x) + \Delta\rho(x)$. Then the equation for $\Delta\rho$ is

$$\Delta\rho(x) + \int_{-B}^{B} a(x-y,2)\Delta\rho(y)dy = -(a(x-B,2) + a(x+B,2))\rho(B)\Delta B. \tag{4.45}$$

Thus the change of energy becomes

$$\Delta e = 2h\left[2\Delta B\rho(B) + \int_{-B}^{B} \Delta\rho(x)dx\right]$$
$$- \frac{2\pi J \sin\gamma}{\gamma}\left[2\Delta Ba(B,1)\rho(B) + \int_{-B}^{B} a(x,1)\Delta\rho(x)dx\right]. \qquad (4.46)$$

This is written as follows:

$$\frac{\Delta e}{4\Delta B\rho(B)} = h\left(1 - \int_{-B}^{B} D(x)dx\right)$$
$$- \frac{\pi J \sin\gamma}{\gamma}\left(a(B,1) - \int_{-B}^{B} a(x,1)D(x)dx\right), \qquad (4.47)$$

where $D(x)$ is defined as the solution of the equation

$$D(x) + \int_{-B}^{B} a(x-y,2)D(y)dy = a(x-B,2). \qquad (4.48)$$

The second parenthesis of the r.h.s. of (4.47) is equal to $\pi\rho(B)$. The magnetic field h for a given B is

$$h = \frac{J\pi \sin\gamma}{\gamma} \frac{\rho(B)}{F(B)}, \qquad (4.49)$$

where $F(x)$ is defined as the solution of the equation

$$F(x) + \int_{-B}^{B} a(x-y,2)F(y)dy = 1. \qquad (4.50)$$

The magnetization is given by (4.41). At $B = 0$ we have $F(B) = 1$, $J \sin\gamma\rho(B)/\gamma = J(1 - \Delta)/(2\pi)$. Then the corresponding magnetic field is $h = J(1 - \Delta)/2$. Above this magnetic field all spins are up and $s_z = 1/2$.

4.8 Susceptibility for $-1 < \Delta < 1$

In the case $-1 < \Delta < 1$, in the limit $B = \infty$, the corresponding magnetic field is zero, because $F(x) = 1/2$, $\rho(\infty) = 0$. The system is magnetized by a weak magnetic field. We can calculate the magnetic susceptibility. In the small magnetic field the energy becomes minimal for large B. The equation (4.40) can be transformed as follows,

$$\rho(x) + \int_{-\infty}^{\infty} a(x-y,2)\rho(y)dy = a(x,1) + \int_{|y|>B} a(x-y,2)\rho(y)dy. \qquad (4.51)$$

By Fourier transformation we have

$$\rho(x) = \rho_0(x) + \int_{|y|>B} R(x-y,\gamma)\rho(y)dy, \quad \rho_0(x) \equiv \frac{1}{4}\text{sech}\left(\frac{\pi x}{2}\right), \qquad (4.52)$$

$$R(x, \gamma) \equiv \frac{1}{2\pi} \int_{-\infty}^{\infty} e^{i\omega x} \frac{\sinh(p_0 - 2)\omega}{2\cosh\omega \sinh(p_0 - 1)\omega} d\omega. \tag{4.53}$$

The integrand of (4.53) has poles at $\frac{\pi n}{p_0 - 1}i$ and $\pi(n + \frac{1}{2})i$. Then for large x,

$$R(x, \gamma) = O(e^{-\frac{\pi}{2}x}) + O(e^{-\frac{\pi}{p_0-1}x}).$$

$\rho_0(x)$ is the solution at $B = \infty$ or $h = 0$. Substituting this into (4.41) we have

$$s_z = \frac{1}{2} - \int_{-\infty}^{\infty} \rho(y)dy + \int_{|y|>B} \rho(y)dy = \frac{\pi}{2(\pi - \gamma)} \int_{|y|>B} \rho(y)dy. \tag{4.54}$$

Here we have used $\int \rho_0(x)dx = 1/2$, $\int R(x, \gamma)dx = (\pi - 2\gamma)/2(\pi - \gamma)$. The energy change is

$$e - e_0 = -hs_z + \frac{2\pi J \sin\gamma}{\gamma} \left[\int_{-\infty}^{\infty} a(x, 1)\rho_0(x)dx - \right.$$
$$\left. \int_{-B}^{B} a(x, 1)\rho(x)dx \right] = -hs_z + \frac{2\pi J \sin\gamma}{\gamma} \int_{|x|>B} \rho_0(x)\rho(x)dx. \tag{4.55}$$

Putting $p(x) = 2\exp(\pi B/2)\rho(x + B)$ we have

$$p(x) = e^{-\pi x/2} + \int_0^{\infty} (R(x - y, \gamma) + R(x + y + 2B, \gamma))p(y)dy,$$

$$s_z = \frac{\pi}{\pi - \gamma} \exp(-\pi B/2)a(B),$$

$$e - e_0 = -hs_z + \left(\frac{J \sin\gamma}{2\gamma} \right) \exp(-\pi B)b(B),$$

$$a(B) = (1/2) \int_0^{\infty} p(x)dx,$$

$$b(B) = \pi \int_0^{\infty} e^{-\pi x/2} p(x)dx. \tag{4.56}$$

The functions $a(B)$ and $b(B)$ behave for $B \gg \gamma$ as follows:

$$a(B) = a_0[1 + O(e^{-\frac{2B\pi}{p_0-1}}) + O(e^{-B\pi})],$$
$$b(B) = b_0[1 + O(e^{-\frac{2B\pi}{p_0-1}}) + O(e^{-B\pi})]. \tag{4.57}$$

We put

$$p(x) = p_1(x) + p_2(x) + p_3(x) + \dots$$

$$p_1(x) - \int_0^{\infty} R(x - y, \gamma)p_1(y)dy = e^{-\pi x/2},$$

$$p_j(x) - \int_0^{\infty} R(x - y, \gamma)p_j(y)dy = \int_0^{\infty} R(x + y + 2B, \gamma)p_{j-1}(y)dy.$$

The integration kernel is factorized as follows:

$$1 - \tilde{R}(\omega, \gamma) = \frac{\sinh(p_0\omega)}{2\cosh(\omega)\sinh((p_0 - 1)\omega)} = \frac{1}{G_+(\omega)G_+(-\omega)},$$

$$G_+(i\pi z) = \sqrt{2\pi(1 - p_0^{-1})}\left(\frac{(p_0 - 1)^{p_0-1}}{p_0^{p_0}}\right)^z \frac{\Gamma(p_0 z + 1)}{\Gamma(\frac{1}{2} + z)\Gamma((p_0 - 1)z + 1)}.$$

Using this function we can obtain a_0 and b_0,

$$a_0 = \pi^{-1}G_+(0)G_+\left(i\frac{\pi}{2}\right), \quad b_0 = G_+\left(i\frac{\pi}{2}\right),$$

$$\frac{a_0^2}{b_0} = (G_+(0)/\pi)^2 = \frac{2(\pi - \gamma)}{\pi^3}.$$

The higher order corrections are $O(\exp(-\pi B))$ and $O(\exp(-2\pi B/(p_0 - 1)))$, because $R(x+y+2B, \gamma)$ decays in this way. In the thermodynamic equilibrium $\partial e/\partial B$ should be zero. Thus,

$$h = \frac{b_0}{a_0}\frac{J(\pi - \gamma)\sin\gamma}{\pi\gamma}e^{-\pi B/2}[1 + O(e^{-\frac{2B\pi}{p_0-1}}) + O(e^{-B\pi})],$$

$$\frac{2s_z}{h} = \chi = 2\frac{\gamma\pi^2}{J(\pi - \gamma)^2\sin\gamma}\frac{a_0^2}{b_0}[1 + O(h^2) + O(h^{\frac{4\gamma}{\pi-\gamma}})]$$

$$= \frac{4\gamma}{J\pi(\pi - \gamma)\sin\gamma}[1 + O(h^2) + O(h^{\frac{4\gamma}{\pi-\gamma}})]. \tag{4.58}$$

There is an algebraic singularity as a function of the magnetic field. χ has an infinite slope for $-1 < \Delta < -0.8$.

4.9 The long range order of the XXZ model

At $\Delta > 1$ and $h = 0$ the ground state is doubly degenerate:

$$|\uparrow\uparrow\uparrow \cdots \uparrow\rangle, \quad |\downarrow\downarrow\downarrow \cdots \downarrow\rangle.$$

Thus the two-point function of the ground state is always 1/4.

$$\langle S_i^z S_j^z \rangle = \frac{1}{4}.$$

In the limit $-\Delta \gg 1$ the system is equivalent to the Ising antiferromagnet. For the case of $N = $ even, the ground states in the limit $\Delta \to -\infty$ are the following two wave functions

$$|\uparrow\downarrow\uparrow \cdots \downarrow\rangle, \quad |\downarrow\uparrow\downarrow \cdots \uparrow\rangle.$$

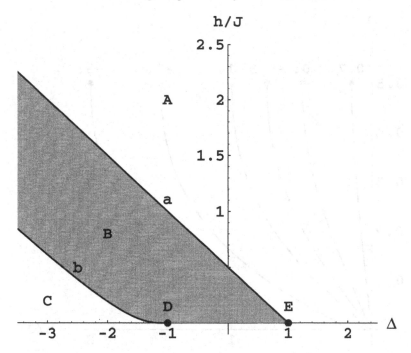

Fig. 4.2. Phase diagram of the XXZ model in the magnetic field. In regions A and C the ground state has an energy gap. In region B the system becomes gapless. Point E is the ferromagnetic XXX point. Point D corresponds to the antiferromagnetic XXX point.

The energy is $-NJ|\Delta|/4$. Then in the Ising limit the two-point correlation function of the ground state is

$$\langle S_i^z S_j^z \rangle = \frac{(-1)^{i-j}}{4}.$$

For finite Δ, $\Delta < -1$, the two-point correlation function of the ground state is not constant but remains finite. The two levels degenerate in the limit $\Delta = \infty$ become slightly different, though this difference is very small. These two states are called nearly degenerate. The third lowest energy states are at $-NJ|\Delta|/4 + J|\Delta|$. Then the third states have a considerable energy gap from the nearly degenerate ground state.

The calculation of the two-point function at the ground state is very difficult. Direct calculation from the wave function was not successful. But it is expected that the two-point function in the z-direction will approach to a finite value,

$$\lim_{i-j\to\infty} 4(-1)^{i-j}\langle S_i^z S_j^z \rangle = P_0^2.$$

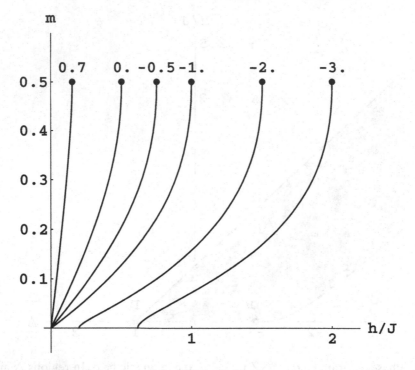

Fig. 4.3. Magnetization curve of the XXZ model at zero temperature for a magnetic field in the z-direction. We take $\Delta = -3, -2, -1, -0.5, 0, 0.7$. At $\Delta < -1$ the system is not magnetized at $h < h_c$.

As will be shown in the next chapter, the ground state of the XXZ model relates to the thermodynamics of the six-vertex model. Especially, the XXZ ground state at $\Delta > 1$ is related to the F model below the transition temperature. P_0 is the spontaneous order of this model. Baxter[10] obtained that

$$P_0 = \Big(\prod_1^\infty \frac{1 - t^{2n}}{1 + t^{2n}}\Big)^2, \quad t = \exp(-\phi). \tag{4.59}$$

The result is very simple and coincides with the perturbation expansion from the Ising limit. This is transformed in various forms,

$$P_0 = \prod_{n=1}^\infty \tanh^2 n\phi = \frac{2(u')^{1/2}K(u')}{\phi} = \frac{2\pi}{\phi}\Big[\sum_{n=0}^\infty \exp(-(n + \tfrac{1}{2})^2\pi^2/\phi)\Big]^2. \tag{4.60}$$

The last formula converges rapidly for small ϕ.

4.10 Excitations from the ground state

4.10.1 $\Delta < -1$ *and the energy gap*

In the case of even N, the true ground state at $h = 0$ has $M = N/2$ and quantum numbers I_j given by (4.37),

$$I_j = \frac{N/2 - 1}{2}, \frac{N/2 - 3}{2}, ..., -\frac{N/2 - 1}{2}. \tag{4.61}$$

This is an Ising-like antiferromagnet. We have a nearly degenerate state with this ground state,

$$I_j = \frac{N/2 + 1}{2}, \frac{N/2 - 1}{2}, ..., -\frac{N/2 - 3}{2}. \tag{4.62}$$

The former has total momentum zero, the latter has total momentum π. The latter has slightly higher energy than the former but the difference is $o(1/N)$. In this state k_1 is always π. In the Ising limit ($\Delta \to -\infty$), these two states correspond to the following wave functions:

$$| \uparrow\downarrow\uparrow\downarrow \,...\rangle \pm | \downarrow\uparrow\downarrow\uparrow \,...\rangle.$$

Next we consider the two-holes state, with $M = N/2 - 1$

$$I'_j = \frac{N'}{2}, \frac{N'}{2} - 1, ..., \frac{N'}{2} - r + 1, \frac{N'}{2} - r - 1,$$
$$..., \frac{N'}{2} - s + 1, \frac{N'}{2} - s - 1, -\frac{N'}{2}, \tag{4.63}$$

with $0 \le r < s \le N'$. The total momentum is $K = 2\pi(r + s)/N + \pi$. From equation (4.22) we have

$$N\theta_1(x'_j, \phi) = 2\pi I'_j + \sum_{l \ne r,s} \theta_2(x'_j - x'_l, \phi). \tag{4.64}$$

Subtracting (4.22) from (4.61) we have

$$2\pi N a_1(x_j)\Delta x_j - \sum_{l=1}^{N'-1} 2\pi a_2(x_j - x_l)(\Delta x_j - \Delta x_l)$$
$$= \pi(\text{sign}(x_j - x_r) + \text{sign}(x_j - x_s) - 1) - \theta_2(x_j - x_{N'}, \phi). \tag{4.65}$$

Putting $J(x_j) = N\Delta x_j \rho(x_j)$ yields

$$J(x) + \int_{-Q}^{Q} a_2(x - y)J(y)dy$$
$$= \frac{1}{2}(\text{sign}(x - x_r) + \text{sign}(x - x_s)) - 1 - \frac{1}{\pi}\tan^{-1}(\tanh\phi\tan(x/2)). \tag{4.66}$$

The Fourier transform of this yields

$$(1 + e^{-2|n|\phi})\tilde{J}(n)$$
$$= \begin{cases} \frac{1}{in}[2(-1)^n - e^{inx_r} - e^{inx_s} - (-1)^n(1 - e^{-2|n|\phi})] & \text{for } n \neq 0 \\ -x_r - x_s - 2\pi & \text{for } n = 0. \end{cases}$$

$$(4.67)$$

The excitation energy ΔE becomes:

$$-2h + \frac{2\pi J \sinh \phi}{\phi}\left[\mathbf{a}_1(x_{N'}) - \int_{-Q}^{Q} \frac{d}{dx}\mathbf{a}_1(x)J(x)dx\right]$$

$$= -2h + J \sinh \phi \left[\tanh \frac{\phi}{2} - \sum_{n \neq 0} \frac{e^{-|n|\phi}}{1 + e^{-2|n|\phi}}[-e^{inx_r} - e^{inx_s} +\right.$$

$$\left. (-1)^n(1 + e^{-2|n|\phi})]\right] = -2h + J \sinh \phi \left[1 + \sum_{n \neq 0} \frac{e^{inx_r} + e^{inx_s}}{2\cosh n\phi}\right]$$

$$= -2h + 2\pi J \frac{\sinh \phi}{\phi}[\rho_0(x_r) + \rho_0(x_s)]$$

$$= -2h + \frac{J \sinh \phi K(u)}{\pi}\left[\text{dn}\left(\frac{K(u)x_r}{\pi}, u\right) + \text{dn}\left(\frac{K(u)x_s}{\pi}, u\right)\right].$$

$$(4.68)$$

Here the modulus u and the elliptic integrals $K(u)$ and $K' = K(u')$ are determined from the parameter ϕ by (4.27). The momentum is given by

$$q_r = \frac{2\pi r}{N} = 2\pi \int_{x_r}^{Q} \rho_0(x)dx = \sin^{-1} \text{sn}\left(\frac{K(u)x}{\pi}, u\right)\Big|_{x_r}^{Q}$$

$$= \frac{\pi}{2} - \sin^{-1} \text{sn}\left(\frac{K(u)x_r}{\pi}, u\right).$$

$$(4.69)$$

Thus we have

$$\Delta E = -2h + \frac{JK(u')\sinh \phi}{\phi}[\sqrt{1 - u^2 \cos^2 q_r} + \sqrt{1 - u^2 \cos^2 q_s}],$$

$$\Delta K = q_r + q_s - \pi.$$

$$(4.70)$$

This excitation has $S_{\text{total}}^z = 1$. By the unitary transformation U_1 we have the following excitation at $S_{\text{total}}^z = -1$:

$$\Delta E = 2h + \frac{JK(u')\sinh \phi}{\phi}[\sqrt{1 - u^2 \cos^2 q_r} + \sqrt{1 - u^2 \cos^2 q_s}].$$

$$(4.71)$$

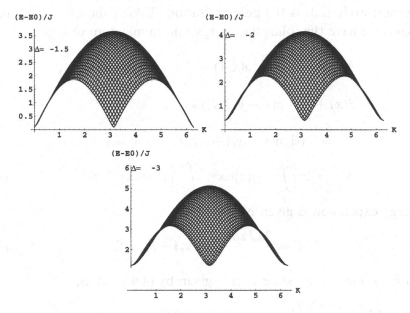

Fig. 4.4. Spinon excitation of the XXZ model at $\Delta = -1.5$, $\Delta = -2$ and $\Delta = -3$. We put $h = 0$.

The spinon spectrum in a zero field is shown in Fig. (4.4). The lowest energy point of (4.71) at $h = 0$ is at $q_r = q_s = 0$,

$$\Delta E = \frac{2JK(u')u' \sinh \phi}{\phi}. \tag{4.72}$$

The energy gap goes to zero as $\Delta \to 1+$. The excitation spectrum were investigated by des Cloizeaux and Gaudin[20], Johnson, Krinsky and McCoy[44]. Ishimura and Shiba[41] calculated the dynamical correlation function of the finite chains.

4.10.2 Excitations for $-1 < \Delta < 1$

As will be shown in part 2, the XXZ chain has many kinds of string solutions which are dependent on the anisotropy parameter $\Delta = -\cos \gamma$. For the ground state, the x_js are on the real axis. For equation (4.39) the ground state is given by the set of half integers

$$\{I_j\} = \left\{ \frac{N/2-1}{2}, \frac{N/2-3}{2}, ..., -\frac{N/2-1}{2} \right\}. \tag{4.73}$$

The simplest excitation is the hole excitation. Taking the quantum number I_j as (4.63) we have the following energy and momentum change:

$$\Delta E = \frac{2\pi J \sin \gamma}{\gamma} \left[a(x, 1) - \int_{-\infty}^{\infty} \frac{d}{dx} a(x, 1) J(x) dx \right],$$

$$J(x) + \int_{-\infty}^{\infty} a(x - y, 2) J(y) dy$$

$$= \frac{1}{2} (\text{sign}(x - x_r) + \text{sign}(x - x_s)) - 1 - \frac{\pi - \gamma}{\pi},$$

$$\Delta K = 2\pi \left[\int_{x_r}^{\infty} \rho_0(x) dx + \int_{x_s}^{\infty} \rho_0(x) dx \right] - \pi. \qquad (4.74)$$

The energy expression is given by

$$\Delta E = \frac{2\pi J \sin \gamma}{\gamma} [\rho_0(x_r) + \rho_0(x_s)]. \qquad (4.75)$$

At zero field, the ground state $\rho(x)$ is given by (4.43). Thus,

$$\Delta E = \frac{J\pi \sin \gamma}{2\gamma} [\sin q_r + \sin q_s], \quad \Delta K = q_r + q_s - \pi. \qquad (4.76)$$

The hole state has almost the same spectrum as the XXX antiferromagnet except for the velocity coefficient. The spinon velocity is given by

$$v_s = \frac{J\pi \sin \gamma}{2\gamma}. \qquad (4.77)$$

This velocity goes to zero as Δ approaches to 1. In this limit the dispersion becomes completely flat. This dispersion also resembles (4.70). The main difference between the excitations of the Ising-like XXZ model and the XY-like XXZ model is that the former has a finite energy gap but the latter is gapless.

4.10.3 Spin-wave like excitations for $0 < \Delta < 1$

We assume that there are $N/2 - 1$ particles on the real axis and 1 particle on the $\Im x = \pi/\gamma$ axis, with γ at $\pi > \gamma > \pi/2$. Then the Bethe-ansatz equation becomes

$$-N\theta(y, p_0 - 1) = 2\pi J - \sum_{l=1}^{N'-1} \theta(y - x_l, p_0 - 2), \qquad (4.78)$$

$$N\theta(x_j, 1) = 2\pi I_j - \theta(x_j - y, p_0 - 2) + \sum_{l=1}^{N'-1} \theta(x_j - x_l, 2).$$

$$(4.79)$$

Here $y = \Re x$, J is an integer and I_j are integers (half-odd integers) for odd (even) $N' - 1$ and

$$\theta(x, \alpha) \equiv 2 \tan^{-1} \left(\frac{\tanh \gamma x/2}{\tan \alpha \gamma/2} \right).$$

We assume a symmetric distribution of the I_j,

$$\frac{N' - 2}{2}, \frac{N' - 4}{2}, \cdots, -\frac{N' - 2}{2}.$$

Then the total momentum is

$$K = \theta(y, p_0 - 1) - \int \theta'(x, 1) J(x) dx. \qquad (4.80)$$

The energy change is

$$\epsilon(y) = \frac{2\pi J \sin \gamma}{\gamma} \left[a(y, p_0 - 1) - \int_{-\infty}^{\infty} a'(x, 1) J(x) dx \right]. \qquad (4.81)$$

Here

$$a(x, \alpha) \equiv (2\pi)^{-1} \frac{\partial \theta(x, \alpha)}{\partial x},$$

and $J(x)$ is the back flow for this excitation,

$$J(x) + \int a(x - x', 2) J(x') dx' = -\frac{1}{p_0} + \frac{1}{2\pi} \theta(x - y, 2 - p_0). \qquad (4.82)$$

The solution of this equation is

$$J(x) = -\frac{1}{2(p_0 - 1)} + \int_0^{x-y} dt \int_{-\infty}^{\infty} \frac{d\omega}{2\pi} \frac{\cosh(p_0 - 1)\omega}{\cosh \omega} e^{i\omega t}. \qquad (4.83)$$

Then we have

$$\epsilon(y) \frac{\gamma}{2\pi J \sin \gamma} = \frac{1}{2\pi} \frac{d}{dy} K(y) = \int_{-\infty}^{\infty} d\omega e^{i\omega y} \frac{\cosh(2 - p_0)\omega}{\cosh \omega}$$

$$= \frac{1}{4} \left[\operatorname{sech} \frac{\pi}{2} (y - i(2 - p_0)) + \operatorname{sech} \frac{\pi}{2} (y + i(2 - p_0)) \right]. \qquad (4.84)$$

Integrating with respect to y yields

$$K = 2[\tan^{-1} e^{\frac{\pi}{2}(y - i(2 - p_0))} + \tan^{-1} e^{\frac{\pi}{2}(y + i(2 - p_0))}],$$

$$\tan \frac{K}{2} = \frac{\sin(p_0 - 1)\pi/2}{\sinh \pi y/2}. \qquad (4.85)$$

We can eliminate y and obtain

$$\epsilon(K) = \frac{\pi J \sin \gamma}{\gamma} \left| \sin \frac{K}{2} \right| \sqrt{1 + \cot^2 \frac{(p_0 - 1)\pi}{2} \sin^2 \frac{K}{2}}. \qquad (4.86)$$

At low momentum this excitation has the same velocity as a spinon excitation. The spinon excitation has periodicity π. But this excitation has periodicity 2π. As Δ goes to 1, $\gamma \to \pi$ and $p_0 \to 1$. Then this excitation becomes

$$\epsilon(K) = 2J \sin^2 \frac{K}{2} = J(1 - \cos K). \tag{4.87}$$

This is the spin wave excitation from the ferromagnetic ground state of the XXX ferromagnet. It should be noted that this excitation exists only for $0 < \Delta < 1$. If we calculate energy and momentum for $-1 < \Delta \leq 0$, we find that the bare energy and momentum are cancelled by the back flow at the ground state. At finite temperature or at finite magnetic field this cancellation is not complete and we can calculate the excitation spectrum.

4.10.4 The spin-wave bound state

A length n string can exist in the ground state of XXZ model, if $(p_0 - 1)n < 1$. In the other region the bare momentum and energy cancelled with the back flow. The positions of rapidities are

$$x_j = y + [(-1)^n p_0 + (n + 1 - 2j)]i. \tag{4.88}$$

The bare momentum and energy of this state are

$$\theta(y, n(p_0 - 1)), \quad J \sin \gamma / \gamma \theta'(y, n(p_0 - 1)). \tag{4.89}$$

The back flow density should satisfy

$$J(x) + \int a(x - x', \gamma)J(x')dx' = -n/p_0 + (2\pi)^{-1}\Theta(x - y),$$
$$\Theta(x) = \theta(x, p_0 - (n - 1)(p_0 - 1)) + \theta(x, p_0 - (n + 1)(p_0 - 1)). \tag{4.90}$$

The solution of this equation is

$$J(x) = -\frac{n}{2(p_0 - 1)} + \int_0^{x-y} dt \int_{-\infty}^{\infty} \frac{d\omega}{2\pi} \frac{\cosh n(p_0 - 1)\omega}{\cosh \omega} e^{i\omega t}. \tag{4.91}$$

Thus we have

$$\epsilon(y)\frac{\gamma}{2\pi J \sin \gamma} = \frac{1}{2\pi}\frac{d}{dy}K(y) = \int d\omega e^{i\omega y}\frac{\cosh(1 - n(p_0 - 1))\omega}{\cosh \omega}$$
$$= \frac{1}{4}\left[\operatorname{sech}\frac{\pi}{2}(y - i(1 - n(p_0 - 1))) + \operatorname{sech}\frac{\pi}{2}(y + i(1 - n(p_0 - 1)))\right]. \tag{4.92}$$

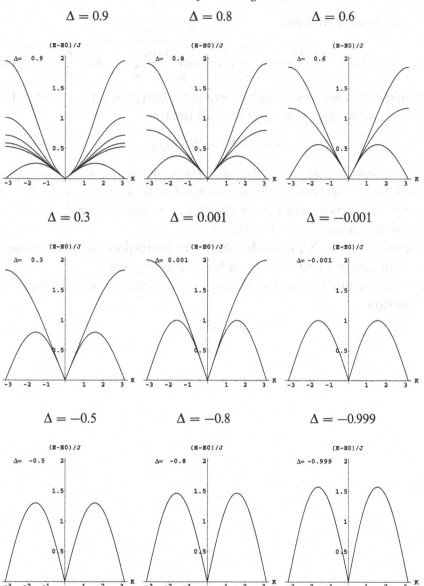

Fig. 4.5. Elementary excitations for various values of Δ from the zero field ground state. If the system is magnetized or the temperature is not zero, the other excitations have finite momentum region also. We put Δ = 0.9, 0.8, 0.6, 0.3, 0.001, −0.001, −0.5, −0.8, −0.999.

We can eliminate y and get

$$\epsilon(K) = \frac{\pi J \sin \gamma}{\gamma} \left| \sin \frac{K}{2} \right| \sqrt{1 + \cot^2 \frac{n(p_0 - 1)\pi}{2} \sin^2 \frac{K}{2}}. \tag{4.93}$$

One should note that this excitation exists for $\cos(\pi/(n+1)) < \Delta < 1$. In the limit $\Delta \to 1$ we have $\gamma \to \pi$, $p_0 \to 1$ and obtain

$$\epsilon(K) = J(1 - \cos k)/n. \tag{4.94}$$

Thus we have the spinon excitation, spin-wave excitation and bound state excitation for the ground state of the XXZ model for $|\Delta| < 1$. All these excitations are gapless and have a common low-momentum group velocity. These excitations are shown in Figs. (4.5).

In the case of the XYZ model there are corresponding excitations, but these are all massive and the system has the energy gap.

The selection rule for the length and parity of strings will be discussed in a later section.

5

XYZ and eight-vertex models

5.1 Transfer matrix of the eight-vertex model

5.1.1 Relation between the six-vertex model and the XXZ model

Here we should consider the partition function of a two-dimensional classical model. Consider a square lattice with $N \times M$ sites and six kinds of vertices as shown in Fig. (5.1) The Boltzmann weights for a vertex are assumed to be $w_1, w_2, w_3, w_4, w_5, w_6$. The Boltzmann weight of one configuration is $W = w_1^{N_1} w_2^{N_2} w_3^{N_3} w_4^{N_4} w_5^{N_5} w_6^{N_6}$. The partition function is given by the sum of Boltzmann weights of all configurations,

$$Z = \sum_{\text{conf}} W_{\text{conf}}. \tag{5.1}$$

This partition function is the trace of the product of M transfer matrices,

$$Z = \text{Tr}\mathbf{T}^M. \tag{5.2}$$

Hereafter we put $w_1 = w_2 = a$, $w_3 = w_4 = b$ and $w_5 = w_6 = c$ for simplicity. The transfer matrix $\mathbf{T}$ is a $2^N \times 2^N$ matrix given by

$$
\begin{aligned}
&\mathbf{T}(\sigma_1, \sigma_2, ..., \sigma_N | \sigma_1', \sigma_2', ..., \sigma_N') \\
&= \sum_{\tau_1} \sum_{\tau_2} \cdots \sum_{\tau_N} R(\sigma_1, \sigma_1')_{\tau_N, \tau_1} R(\sigma_2, \sigma_2')_{\tau_1, \tau_2} ... R(\sigma_N, \sigma_N')_{\tau_{N-1}, \tau_N}.
\end{aligned} \tag{5.3}
$$

This matrix element can be written as the trace of products of 2×2 matrices,

$$\mathbf{T}(\sigma_1, \sigma_2, ..., \sigma_N | \sigma_1', \sigma_2', ..., \sigma_N') = \text{Tr}R(\sigma_1, \sigma_1')R(\sigma_2, \sigma_2')...R(\sigma_N, \sigma_N'). \tag{5.4}$$

Here we have four kinds of matrices,

$$\mathbf{R}(++) = \begin{pmatrix} a & 0 \\ 0 & b \end{pmatrix}, \quad \mathbf{R}(+-) = \begin{pmatrix} 0 & 0 \\ c & 0 \end{pmatrix},$$

$$\mathbf{R}(-+) = \begin{pmatrix} 0 & c \\ 0 & 0 \end{pmatrix}, \quad \mathbf{R}(--) = \begin{pmatrix} b & 0 \\ 0 & a \end{pmatrix}. \tag{5.5}$$

67

Fig. 5.1. Vertices and their Boltzman weights of the six-vertex model.

The problem of the thermodynamics of an infinite M system is to calculate the largest eigenvalue of the transfer matrix $\mathbf{T}$,

$$Z = \sum_{l=1}^{2^N} \lambda_l^M. \tag{5.6}$$

The free energy per site is

$$f = -\lim_{M \to \infty} \frac{T}{NM} \ln Z = -\frac{T}{N} \ln \lambda_1. \tag{5.7}$$

Here T is the temperature.

One trivial eigenstate of this matrix is $| \uparrow, ..., \uparrow \rangle$ with eigenvalue $a^N + b^N$. The next simplest solution is the state with one down spin. Two transfer matrices $\mathbf{T}(a, b, c)$ and $\mathbf{T}(a', b', c')$ commute if

$$\frac{a^2 + b^2 - c^2}{2ab} = \frac{a'^2 + b'^2 - c'^2}{2a'b'}. \tag{5.8}$$

To ensure the commutation, a, b and c are parametrized as follows,

$$a = \rho \sin(v + \eta), \quad b = \rho \sin(v - \eta), \quad c = \rho \sin 2\eta. \tag{5.9}$$

ρ and v might be different but $\cos 2\eta$ must be common for a, b, c and a', b', c' to commute $\mathbf{T}(a, b, c)$ and $\mathbf{T}(a', b', c')$. Then these transfer matrices $\mathbf{T}(v)$ have the common set of eigenvectors. One can construct another series of matrices $\mathbf{Q}(v)$ which satisfy

$$\mathbf{T}(v)\mathbf{Q}(v) = \phi(v - \eta)\mathbf{Q}(v + 2\eta) + \phi(v + \eta)\mathbf{Q}(v + 2\eta), \tag{5.10}$$

$$[\mathbf{T}(v), \mathbf{Q}(v')] = [\mathbf{Q}(v), \mathbf{Q}(v')] = 0, \quad \phi(v) = (\rho \sin(v))^N. \tag{5.11}$$

As $\mathbf{T}$ and $\mathbf{Q}$ always commute, they are diagonalized by a certain linear transformation which is independent of the parameter v. As all elements of $\mathbf{T}(v)$ and $\mathbf{Q}(v)$ are entire functions of v on the complex plane, diagonal elements of $\mathbf{T}$ and $\mathbf{Q}$ must be so as well. Thus we have an equation for one diagonal element in the expression:

$$T(v)Q(v) = \phi(v - \eta)Q(v + 2\eta) + \phi(v + \eta)Q(v + 2\eta). \tag{5.12}$$

Moreover they must be periodic functions with period 2π. Assume that $Q(v)$ has n zeros in the region $-\pi < \Re v < \pi$,

$$Q(v_j) = 0, j = 1, 2, ..., n, \quad Q(v) = C \prod_{l=1}^{n} \sin(v - v_l). \tag{5.13}$$

Putting $v = v_j$ in (5.12) we have

$$\left[\frac{\sin(v_j + \eta)}{\sin(v_j - \eta)}\right]^N = -\prod_{l=1}^{n} \frac{\sin(v_j - v_l + 2\eta)}{\sin(v_j - v_l - 2\eta)}, \quad j = 1, 2, ..., n. \tag{5.14}$$

The corresponding eigenvalue is given by

$$T(v) = \rho^N \left[\sin^N(v - \eta) \prod_{j=1}^{n} \frac{\sin(v - v_j + 2\eta)}{\sin(v - v_j)}\right.$$

$$\left. + \sin^N(v + \eta) \prod_{j=1}^{n} \frac{\sin(v - v_j - 2\eta)}{\sin(v - v_j)}\right]. \tag{5.15}$$

The **R** matrix is represented as follows:

$$\mathbf{R}(\sigma_l, \sigma_l')_{\tau_{l-1}, \tau_l} =$$
$$a(\delta_{\sigma_l +}\delta_{\sigma_l' +}\delta_{\tau_{l-1} +}\delta_{\tau_l +} + \delta_{\sigma_l -}\delta_{\sigma_l' -}\delta_{\tau_{l-1} -}\delta_{\tau_l -})$$
$$+ b(\delta_{\sigma_l +}\delta_{\sigma_l' +}\delta_{\tau_{l-1} -}\delta_{\tau_l -} + \delta_{\sigma_l -}\delta_{\sigma_l' -}\delta_{\tau_{l-1} +}\delta_{\tau_l +})$$
$$+ c(\delta_{\sigma_l +}\delta_{\sigma_l' -}\delta_{\tau_{l-1} +}\delta_{\tau_l -} + \delta_{\sigma_l -}\delta_{\sigma_l' +}\delta_{\tau_{l-1} -}\delta_{\tau_l +}). \tag{5.16}$$

Let us consider the transfer matrix at $v = \eta$. Here we have $a = c, b = 0$ and $\mathbf{R}(\sigma_l, \sigma_l')_{\tau_{l-1}, \tau_l} = a\delta_{\sigma_l, \tau_{l-1}}\delta_{\sigma_l', \tau_l}$. In this case

$$\langle \sigma_1, \sigma_2, ..., \sigma_N | \mathbf{T}(\eta) | \sigma_1', \sigma_2', ..., \sigma_N' \rangle = a^N \delta_{\sigma_1' \sigma_2} \delta_{\sigma_2' \sigma_3} ... \delta_{\sigma_N' \sigma_1}. \tag{5.17}$$

This means that the transfer matrix is merely the right-ward shift operator times a^N. Then the total momentum K of an eigenstate is given by

$$\exp iK = a^{-n} T(\eta) = \prod_{j=1}^{n} \frac{\sin(v_j + \eta)}{\sin(v_j - \eta)}. \tag{5.18}$$

Now let us consider $\mathbf{T}'(\eta)\mathbf{T}^{-1}(\eta)$,

$$\langle \sigma | \frac{d}{dv} \mathbf{T}(v) \Big|_{v=\eta} | \sigma' \rangle = \sum_l \rho^N (\sin 2\eta)^{N-1} \prod_{i \neq l-1, l} \delta_{\sigma_i', \sigma_{i+1}} \Big[\cos 2\eta$$

$$(\delta_{\sigma_l +}\delta_{\sigma_l' +}\delta_{\sigma_{l-1}' +}\delta_{\sigma_{l+1} +} + \delta_{\sigma_l -}\delta_{\sigma_l' -}\delta_{\sigma_{l-1}' -}\delta_{\sigma_{l+1} -})$$

$$+ (\delta_{\sigma_l +}\delta_{\sigma_l' +}\delta_{\sigma_{l-1}' -}\delta_{\sigma_{l+1} -} + \delta_{\sigma_l -}\delta_{\sigma_l' -}\delta_{\sigma_{l-1}' +}\delta_{\sigma_{l+1} +})\Big], \tag{5.19}$$

$$\langle \sigma' | \mathbf{T}^{-1}(\eta) | \sigma'' \rangle = (\rho \sin 2\eta)^{-N} \prod_i \delta_{\sigma_i', \sigma_{i+1}''}. \tag{5.20}$$

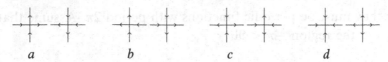

Fig. 5.2. Vertices and Boltzmann weights of the eight-vertex model.

Thus we have

$$\langle\sigma|\frac{d}{dv}T(v)\Big|_{v=\eta}T^{-1}(\eta)|\sigma''\rangle = (\sin 2\eta)^{-1}\sum_l\prod_{i\neq l,l+1}\delta_{\sigma_i,\sigma_i''}$$

$$\times\Big[\cos 2\eta(\delta_{\sigma_l,+}\delta_{\sigma_{l+1}''}+\delta_{\sigma_l''+}\delta_{\sigma_{l+1}+}+\delta_{\sigma_l-}\delta_{\sigma_{l+1}''}-\delta_{\sigma_l''-}\delta_{\sigma_{l+1},-})$$

$$+(\delta_{\sigma_l+}\delta_{\sigma_{l+1}''}+\delta_{\sigma_l''-}\delta_{\sigma_{l+1}-}+\delta_{\sigma_l-}\delta_{\sigma_{l+1}''}-\delta_{\sigma_l''+}\delta_{\sigma_{l+1}+})\Big]$$

$$=\frac{1}{2}(\sin 2\eta)^{-1}\sum_l\Big[\cos 2\eta(1+\sigma_l^z\sigma_{l+1}^z)+\sigma_l^x\sigma_{l+1}^x+\sigma_l^y\sigma_{l+1}^y\Big]. \quad (5.21)$$

This is just the Hamiltonian of the XXZ model plus a constant. Thus the energy eigenvalue of the Hamiltonian

$$\mathscr{H} = -J\sum_{l=1}^{N} S_l^x S_{l+1}^x + S_l^y S_{l+1}^y + \Delta S_l^z S_{l+1}^z \quad (5.22)$$

is given by

$$E = -\frac{J}{2}\sin 2\eta\frac{d}{dv}\ln T(v)\Big|_{v=\eta}-\frac{JN\Delta}{4}$$

$$= -\frac{JN\Delta}{4} - J\sin 2\eta\sum_{j=1}^{n}\frac{\sin 2\eta}{\cos 2\eta - \sin 2v}. \quad (5.23)$$

If we put $\eta = i\phi/2, v_j = \phi x_j/2$, equation (5.14) coincides with (4.11). The energy eigenvalue also coincides with (4.6) at $h=0$.

5.1.2   *The eight-vertex model and the XYZ model*

The exact solution of the eight-vertex model was found by Baxter[8,9]. In eight-vertex model vertices shown in Fig. (5.2) are possible. Here we have four kinds of matrices,

$$\mathbf{R}(++) = \begin{pmatrix} a & 0 \\ 0 & b \end{pmatrix}, \quad \mathbf{R}(+-) = \begin{pmatrix} 0 & d \\ c & 0 \end{pmatrix},$$

$$\mathbf{R}(-+) = \begin{pmatrix} 0 & c \\ d & 0 \end{pmatrix}, \quad \mathbf{R}(--) = \begin{pmatrix} b & 0 \\ 0 & a \end{pmatrix}. \quad (5.24)$$

In this case two transfer matrices, $\mathbf{T}(a, b, c, d)$ and $\mathbf{T}(a', b', c', d')$, commute with each other if

$$\frac{a^2 + b^2 - c^2 - d^2}{2(ab - cd)} = \frac{a'^2 + b'^2 - c'^2 - d'^2}{2(a'b' - c'd')}, \tag{5.25}$$

$$ab/cd = a'b'/c'd'. \tag{5.26}$$

Details are in Appendix C. This condition is parametrized by Jacobian elliptic theta functions,

$$\begin{aligned}
a &= \rho\Theta(2\eta)\Theta(v - \eta)H(v + \eta), \\
b &= \rho\Theta(2\eta)H(v - \eta)\Theta(v + \eta), \\
c &= \rho H(2\eta)\Theta(v - \eta)\Theta(v + \eta), \\
d &= \rho H(2\eta)H(v - \eta)H(v + \eta).
\end{aligned} \tag{5.27}$$

Here $H(x)$ and $\Theta(x)$ are

$$H(x) \equiv 2\sum_{n=1}^{\infty}(-1)^{n+1}q^{n(n-1)+1/4}\sin(2n - 1)\frac{\pi x}{2K},$$

$$\Theta(x) \equiv 1 + 2\sum_{n=1}^{\infty}(-1)^n q^{n^2}\cos\frac{n\pi x}{K}, \quad q \equiv \exp\left(-\frac{\pi K'}{K}\right). \tag{5.28}$$

These functions have the following properties and are doubly quasi-periodic on the complex plane,

$$\begin{aligned}
H(x) &= -H(x + 2K) = -iq^{1/4}e^{i\pi x/2K}\Theta(x + iK') \\
&= -qe^{i\pi x/K}H(x + 2iK'), \\
\Theta(x) &= \Theta(x + 2K) = -iq^{1/4}e^{i\pi x/2K}H(x + iK') \\
&= -qe^{i\pi x/K}\Theta(x + 2iK').
\end{aligned} \tag{5.29}$$

The modulus k and parameter η are determined by

$$k = \frac{1 - l}{1 + l}, \quad l \equiv \sqrt{\frac{A_z^2 - A_y^2}{A_z^2 - A_x^2}},$$

$$\text{sn}^2(2\eta) = -\left(\frac{\sqrt{A_z^2 - A_x^2} + \sqrt{A_z^2 - A_y^2}}{A_x + A_y}\right)^2,$$

$$A_z \equiv a^2 + b^2 - c^2 - d^2, A_y \equiv 2(ab + cd), A_x \equiv 2(ab - cd). \tag{5.30}$$

One can construct the matrix $\mathbf{Q}(v)$ which satisfies

$$\begin{aligned}
&\mathbf{T}(v)\mathbf{Q}(v) = \phi(v + \eta)\mathbf{Q}(v - 2\eta) + \phi(v - \eta)\mathbf{Q}(v + 2\eta), \\
&\phi(v) = (\rho\Theta(0)H(v)\Theta(v))^N, \\
&[\mathbf{T}(v), \mathbf{Q}(v)] = [\mathbf{Q}(v), \mathbf{Q}(v')] = 0.
\end{aligned} \tag{5.31}$$

Then $\mathbf{T}(v)$ and $\mathbf{Q}(v)$ are diagonalized at the same time. Details are in Appendix D. An eigenvalue of $\mathbf{Q}(v)$ is factorized as follows,

$$Q(v) = \exp\left(-\frac{iv\pi v}{2K}\right) \prod_{j=1}^{N/2} \mathsf{h}(v - w_j), \quad \mathsf{h}(x) \equiv H(x)\Theta(x). \tag{5.32}$$

The sum of the w_js should satisfy

$$\sum_{j=1}^{N/2} w_j = -K(v'' + N/2 + \text{even integer}) + ivK'/2. \tag{5.33}$$

Here v is an even (odd) integer if $N/2$ minus the number of down spins is even (odd). v'' is an even (odd) integer if the eigenstate is symmetric (anti-symmetric) for reversing up-spin and down-spin. We obtain transcendental equations for the w_js,

$$0 = \phi(w_j + \eta)Q(w_j - 2\eta) + \phi(w_j - \eta)Q(w_j + 2\eta),$$

or

$$\left(\frac{\mathsf{h}(w_l + \eta)}{\mathsf{h}(w_l - \eta)}\right)^N = -e^{-2iv\pi\eta/K} \prod_{j=1}^{N/2} \frac{\mathsf{h}(w_l - w_j + 2\eta)}{\mathsf{h}(w_l - w_j - 2\eta)}. \tag{5.34}$$

The corresponding eigenvalue of $\mathrm{T}(v)$ is given by

$$T(v) = \phi(v + \eta)e^{i\pi v\eta/K} \prod_{l=1}^{N/2} \frac{\mathsf{h}(v - w_l - 2\eta)}{\mathsf{h}(v - w_l)}$$

$$+ \phi(v - \eta)e^{-i\pi v\eta/K} \prod_{l=1}^{N/2} \frac{\mathsf{h}(v - w_l + 2\eta)}{\mathsf{h}(v - w_l)}. \tag{5.35}$$

At $v = \eta$ we have $b = d = 0$, $a = c$ and therefore

$$\langle \sigma_1, \sigma_2, ..., \sigma_N | \mathbf{T}(\eta) | \sigma'_1, \sigma'_2, ..., \sigma'_N \rangle$$
$$= (\rho\Theta(0)\mathsf{h}(2\eta))^N \delta_{\sigma'_1\sigma_2}\delta_{\sigma'_2\sigma_3}...\delta_{\sigma'_N\sigma_1}. \tag{5.36}$$

The logarithmic derivative is

$$\langle \sigma | \frac{\mathrm{d}}{\mathrm{d}v} \mathbf{T}(v) \Big|_{v=\eta} \mathbf{T}^{-1}(\eta) | \sigma'' \rangle = \sum_l \prod_{i \neq l, l+1} \delta_{\sigma_i, \sigma''_i}$$

$$\times \Big[p_1(\delta_{\sigma_l +}\delta_{\sigma''_{l+1} +} + \delta_{\sigma''_l +}\delta_{\sigma_{l+1} +} + \delta_{\sigma_l -}\delta_{\sigma''_{l+1} -} - \delta_{\sigma''_l -}\delta_{\sigma_{l+1} -})$$
$$+ p_2(\delta_{\sigma_l +}\delta_{\sigma''_{l+1} +} + \delta_{\sigma''_l -}\delta_{\sigma_{l+1} -} + \delta_{\sigma_l -}\delta_{\sigma''_{l+1} -} - \delta_{\sigma''_l +}\delta_{\sigma_{l+1} +})$$
$$+ p_3(\delta_{\sigma_l +}\delta_{\sigma''_{l+1} -} - \delta_{\sigma''_l +}\delta_{\sigma_{l+1} -} + \delta_{\sigma_l -}\delta_{\sigma''_{l+1} +} - \delta_{\sigma''_l -}\delta_{\sigma_{l+1} +})$$
$$+ p_4(\delta_{\sigma_l +}\delta_{\sigma''_{l+1} -} - \delta_{\sigma''_l -}\delta_{\sigma_{l+1} +} + \delta_{\sigma_l -}\delta_{\sigma''_{l+1} +} + \delta_{\sigma''_l +}\delta_{\sigma_{l+1} -}) \Big], \tag{5.37}$$

where p_1, p_2, p_3, p_4 are the logarithmic derivatives of a, b, c, d,

$$p_1 = \frac{H'(2\eta)}{H(2\eta)}, \quad p_2 = \frac{H'(0)\Theta(2\eta)}{H(2\eta)\Theta(0)}, \quad p_3 = \frac{\Theta'(2\eta)}{\Theta(2\eta)}, \quad p_4 = \frac{H'(0)H(2\eta)}{\Theta(2\eta)\Theta(0)}. \quad (5.38)$$

In Pauli matrix representation (5.37) becomes

$$\sum_l \left[\frac{p_1 + p_3}{2} + \frac{p_1 - p_3}{2}\sigma_l^z \sigma_{l+1}^z + \frac{p_2 - p_4}{2}\sigma_l^y \sigma_{l+1}^y + \frac{p_2 + p_4}{2}\sigma_l^x \sigma_{l+1}^x \right]. \quad (5.39)$$

Moreover from (5.38) we have

$$p_1 + p_3 = \frac{h'(2\eta)}{h(2\eta)}, \quad p_1 - p_3 = \frac{cn(2\eta)dn(2\eta)}{sn(2\eta)}, \quad p_2 \pm p_4 = \frac{1}{sn(2\eta)} \pm ksn(2\eta). \quad (5.40)$$

Then the eigenvalue of the XYZ Hamiltonian

$$\mathcal{H} = -\sum_l J_z \left[S_l^z S_{l+1}^z + \frac{1 - ksn^2 2\eta}{cn2\eta dn2\eta} S_l^y S_{l+1}^y + \frac{1 + ksn^2 2\eta}{cn2\eta dn2\eta} S_l^x S_{l+1}^x \right], \quad (5.41)$$

is given by

$$\begin{aligned} E &= -\frac{J_z}{2} \frac{sn2\eta}{cn2\eta dn2\eta} \frac{d}{dv} \ln T(v) \Big|_{v=\eta} \\ &= -\frac{J_z}{2} \frac{sn2\eta}{cn2\eta dn2\eta} \sum_{j=1}^{N/2} \left[\frac{h'(w_j - \eta)}{h(w_j - \eta)} - \frac{h'(w_j + \eta)}{h(w_j + \eta)} + \frac{h'(2\eta)}{h(2\eta)} \right]. \end{aligned}$$
$$(5.42)$$

From (5.36) we find the shift operator is given by

$$\delta_{\sigma_1' \sigma_2} \delta_{\sigma_2' \sigma_3} \ldots \delta_{\sigma_N' \sigma_1} = \frac{1}{\phi(2\eta)} T(\eta). \quad (5.43)$$

Using (5.35) we find that the total momentum K is given by

$$e^{iK} = e^{i\pi v\eta/K(k)} \prod_{l=1}^{N/2} \frac{h(w_l + \eta)}{h(w_l - \eta)}. \quad (5.44)$$

Thus one can calculate the energy of the XYZ model in the same way as the XXZ Hamiltonian. But the solution of the XYZ model with a non-zero magnetic field is not known.

5.2 The symmetry of the XYZ model

Here we consider the symmetry of the following Hamiltonian,

$$\mathcal{H} = -\sum_{l=1}^{N} J_x S_l^x S_{l+1}^x + J_y S_l^y S_{l+1}^y + J_z S_l^z S_{l+1}^z. \quad (5.45)$$

We assume N is even. By the transformation

$$U_2 \mathscr{H} U_2^{-1}, \quad U_2 = \prod_{l=even} 2S_l^z,$$

we have $\mathscr{H}(J_x, J_y, J_z) \to \mathscr{H}(-J_x, -J_y, J_z)$. In the same way we have $\mathscr{H}(J_x, J_y, J_z) \to \mathscr{H}(J_x, -J_y, -J_z)$ and $\mathscr{H}(J_x, J_y, J_z) \to \mathscr{H}(J_x, -J_y, -J_z)$. Specifically the energy spectrum of this Hamiltonian is unchanged on reversing the signs of two J_αs. It is evident that the spectrum is unchanged under exchanging J_αs. Then it is sufficient to treat only the case $J_z \geq J_y \geq |J_x| \geq 0$. Comparing (5.41) and (5.45) we can determine k and η,

$$k = \frac{1-l}{1+l}, \quad l = \sqrt{\frac{J_z^2 - J_y^2}{J_z^2 - J_x^2}}, \tag{5.46}$$

$$\text{sn}^2 2\eta = -\frac{1}{k}\frac{J_y - J_x}{J_y + J_x} = -\left(\frac{\sqrt{J_z^2 - J_x^2} + \sqrt{J_z^2 - J_y^2}}{J_y + J_x}\right)^2. \tag{5.47}$$

In the case of the Ising-like XXZ model ($|\Delta| > 1$), $J_z = J\Delta$ and $J_y = J_x = J$. Then we have $k = 0, 2\eta = i\cosh^{-1}\Delta$. In the case of the XY-like XXZ model ($|\Delta| < 1$), we have $k = 1, 2\eta = i\cos^{-1}\Delta$. Thus the zero field XXZ model is included as a special case of the XYZ model.

5.3 Modulus l and modulus k

In previous sections we used modulus k. But one can use l, defined in (5.46). We have the following relation:

$$\frac{K_k'}{K_k} = 2\frac{K_l}{K_l'}, \quad K_k = \frac{1+l}{2}K_l', \quad K_k' = (1+l)K_l. \tag{5.48}$$

The function $\text{h}(x) = H(x)\Theta(x)$ has zeros at $x = 2mK_k + inK_k'$. On the other hand $H_{l'}(x)$ has zeros at $x = 2mK_l' + 2inK_l$. Then $H_{l'}(\frac{2x}{1+l})$ has common zeros with $\text{h}(x)$. One can show that

$$\text{h}(x) = \sqrt{\frac{l'K_{l'}}{\pi}}H_{l'}\left(\frac{2x}{1+l}\right)$$

$$= -i\sqrt{\frac{l'K_l}{\pi}}\exp\left(-\frac{\pi x^2}{K_l K_l'(1+l)^2}\right)H_l\left(\frac{2ix}{1+l}\right). \tag{5.49}$$

Next we consider the function $\text{sn}(x)/(\text{cn}(x)\text{dn}(x))$. This function has poles at $(2m-1)K + inK'$ with residue $(-1)^m$. On the other hand the function $\text{sn}(x)$

has poles at $2mK + i(2n + 1)K'$ with residue $(-1)^m/k$. Thus we have

$$\frac{\text{sn}(x,k)}{\text{cn}(x,k)\text{dn}(x,k)} = \frac{1+l}{2i}\text{sn}\left(\frac{2ix}{1+l}, l\right). \tag{5.50}$$

Using $\text{cn}(x,l) = \sqrt{1 - \text{sn}^2(x,l)}, \text{dn}(x,l) = \sqrt{1 - l^2\text{sn}^2(x,l)}$ we have

$$\text{cn}\left(\frac{2ix}{1+l}, l\right) = \frac{1 + k\text{sn}^2(x,k)}{\text{cn}(x,k)\text{dn}(x,k)}, \text{dn}\left(\frac{2ix}{1+l}, l\right) = \frac{1 - k\text{sn}^2(x,k)}{\text{cn}(x,k)\text{dn}(x,k)}. \tag{5.51}$$

We define the following parameters:

$$\zeta = K_l + \frac{2i\eta}{1+l}, \quad x_j = \frac{2w_j}{\zeta(1+l)}, \quad Q = K_l'/\zeta. \tag{5.52}$$

Then equations (5.41) and (5.34) become

$$\mathcal{H} = -\sum_l J_z\left[S_l^z S_{l+1}^z + \text{dn}2\zeta S_l^y S_{l+1}^y - \text{cn}2\zeta S_l^x S_{l+1}^x\right], \tag{5.53}$$

$$\left(\frac{H_l(i\zeta(x_l + i))}{H_l(i\zeta(x_l - i))}\right)^N = -e^{-2\pi i v'/p_0} \prod_{j=1}^{N/2} \frac{H_l(i\zeta(x_l - x_j + 2i))}{H_l(i\zeta(x_l - x_j - 2i))}, \tag{5.54}$$

$$\sum_{\alpha=1}^{N/2} \zeta x_\alpha = K_l'v' + iK_lv. \tag{5.55}$$

Substituting (5.49) into (5.44), we have the total momentum

$$K = \frac{\pi v'}{p_0} + \sum_{\alpha=1}^{N/2}\left(\frac{H_l(i(x_\alpha + i))}{H_l(i(x_\alpha - i))}\right). \tag{5.56}$$

Equation (5.42) is

$$E = -\frac{J_z\text{sn}(2\zeta)}{2} \sum_{j=1}^{N/2} g(x_j), \tag{5.57}$$

$$g(x) \equiv \frac{H'(i\zeta(x - i))}{H(i\zeta(x - i))} - \frac{H'(i\zeta(x + i))}{H(i\zeta(x + i))} - \frac{H'(2\zeta)}{H(2\zeta)}. \tag{5.58}$$

The function $g(x)$ has poles at $x = \pm i + (2mK_l' + 2inK_l)/\zeta$ with residue $\mp i\zeta^{-1}$ and is doubly periodic. On the other hand the function

$$b(x) \equiv \frac{2\text{sn}\zeta\text{cn}\zeta\text{dn}\zeta}{\text{sn}^2\zeta - \text{sn}^2(i\zeta x)}, \tag{5.59}$$

is also such a function. Then $g(x) - b(x)$ is a doubly periodic function with no singularity. By the Liouville theorem this must be a constant as a function of x. Putting $x = 0$ this constant is again a function of ζ,

$$2\frac{H'(\zeta)}{H(\zeta)} - \frac{H'(2\zeta)}{H(2\zeta)} - 2\frac{\text{cn}\zeta\,\text{dn}\zeta}{\text{sn}\zeta}. \tag{5.60}$$

This has periodicity $2K, i2K'$ and poles at $0, K, iK', K + iK'$ with residues $-\frac{1}{2}, -\frac{1}{2}, \frac{3}{2}, -\frac{1}{2}$, respectively. On the other hand the poles and residues of

$$(\ln(\text{sn}\zeta))' = \frac{\text{cn}\zeta\,\text{dn}\zeta}{\text{sn}\zeta}$$

are $0, iK'$ and $1, -1$. The poles and residues of

$$(\ln(\text{cn}\zeta))' = -\frac{\text{sn}\zeta\,\text{dn}\zeta}{\text{cn}\zeta}$$

are K, iK' and $1, -1$. The poles and residues of

$$(\ln(\text{dn}\zeta))' = -l^2\frac{\text{sn}\zeta\,\text{cn}\zeta}{\text{dn}\zeta}$$

are $iK', K + iK'$ and $-1, 1$. Then we find that the following quantity is a constant:

$$(5.60) + \frac{1}{2}\left(\frac{\text{cn}\zeta\,\text{dn}\zeta}{\text{sn}\zeta} - \frac{\text{sn}\zeta\,\text{dn}\zeta}{\text{cn}\zeta} - l^2\frac{\text{sn}\zeta\,\text{cn}\zeta}{\text{dn}\zeta}\right). \tag{5.61}$$

This must be an odd function of ζ and therefore must be zero. Then we have finally

$$g(x) = \frac{2\text{sn}\zeta\,\text{cn}\zeta\,\text{dn}\zeta}{\text{sn}^2\zeta - \text{sn}^2(i\zeta x)} - \frac{1}{2}\left(\frac{\text{cn}\zeta\,\text{dn}\zeta}{\text{sn}\zeta} - \frac{\text{sn}\zeta\,\text{dn}\zeta}{\text{cn}\zeta} - l^2\frac{\text{sn}\zeta\,\text{cn}\zeta}{\text{dn}\zeta}\right)$$

$$= b(x) + \frac{1}{\text{sn}2\zeta} - \frac{\text{cn}\zeta\,\text{dn}\zeta}{\text{sn}\zeta} = b(x) + \frac{1}{\text{sn}2\zeta} - \frac{1}{2}(b(0) + b(Q)). \tag{5.62}$$

Then (5.57) becomes

$$E = -\frac{NJ_z}{4}\left[1 - \frac{\text{sn}2\zeta}{2}(b(0) + b(Q))\right] - \frac{J_z\text{sn}2\zeta}{2}\sum_{l=1}^{N/2} b(x_l). \tag{5.63}$$

5.4 The case $J_x = 0$

In this case we have $\text{cn}2\zeta = 0$ and $2\zeta = K_l$. The r.h.s. of (5.54) becomes

$$H_l(i\zeta(x + 2i))/H_l(i\zeta(x - 2i)) = -1.$$

Then equation (5.54) becomes very simple,

$$\left(\frac{H_l(i\frac{K}{2}(x_\alpha + i))}{H_l(i\frac{K}{2}(x_\alpha - i))}\right)^N = -(-1)^{\nu''}. \tag{5.64}$$

This is equivalent to

$$\left(\sqrt{l'}\mathrm{tn}(\frac{K}{2}(1 - ix_\alpha, l))\right)^N = -(-1)^{v''}. \tag{5.65}$$

Thus we have

$$\mathrm{tn}(\frac{K}{2}(1 - ix_\alpha)) = (l')^{-1/2}e^{-iq_\alpha}, \quad q_\alpha \equiv \frac{\pi}{N}(1 + v'' + 2I_\alpha), \tag{5.66}$$

where I_α is some integer. This yields

$$\mathrm{sn}(iKx_\alpha/2) = \frac{\sqrt{l' + e^{2iq_\alpha}} - \mathrm{sign}(\cos q_\alpha)\sqrt{l' + e^{-2iq_\alpha}}}{2\sqrt{1 + l'}\cos q_\alpha}. \tag{5.67}$$

The r.h.s. is always real or pure imaginary. This means that x_α must be on the real axis or on $\Im x_\alpha = 2$. Substituting (5.67) into (5.59) we have

$$g(x_\alpha) = \mathrm{sign}(\cos q_\alpha)\sqrt{(1 - l')^2 + 4l'\cos^2 q_\alpha} + 1 - l'. \tag{5.68}$$

Substituting this into (5.63) we have the lowest energy, or the ground state energy,

$$E = -\frac{J_z}{2}\sum_{j=1}^{N/2}\sqrt{(1 - l')^2 + 4l'\cos^2\frac{\pi}{N}(N/2 + 1 - 2j)}. \tag{5.69}$$

For $J_z > 0$, the ground state x_αs are on the real axis and for $J_z < 0$ they are on the $p_0 i$ axis.

5.5 The ground state for $J_z > 0$

Taking the logarithm of equation (5.54) we have

$$N\theta(x_\alpha, 1) = 2\pi I_\alpha - \frac{2\pi v'}{p_0} + \sum_{\beta=1}^{N/2}\theta(x_\alpha - x_\beta, 2),$$

$$\theta(x, n) \equiv \frac{1}{i}\ln\left(-\frac{H_l(i\zeta(x + ni))}{H_l(i\zeta(x - ni))}\right) + 2\pi\left(\frac{p_0 - n}{p_0}\right)\left[\frac{x + Q}{2Q}\right],$$

$$n < 2p_0. \tag{5.70}$$

The function $\theta(x, n)$ has the following properties,

$$\theta(x, n) = -\theta(-x, n), \quad \theta(jQ, n) = j\pi\left(\frac{p_0 - n}{p_0}\right).$$

The ground state should be characterized by the following set of integers,

$$v' = 0, \quad I_\alpha = \frac{N/2 - 1}{2}, \frac{N/2 - 3}{2}, ..., -\frac{N/2 - 1}{2}. \tag{5.71}$$

Thus the distribution function $\rho(x)$ of x_αs on the real axis for $-Q < x < Q$ should satisfy

$$\mathbf{a}(x, 1) = \rho(x) + \int_{-Q}^{Q} \mathbf{a}(x - y, 2)\rho(y)dy, \tag{5.72}$$

$$\mathbf{a}(x, n) \equiv \frac{1}{2\pi}\frac{d}{dx}\theta(x, n) = \sum_{j=-\infty}^{\infty} \frac{1}{2p_0} \frac{\sin\frac{\pi n}{p_0}}{\cosh\frac{\pi(x-2Qj)}{p_0} - \cos\frac{\pi n}{p_0}}, \tag{5.73}$$

where $p_0 \equiv K_l/\zeta$. The energy (5.63) and momentum in (5.56) are rewritten as follows,

$$E = -\frac{NJ_z}{4}\left[1 - \frac{\pi \mathrm{sn}2\zeta}{\zeta}(\mathbf{a}(0, 1) + \mathbf{a}(Q, 1))\right] - \frac{J_z\pi \mathrm{sn}2\zeta}{\zeta}\sum_{l=1}^{N/2}\mathbf{a}(x_l, 1), \tag{5.74}$$

$$K = \sum_{\alpha=1}^{N/2}\theta(x_\alpha, 1) + \frac{\pi v'}{p_0}. \tag{5.75}$$

The Fourier series $\tilde{f}(m)$ for a function $f(x)$ in the region $[Q, -Q]$ is defined as follows,

$$\tilde{f}(m) \equiv \int_{-Q}^{Q} f(x)\exp(-i\pi mx/Q)dx,$$

$$f(x) = \frac{1}{2Q}\sum_{m=-\infty}^{\infty}\tilde{f}(m)\exp(i\pi mx/Q). \tag{5.76}$$

The functions $\mathbf{a}(x, n)$ are transformed as follows,

$$\tilde{\mathbf{a}}(m, n) = \begin{cases} \dfrac{\sinh(\pi m(p_0 - n)/Q)}{\sinh(\pi mp_0/Q)}, & \text{for } m \neq 0 \\[3mm] \dfrac{p_0 - n}{p_0}, & \text{for } m = 0. \end{cases} \tag{5.77}$$

The equation (5.72) becomes

$$\tilde{\mathbf{a}}(n, 1) = \tilde{\rho}(n) + \tilde{\mathbf{a}}(n, 2)\tilde{\rho}(n). \tag{5.78}$$

Thus we obtain

$$\tilde{\rho}(m) = \frac{1}{2\cosh(\pi m/Q)},$$

$$\rho(x) = \sum_{j=-\infty}^{\infty}\frac{1}{4}\mathrm{sech}\left(\frac{\pi}{2}(x - 2jQ)\right) = \frac{K(u')}{2\pi}\mathrm{dn}(K(u')x, u). \tag{5.79}$$

Here the modulus u is determined by

$$\frac{K(u')}{K(u)} = \frac{1}{Q}.$$

The energy per site is given by

$$\frac{E}{NJ_z} = -\frac{1}{4}\left[1 - \frac{\pi \operatorname{sn}2\zeta}{\zeta}(\mathbf{a}(0,1) + \mathbf{a}(Q,1))\right]$$

$$-\frac{\pi \operatorname{sn}2\zeta}{\zeta}\frac{1}{2Q}\sum_{m=-\infty}^{\infty}\tilde{p}(m)\tilde{\mathbf{a}}(n,1)$$

$$= -\frac{1}{4} - \frac{\tau \operatorname{sn}2\zeta}{2\zeta}\sum_{n=1}^{\infty}\left[\frac{\sinh(p_0-1)\tau n}{\cosh(\tau n)\sinh(\tau n p_0)} - \frac{\sinh 2(p_0-1)\tau n}{\sinh 2p_0 \tau n}\right], \tag{5.80}$$

where $\tau \equiv \pi/Q$.

5.6 Long range order

In the case $J_z > J_y \geq |J_x| \geq 0$, the two-point function of the ground state of the system as described by (5.45) is

$$\lim_{n\to\infty}\lim_{N\to\infty}\langle S_l^z S_{l+n}^z\rangle = \frac{1}{4}P_0^2. \tag{5.81}$$

P_0 is called the spontaneous polarization. This is equivalent to the order parameter of the eight-vertex model below the transition point. For the case $J_y = J_x$, the system is the ferromagnetic Ising-like XXZ model. In this case it is evident that $P_0 = 1$. At $J_y = -J_x$ this is equivalent to the antiferromagnetic Ising-like XXZ model, discussed in §4.9. Baxter and Kelland[12] conjectured that

$$P_0 = \prod_{n=1}^{\infty}\left(\frac{1+q^n}{1-q^n}\times\frac{1-t^{2n}}{1+t^{2n}}\right)^2,$$

$$t \equiv \exp\left(-\frac{\pi\zeta}{2K(l')}\right),$$

$$q \equiv \exp\left(-\frac{\pi K(k')}{K(k)}\right) = \exp\left(-\frac{2\pi K(l)}{K(l')}\right). \tag{5.82}$$

Later, Jimbo, Miwa and Nakayashiki[43] derived this formula using the corner transfer matrix.

5.7 Elementary excitations

5.7.1 The nearly degenerate ground state

For the case $N =$ even, it is possible to consider the state with the following quantum numbers:

$$v' = 1, \ I_j = \frac{N'+1}{2}, \frac{N'-1}{2}, ..., -\frac{N'-3}{2}.$$

For this state the total momentum is π. The true ground state has zero total momentum. In this case all zeros shift slightly to the right,

$$N\Delta x_j \mathbf{a}(x_j, 1) = 1 - \frac{1}{p_0} + \sum_l \mathbf{a}(x_j - x_l, 2)(\Delta x_j - \Delta x_l). \quad (5.83)$$

In the $N \to \infty$ limit we introduce the flow-function $J(x) = N\Delta x_j \rho(x)$. We have the following equation for $J(x)$:

$$J(x) + \int_{-Q}^{Q} \mathbf{a}(x - y, 2)J(y)\mathrm{d}y = 1 - \frac{1}{p_0}. \quad (5.84)$$

This has a very simple solution: $J(x) = 1/2$. We have energy and momentum change as follows:

$$\Delta E = -\frac{J_z \mathrm{sn}_l 2\zeta}{\zeta} \int_{-Q}^{Q} \mathbf{a}'(x, 1)J(x)\mathrm{d}x = 0,$$

$$K = 2\pi \int \mathbf{a}(x, 1)J(x)\mathrm{d}x + \frac{\pi}{p_0} = \pi. \quad (5.85)$$

Actually, for the finite system this state has a slightly higher energy than the true ground state. But the energy difference decreases rapidly as N becomes large. This state is unique and has no dispersion, but it is important that the ground state is actually degenerate.

5.7.2 Spinon excitations

For the ground state all zeros are on the real axis. We consider one simple excitation where we have one string with length 2 and several holes in the distribution of zeros on the real axis. From equation (5.74) we have

$$\Delta E = -\frac{J_z \pi \mathrm{sn}_l 2\zeta}{\zeta} \left[\int \mathbf{a}'(x, 1)J(x)\mathrm{d}x \right.$$

$$\left. -\mathbf{a}(x_r, 1) - \mathbf{a}(x_s, 1) + \mathbf{a}(x_0, 2) \right], \quad (5.86)$$

$$K = 2\pi \left[\int_{x_r}^{Q} \rho_0(x)\mathrm{d}x + \int_{x_s}^{Q} \rho_0(x)\mathrm{d}x \right] - \pi. \quad (5.87)$$

Here x_r and x_s are the positions of two holes on the real axis. x_0 is the real part of the length 2 string. The equation for the back flow $J(x)$ is

$$J(x) + \int_{-Q}^{Q} \mathbf{a}(x - y, 2)J(y)\mathrm{d}y = -\theta(x - x_r, 2) - \theta(x - x_s, 2)$$

$$+ \Theta_{12}(x - x_0), \tag{5.88}$$

$$\Theta_{12}(x) \equiv \theta(x, 1) + \theta(x, 3). \tag{5.89}$$

We can calculate $J(x)$ analytically using the Fourier transform. The terms in (5.86) which contain x_0 cancel each other and we get

$$\Delta E = \frac{J_z \pi \mathrm{sn}_l 2\zeta}{\zeta} \Big[\rho_0(x_r) + \rho_0(x_s) \Big],$$

$$K = \mathrm{am}(K'x_r) + \mathrm{am}(K'x_s). \tag{5.90}$$

Here $\mathrm{am}(x) = \sin^{-1}\mathrm{sn}(x)$. Then this excitation is represented by two spinons

$$\Delta E = \frac{J_z K(u')\mathrm{sn}_l 2\zeta}{2\zeta} \Big[\sqrt{1 - u^2 \cos^2 q_r} + \sqrt{1 - u^2 \cos^2 q_s} \Big],$$

$$K = q_r + q_s - \pi. \tag{5.91}$$

The energy gap of one spinon is

$$\Delta_{\mathrm{spinon}} = \frac{J_z \mathrm{sn}_l 2\zeta K(u')u'}{2\zeta}. \tag{5.92}$$

The energy gap from the ground state to the excited state is $2\Delta_{\mathrm{spinon}}$, because the spinon number is even for a ring. One should note the low-temperature specific heat behaves as $\exp(-\Delta_{\mathrm{spinon}}/T)$ and not as $\exp(-2\Delta_{\mathrm{spinon}}/T)$, as will be shown later. This spinon excitation exists for any set of parameters of $|J_x| \leq J_y \leq J_z$.

5.7.3 Spin-wave excitations

We consider the case where $J_x > 0$ and $1 < \rho_0 (\equiv K(l')/\zeta) < 2$. Assume that $N' - 1$ zeros are on the real axis and that one zero is on the $\rho_0 i$ axis. Then,

$$\Delta E = \frac{J_z \pi \mathrm{sn}_l 2\zeta}{\zeta} \Big[-\int_{-Q}^{Q} \mathbf{a}'(x, 1)J(x)\mathrm{d}x - \mathbf{a}(-Q, 1) + \mathbf{a}(y, \rho_0 - 1) \Big],$$

$$K = \theta(y, \rho_0 - 1) - \theta(-Q, 1) - \int_{-Q}^{Q} \theta'(x, 1)J(x)\mathrm{d}x. \tag{5.93}$$

Here

$$\mathbf{a}(x, \alpha) \equiv (2\pi)^{-1} \frac{\partial \theta(x, \alpha)}{\partial x},$$

and $J(x)$ must satisfy

$$J(x) + \int_{-Q}^{Q} a(x - x', 2) J(x') dx' = \frac{1}{2\pi} [-\pi + \theta(x + Q, 2) + \theta(x - y, 2 - p_0)]. \quad (5.94)$$

The solution of this equation is

$$J(x) = \sum_m \int_Q^{Q+x} dt e^{i\tau mt} \frac{\sinh(p_0 - 2)\tau m}{2 \cosh \tau m \sinh(p_0 - 1)\tau m}$$

$$+ \int_0^{x-y} dt e^{i\tau mt} \frac{\cosh(p_0 - 1)\tau m}{\cosh \tau m}. \quad (5.95)$$

Then we get

$$\Delta E = \frac{J_z K(u') \mathrm{sn}_l 2\zeta}{2\zeta} \Big[\mathrm{dn}(K(u')(y - i(p_0 - 2)))$$

$$+ \mathrm{dn}(K(u')(y + i(p_0 - 2))) \Big], \quad (5.96)$$

$$K = \pi - \mathrm{am}(K(u')(y - i(p_0 - 2)))$$

$$- \mathrm{am}(K(u')(y + i(p_0 - 2))). \quad (5.97)$$

Taking the cosine of (5.97) we have

$$\cos K = -\frac{1 - 2\mathrm{sn}^2 z + \mathrm{sn}^4 z (2u^2 \mathrm{sn}^2 \alpha_1 - u^4 \mathrm{sn}^4 \alpha_1)}{(1 - u^2 \mathrm{sn}^2 z \mathrm{sn}^2 \alpha_1)^2},$$

$$x \equiv K(u')y, \quad \alpha_1 \equiv K(u') i(2 - p_0). \quad (5.98)$$

This is a quadratic equation for $\mathrm{sn}^2 z$. The solution is

$$\mathrm{sn}^2 z = \frac{1 + \cos K}{2 - u^2 \mathrm{sn}^2 \alpha_1 (1 - \cos K)}. \quad (5.99)$$

By the addition theorem of elliptic functions (5.96) is transformed,

$$\Delta E = \frac{J_z \mathrm{sn}_l 2\zeta K(u')}{2\zeta} \frac{4 \mathrm{dn} \alpha_1 \mathrm{dn} z}{1 - u^2 \mathrm{sn}^2 z \mathrm{sn}^2 \alpha_1}. \quad (5.100)$$

Substituting (5.99) we have the spin-wave dispersion

$$\Delta E = \frac{J_z \mathrm{sn}_l 2\zeta K(u')}{\zeta} \sqrt{\left(1 + \mathrm{ctn}_{u'}^2(\beta) \sin^2 \frac{K}{2}\right)\left(1 - \mathrm{dn}_{u'}^2(\beta) \cos^2 \frac{K}{2}\right)}, \quad (5.101)$$

where $\beta \equiv K(u')(p_0 - 1)$. This state corresponds to (4.86) of the XXZ model. This excitation has a minimum at $K = 0$. The energy gap for one spin wave is

$$\Delta_{\text{spin-wave}} = \frac{J_z \mathrm{sn}_l 2\zeta K(u') u'}{\zeta} \mathrm{sn}_{u'}(K(u')(p_0 - 1)). \quad (5.102)$$

This excitation exists only at $J_x > 0$ and not at $J_x \leq 0$. This energy gap is competitive with the spinon energy gap (5.92). The condition that the spin-wave gap is lower than the spinon gap is $\mathrm{sn}_{u'}(K(u')(p_0 - 1)) < 1/2$. In region Y of figure (11.2) the spin-wave gap is dominant.

5.7.4 Spin-wave bound states

Assume that $1 < p_0 < 1 + 1/n$. Consider the condition that $N' - n$ zeros are on the real axis and n zeros form a string,

$$x_l = y + i\{p_0 + (p_0 - 1)(n + 1 - 2l)\} + \text{deviation}, \quad l = 1, 2, ..., n. \quad (5.103)$$

The energy and momentum difference from the ground state are

$$\Delta E = \frac{J_z \pi \mathrm{sn}_l 2\zeta}{\zeta}\left[-\int_{-Q}^{Q} \mathbf{a}'(x, 1)J(x)\mathrm{d}x\right.$$
$$\left. -n\mathbf{a}(-Q, 1) + \mathbf{a}(y, n(p_0 - 1))\right],$$
$$K = \theta(y, n(p_0 - 1)) - n\theta(-Q, 1) - \int_{-Q}^{Q} \theta'(x, 1)J(x)\mathrm{d}x.$$
$$(5.104)$$

$J(x)$ satisfies

$$J(x) + \int_{-Q}^{Q} \mathbf{a}(x - x', 2)J(x')\mathrm{d}x' =$$
$$\frac{1}{2\pi}[-n\pi + n\theta(x + Q, 2) + \Theta(x - y)],$$
$$\Theta(x) \equiv \theta(x, p_0 - (n - 1)(p_0 - 1)) + \theta(x, p_0 - (n + 1)(p_0 - 1)).$$
$$(5.105)$$

The solution of this equation is

$$J(x) = n\sum_m \int_{Q}^{Q+x} \mathrm{d}t e^{i\tau m t} \frac{\sinh(p_0 - 2)\tau m}{2\cosh \tau m \sinh(p_0 - 1)\tau m}$$
$$+ \int_{0}^{x-y} \mathrm{d}t e^{i\tau m t} \frac{\cosh n(p_0 - 1)\tau m}{\cosh \tau m}. \quad (5.106)$$

Substituting this into (5.104) we get

$$\Delta E = \frac{J_z K(u')\mathrm{sn}_l 2\zeta}{2\zeta}\left[\mathrm{dn}(K(u')(y + i(1 - n(p_0 - 1))))\right.$$
$$\left. + \mathrm{dn}(K(u')(y - i(1 - n(p_0 - 1))))\right], \quad (5.107)$$
$$K = \pi - \mathrm{am}(K(u')(y + i(1 - n(p_0 - 1))))$$
$$- \mathrm{am}(K(u')(y - i(1 - n(p_0 - 1)))). \quad (5.108)$$

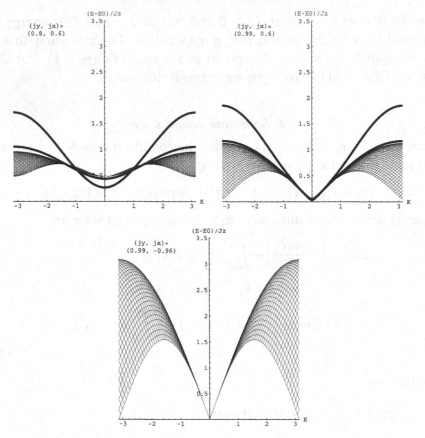

Fig. 5.3. Elementary excitations for various values of $jy = J_y/J_z$ and $jx = J_x/J_z$ from the ground state.

Eliminating y we get the dispersion relation of this excitation at $p_0 < 1 + 1/n$,

$$\Delta E = \frac{J_z \mathrm{sn}_l 2\zeta K(u')}{\zeta} \sqrt{\left(1 + \mathrm{ctn}_{u'}^2(n\beta)\sin^2\frac{K}{2}\right)\left(1 - \mathrm{dn}_{u'}^2(n\beta)\cos^2\frac{K}{2}\right)}. \quad (5.109)$$

This excitation corresponds to (4.93). The minimum energy is at $K = 0$

$$\frac{J_z \mathrm{sn}_l 2\zeta K(u')u'}{\zeta} \mathrm{sn}_{u'}(nK(u')(p_0 - 1)). \quad (5.110)$$

This is always bigger than $\Delta_{\mathrm{spin-wave}}$ defined in (5.102). This excitation cannot be dominant at low temperature. These spinon, spinwave and spin-wave bound state excitations are shown in Fig. (5.3)

6

The Hubbard model

6.1 Symmetry of the Hamiltonian

6.1.1 Particle–hole symmetry

In the problem of electrons with a narrow band, the Hubbard model is the most famous and has been well investigated by many authors. It relates to the metal–insulator transition (or Mott transition). In the half-filled state with strong repulsion, the system becomes insulating and the effective Hamiltonian is represented by the XXX antiferromagnet. The Hubbard Hamiltonian is

$$\mathcal{H}(t, U, A, h) = -t \sum_{\langle ij \rangle} \sum_{\sigma} (c_{i\sigma}^{\dagger} c_{j\sigma} + c_{j\sigma}^{\dagger} c_{i\sigma}) + U \sum_{i=1}^{N_a} c_{i\uparrow}^{\dagger} c_{i\uparrow} c_{i\downarrow}^{\dagger} c_{i\downarrow}$$

$$-A \sum_{i=1}^{N_a} (c_{i\uparrow}^{\dagger} c_{i\uparrow} + c_{i\downarrow}^{\dagger} c_{i\downarrow}) - h \sum_{i=1}^{N_a} (c_{i\uparrow}^{\dagger} c_{i\uparrow} - c_{i\downarrow}^{\dagger} c_{i\downarrow}). \qquad (6.1)$$

Here $c_{j\sigma}^{\dagger}$ and $c_{j\sigma}$ are creation and annihilation operators of an electron at site j. $\langle ij \rangle$ means that sites i and j are nearest neighbours. N_a is the number of atoms. These operators satisfy

$$\{c_{i\sigma}, c_{j\sigma'}\} = \{c_{i\sigma}^{\dagger}, c_{j\sigma'}^{\dagger}\} = 0, \quad \{c_{i\sigma}^{\dagger}, c_{j\sigma'}\} = \delta_{ii'} \delta_{\sigma\sigma'}. \qquad (6.2)$$

Each atomic site has four states, namely the vacant state $|0\rangle$, the up-spin state $c_{j\uparrow}^{\dagger}|0\rangle$, the down-spin state $c_{j\downarrow}^{\dagger}|0\rangle$, and the full state $c_{j\uparrow}^{\dagger} c_{j\downarrow}^{\dagger}|0\rangle$. Thus the total number of states is 4^{N_a}. The following quantities commute with the Hamiltonian and are therefore conserved quantities,

$$N_{\uparrow} = \sum_{j} c_{j\uparrow}^{\dagger} c_{j\uparrow}, \quad N_{\downarrow} = \sum_{j} c_{j\downarrow}^{\dagger} c_{j\downarrow}.$$

Now we assume that the lattice is divided into two sub-lattices. The nearest neighbour atoms of one atom of the $\mathcal{A}$ sub-lattice belong always to the $\mathcal{B}$

sub-lattice. We call such a lattice a bipartite lattice. In the case of a bipartite lattice there are many symmetries of this Hamiltonian.

Let us rewrite the field operators as follows,

$$
\begin{aligned}
a_{i\downarrow}^{\dagger} &= c_{i\downarrow}, & a_{i\downarrow} &= c_{i\downarrow}^{\dagger}, & a_{i\uparrow}^{\dagger} &= c_{i\uparrow}, & a_{i\uparrow} &= c_{i\uparrow}^{\dagger}, & &\text{for } i \in \mathcal{A}, \\
a_{i\downarrow}^{\dagger} &= -c_{i\downarrow}, & a_{j\downarrow} &= -c_{j\downarrow}^{\dagger}, & a_{j\uparrow}^{\dagger} &= -c_{j\uparrow}, & a_{j\uparrow} &= -c_{j\uparrow}^{\dagger}, & &\text{for } j \in \mathcal{B}.
\end{aligned} \tag{6.3}
$$

By this transformation we have

$$
\mathcal{H} \to -t \sum_{\langle ij \rangle} \sum_{\sigma} a_{i\sigma}^{\dagger} a_{j\sigma} + a_{j\sigma}^{\dagger} a_{i\sigma} + U \sum_{i=1}^{N_a} a_{i\uparrow}^{\dagger} a_{i\uparrow} a_{i\downarrow}^{\dagger} a_{i\downarrow}
$$

$$
+ U N_a - (U - A) \sum_{i=1}^{N_a} (a_{i\uparrow}^{\dagger} a_{i\uparrow} + a_{i\downarrow}^{\dagger} a_{i\downarrow}) + h \sum_{i=1}^{N_a} (a_{i\uparrow}^{\dagger} a_{i\uparrow} - a_{i\downarrow}^{\dagger} a_{i\downarrow}). \tag{6.4}
$$

Thus the Hamiltonian and particle number are changed as follows,

$$
\begin{aligned}
\mathcal{H}(t, U, A, h) &\to U N_a + \mathcal{H}(t, U, U - A, -h), \\
N_{\uparrow} &\to N_a - N_{\uparrow}, \quad N_{\downarrow} \to N_a - N_{\downarrow}.
\end{aligned} \tag{6.5}
$$

It should be noted that the more than half-filled case ($N > N_a$) is transformed to the less than half-filled case ($N < N_a$).

If we exchange particles and holes only for spin-up electrons,

$$
\begin{aligned}
a_{i\uparrow}^{\dagger} &= c_{i\uparrow}, & a_{i\uparrow} &= c_{i\uparrow}^{\dagger}, & &\text{for } i \in \mathcal{A}, \\
a_{j\uparrow}^{\dagger} &= -c_{j\uparrow}, & a_{j\uparrow} &= -c_{j\uparrow}^{\dagger}, & &\text{for } j \in \mathcal{B},
\end{aligned} \tag{6.6}
$$

we have

$$
\mathcal{H} \to -t \sum_{\langle ij \rangle} a_{i\uparrow}^{\dagger} a_{j\uparrow} + a_{j\uparrow}^{\dagger} a_{i\uparrow} + c_{i\downarrow}^{\dagger} c_{j\downarrow} + c_{j\downarrow}^{\dagger} c_{j\downarrow} - U \sum_{i=1}^{N_a} a_{i\uparrow}^{\dagger} a_{i\uparrow} c_{i\downarrow}^{\dagger} c_{i\downarrow}
$$

$$
-(A + h)N_a + \left(h + \frac{U}{2}\right) \sum_{i=1}^{N_a} (a_{i\uparrow}^{\dagger} a_{i\uparrow} + c_{i\downarrow}^{\dagger} c_{i\downarrow}) + \left(A - \frac{U}{2}\right) \sum_{i=1}^{N_a} (a_{i\uparrow}^{\dagger} a_{i\uparrow} - c_{i\downarrow}^{\dagger} c_{i\downarrow}),
$$

and therefore

$$
\begin{aligned}
\mathcal{H}(t, U, A, h) &\to -(A + h)N_a + \mathcal{H}\left(t, -U, -h - \frac{U}{2}, \frac{U}{2} - A\right), \\
N_{\uparrow} &\to N_a - N_{\uparrow}, \quad N_{\downarrow} \to N_{\downarrow}.
\end{aligned} \tag{6.7}
$$

The attractive Hubbard model ($U < 0$) can be obtained by this transformation from the repulsive Hubbard model ($U > 0$).

6.1.2 SU(2) symmetry

The total magnetization operators are defined as follows,

$$S_{tot}^x \equiv \frac{1}{2} \sum_j c_{j\uparrow}^\dagger c_{j\downarrow} + c_{j\downarrow}^\dagger c_{j\uparrow}, \quad S_{tot}^y \equiv \frac{1}{2i} \sum_j c_{j\uparrow}^\dagger c_{j\downarrow} - c_{j\downarrow}^\dagger c_{j\uparrow},$$

$$S_{tot}^z \equiv \frac{1}{2} \sum_j c_{j\uparrow}^\dagger c_{j\uparrow} - c_{j\downarrow}^\dagger c_{j\downarrow}. \tag{6.8}$$

Among these operators and the Hamiltonian (6.1), the following commutation relations exist,

$$[S_{tot}^x, S_{tot}^y] = iS_{tot}^z, \quad [S_{tot}^y, S_{tot}^z] = iS_{tot}^x, \quad [S_{tot}^z, S_{tot}^x] = iS_{tot}^y,$$

$$[\mathcal{H}, S_{tot}^z] = 0, \quad [\mathcal{H}, S_{tot}^x] = -2hiS_{tot}^y, \quad [\mathcal{H}, S_{tot}^y] = 2hiS_{tot}^x. \tag{6.9}$$

Defining

$$S_{tot}^+ = S_{tot}^x + iS_{tot}^y = \sum_j c_{j\uparrow}^\dagger c_{j\downarrow}, \, S_{tot}^- = S_{tot}^x - iS_{tot}^y = \sum_j c_{j\downarrow}^\dagger c_{j\uparrow}, \tag{6.10}$$

we have

$$[S_{tot}^+, S_{tot}^-] = 2S_{tot}^z, \quad [S_{tot}^z, S_{tot}^\pm] = \pm S_{tot}^\pm, \quad [\mathcal{H}, S_{tot}^\pm] = \mp 2hS_{tot}^\pm. \tag{6.11}$$

For an eigenstate of the Hamiltonian and S_{tot}^z, $|\psi\rangle$ with eigenvalue E and l, $S_{tot}^+|\psi\rangle$ and $S_{tot}^-|\psi\rangle$ are also eigenstates with eigenvalues $E \mp 2h$ and $l \pm 1$, if they are not zero. Thus we can generate a class of states from the state with highest l. Assume that $|\psi\rangle$ is an eigenstate of $\mathcal{H}$ and S_{tot}^z, and that $S_{tot}^+|\psi\rangle = 0$. We can show that

$$\langle\psi|(S_{tot}^+)^n(S_{tot}^-)^n|\psi\rangle = n(2l+1-n)\langle\psi|(S_{tot}^+)^{n-1}(S_{tot}^-)^{n-1}|\psi\rangle. \tag{6.12}$$

This series of states must be finite. Then $2l + 1$ should be an integer and therefore l must be an integer or half-odd integer. We have $2l + 1$ states,

$$\left[\prod_{k=1}^n k(2l+1-k)\right]^{-1/2} (S_{tot}^-)^n|\psi\rangle, n = 0, 1, ..., 2l. \tag{6.13}$$

The eigenvalues of S_{tot}^z for these states are $l, l-1, l-2, ..., -l$. This is the $SU(2)$ symmetry of the spin degrees of freedom. This type of symmetry is observed in the XXX model and delta function fermion theory. But the Hubbard model has another hidden $SU(2)$ symmetry. By the transformation (6.6) $S_{tot}^z, S_{tot}^+, S_{tot}^-$ are changed as follows,

$$\eta^z = \frac{1}{2}(N_a - a_{j\uparrow}^\dagger a_{j\uparrow} - a_{j\downarrow}^\dagger a_{j\downarrow}), \quad \eta^+ = (-1)^j a_{j\uparrow} a_{j\downarrow}, \quad \eta^- = (-1)^j a_{j\uparrow}^\dagger a_{j\downarrow}^\dagger. \tag{6.14}$$

These operators satisfy the relations

$$[\eta^+, \eta^-] = 2\eta^z, \quad [\eta^z, \eta^\pm] = \pm\eta^\pm \quad, [\mathcal{H}, \eta^\pm] = \mp(U - 2A)\eta^\pm. \qquad (6.15)$$

Thus we can generate a series of states by successively applying the η^- operator to an eigenstate of the Hamiltonian. Therefore the Hubbard model with nearest neighbour interactions on a bipartite lattice has a $SU(2) \times SU(2)$ symmetry.[118] Then the eigenstates of the Hubbard Hamiltonian (6.1) are characterized by η, η_z, S, S^z. For an eigenstate of Hamiltonian $|\psi\rangle$ which satisfies

$$\mathcal{H}|\psi\rangle = E|\psi\rangle, \eta^+|\psi\rangle = 0, \eta^z|\psi\rangle = l'|\psi\rangle, S_{\text{tot}}^+|\psi\rangle = 0, A_{\text{tot}}^z = l|\psi\rangle, \qquad (6.16)$$

we can obtain the following states:

$$|n, n'\rangle = \frac{(\eta^-)^{n'}(S^-)^n}{\left[\prod_{k=1}^n k(2l + 1 - k) \prod_{k'=1}^{n'} k'(2l' + 1 - k')\right]^{1/2}}|\psi\rangle,$$
$$n = 0, 1, ..., 2l, \quad n' = 0, 1, ..., 2l'. \qquad (6.17)$$

The eigenvalues for the operators $\mathcal{H}, S^z$ and η^z are $E + 2nh + (U - 2A)n'$, $l - n$ and $l' - n'$, respectively.

6.2 The Bethe-ansatz equation for the one-dimensional Hubbard model

6.2.1 The wave function for a finite system

For the one-dimensional Hubbard model with nearest neighbour hopping, Lieb and Wu[62] found that this is solvable by the Bethe-ansatz method, just after the discovery of the solution for spin 1/2 delta-function fermions by Gaudin[29] and Yang[117]. In the first quantization representation, the eigenvalue equation for the Hamiltonian (6.1) is

$$-t \sum_{j=1}^N \sum_{s=\pm1} f(x_1, x_2, ..., x_j + s, x_{j+1}, ..., x_N)$$
$$+(U \sum_{j<l} \delta_{jl} - AN - h(N - 2M) - E)f(x_1, x_2, ..., x_N) = 0.$$

$$(6.18)$$

Here N is the total number of electrons and M is the number of down-spin electrons. We assume $2M \leq N \leq N_a$. The other cases are obtained from the symmetry relations in §6.1.1. x_j is the coordinate on the lattice of the j-th electron and takes integer values $1, 2, ..., N_a$. We assume that the $1, 2, ..., M$-th electrons have down spins and that the others have up spins. Applying the Bethe-ansatz we assume

$$f(x_1, x_2, ..., x_N) = \sum_P [Q, P] \exp\left(i \sum_{j=1}^N k_{Pj} x_{Qj}\right),$$

$$\text{for} \quad x_{Q1} \le x_{Q2} \le x_{Q3} \le ... \le x_{QN}. \tag{6.19}$$

Here Ps and Qs are permutations of the integers $1, 2, ..., N$. The energy eigenvalue and total momentum are given by

$$E = -2t \sum_{j=1}^N \cos k_j - AN - h(N - 2M), \quad K = \sum_{j=1}^N k_j. \tag{6.20}$$

As we treat spin $1/2$ fermions on the lattice we only consider two-body scattering by these particles. From the continuity condition on the boundary of the regions we get

$$[Q, P] = (u_{Pj,P(j+1)} - 1)[Q, P'] + u_{Pj,P(j+1)}[Q', P'],$$

$$\text{for} \quad j = 1, 2, ..., N - 1, \tag{6.21}$$

where

$$u_{nm} \equiv \frac{\sin k_n - \sin k_m}{\sin k_n - \sin k_m + iU/(2t)}, \quad Q' = Q(j, j+1), \quad P' = P(j, j+1). \tag{6.22}$$

We regard the coefficients $[Q, P]$ as an elements of an $N! \times N!$ matrix. Then the above equation can be rewritten as follows:

$$\xi_P = Y_{Pj,P(j+1)}^{j,j+1} \xi_{P'}, \quad Y_{kl}^{ab} = (u_{kl} - 1) + u_{kl} P_{ab}. \tag{6.23}$$

Here ξ_P is the column vector for a permutation P. There are $N! \times (N - 1)$ relations between $N!$ vectors. But by the Yang–Baxter relations

$$Y_{ij}^{ab} Y_{ji}^{ab} = I, \quad Y_{jk}^{ab} Y_{ik}^{bc} Y_{ij}^{ab} = Y_{ij}^{bc} Y_{ik}^{ab} Y_{jk}^{bc}, \tag{6.24}$$

these are mutually consistent with each other. The coefficients $[Q, P]$ must be antisymmetric for the exchange of spin-up electrons and for the exchange of spin-down electrons. Then we can write

$$[Q, P] = \epsilon(Q_1)\epsilon(Q_2)\Phi(y_1, y_2, ..., y_M; P), \quad 1 \le y_1 < y_2 < ... < y_M \le N. \tag{6.25}$$

Here Q_1 is the ordering of the first M electrons and Q_2 is the ordering of the of the $M + 1$-st to N-th electrons. $y_1, y_2, ..., y_M$ are the coordinates of down spins in the lattice with length N. In the same way in §2.2 we have

$$
\begin{aligned}
&\Phi(y_1, y_2, ..., y_M ; P) \\
&= \sum_R A(R) F_P(y_1, \Lambda_{R1}) F_P(y_2, \Lambda_{R2})...F_P(y_M, \Lambda_{RM}),
\end{aligned}
\tag{6.26}
$$

$$
F_P(y, \Lambda) = \prod_{j=1}^{y-1} (\sin k_{Pj} - \Lambda + iU') \prod_{l=y+1}^{N} (\sin k_{Pl} - \Lambda - iU'),
$$

$$
U' \equiv \frac{U}{4t},
\tag{6.27}
$$

$$
A(R) = \epsilon(R) \prod_{j<l} (\Lambda_{Rj} - \Lambda_{Rl} - 2iU').
\tag{6.28}
$$

6.2.2 Periodic boundary condition

The periodic boundary condition is

$$
\exp(ik_j N_a) = \prod_{\beta=1}^{M} \frac{\sin k_j - \Lambda_\beta + iU'}{\sin k_j - \Lambda_\beta - iU'}
\tag{6.29}
$$

$$
\prod_{j=1}^{N} \frac{\Lambda_\alpha - \sin k_j + iU'}{\Lambda_\alpha - \sin k_j - iU'} = - \prod_{\beta=1}^{M} \frac{\Lambda_\alpha - \Lambda_\beta + 2iU'}{\Lambda_\alpha - \Lambda_\beta - 2iU'}.
\tag{6.30}
$$

Taking the logarithm of these equations we have

$$
N_a k_j = 2\pi I_j - 2 \sum_{\beta=1}^{M} \tan^{-1} \frac{\sin k_j - \Lambda_\beta}{U'},
\tag{6.31}
$$

$$
\sum_{j=1}^{N} 2 \tan^{-1} \frac{\Lambda_\alpha - \sin k_j}{U'} = 2\pi J_\alpha + \sum_{\beta=1}^{M} 2 \tan^{-1} \frac{\Lambda_\alpha - \Lambda_\beta}{2U'}.
\tag{6.32}
$$

Here the I_js are integers (half-odd integers) for even (odd) M and the J_αs are integers (half-odd integers) for odd (even) $N - M$. Using these equations and (6.20) we have the total momentum:

$$
K = \frac{2\pi}{N_a} \left[\sum_j I_j + \sum_\alpha J_\alpha \right].
\tag{6.33}
$$

6.2.3 Fredholm type integral equations for the ground state

For the ground state in the case $U > 0$, I_j and J_α are successive numbers centered around the origin. For the lowest energy state at even N and odd

M they are

$$I_j = \frac{N-1}{2}, \frac{N-3}{2}, ..., -\frac{N-1}{2}, \quad J_\alpha = \frac{M-1}{2}, \frac{M-3}{2}, ..., -\frac{M-1}{2}. \quad (6.34)$$

Putting N/N_a and M/N_a as constants, we take the limit $N_a \to \infty$. We put the distribution functions of ks and Λs as $\rho(k)$ and $\sigma(\Lambda)$. From equations (6.31) and (6.32) we get

$$k = 2\pi f(k) - \int_{-B}^{B} 2\tan^{-1}\frac{\sin k - \Lambda}{U'}\sigma(\Lambda)d\Lambda, \quad (6.35)$$

$$\int_{-Q}^{Q} 2\tan^{-1}\frac{\Lambda - \sin k}{U'}\rho(k)dk$$

$$= 2\pi g(\Lambda) + \int_{-B}^{B} 2\tan^{-1}\frac{\Lambda - \Lambda'}{2U'}\sigma(\Lambda')d\Lambda'. \quad (6.36)$$

Here $f(k_j) = I_j/N_a, g(\Lambda_\alpha) = J_\alpha/N_a$ and therefore $f'(k) = \rho(k), g'(\Lambda) = \sigma(\Lambda)$. $\pm Q$ are the upper and lower bounds of the distribution of ks, and $\pm B$ are bounds on the Λs. Differentiating with respect to k and Λ we get

$$\rho(k) = \frac{1}{2\pi} + \cos k \int_{-B}^{B} a_1(\sin k - \Lambda)\sigma(\Lambda)d\Lambda, \quad (6.37)$$

$$\sigma(\Lambda) + \int_{-B}^{B} a_2(\Lambda - \Lambda')\sigma(\Lambda')d\Lambda' = \int_{-Q}^{Q} a_1(\Lambda - \sin k)\rho(k)dk, \quad (6.38)$$

$$a_n(x) \equiv \frac{1}{\pi}\frac{nU'}{(nU')^2 + x^2}. \quad (6.39)$$

Q and B are determined by

$$\int_{-Q}^{Q} \rho(k)dk = N/N_a = n = n_\uparrow + n_\downarrow, \quad (6.40)$$

$$\int_{-B}^{B} \sigma(\Lambda)d\Lambda = M/N_a = n_\downarrow. \quad (6.41)$$

From equation (6.20) we have the energy per site

$$e = \int_{-Q}^{Q} (-2t\cos k - A - h)\rho(k)dk + 2h\int_{-B}^{B} \sigma(\Lambda)d\Lambda. \quad (6.42)$$

Using these equations we can determine the energy at $t > 0, U > 0, N \leq N_a, 2M \leq N$. The energy in other regions can be obtained by the symmetry relations given in the previous subsection.

In the $Q = \pi$ case we can show that $n = 1$, by substituting (6.37) into (6.40). This means that the system is in the half-filled state at $Q = \pi$. In the

$B = \infty$ case equation (6.38) is transformed as follows:

$$\sigma(\Lambda) = \int_{-Q}^{Q} \frac{1}{4U'} \text{sech}\left(\frac{\pi(\Lambda - \sin k)}{2U'}\right) \rho(k) dk. \tag{6.43}$$

Substituting this into (6.41) we have $n_\downarrow = n_\uparrow = n/2$.

6.2.4 Analytic solution for the ground state in the half-filled case

Coupled integral equations (6.37) and (6.38) can be solved analytically at $Q = \pi, B = \infty$. Substituting (6.37) into (6.43) we have

$$\sigma(\Lambda) = \frac{1}{2\pi} \int_{-\pi}^{\pi} \frac{1}{4U'} \text{sech}\left(\frac{\pi(\Lambda - \sin k)}{2U'}\right) dk = \frac{1}{2\pi} \int_{-\infty}^{\infty} \frac{J_0(\omega)e^{i\omega\Lambda} d\omega}{2\cosh U'\omega}. \tag{6.44}$$

Substituting this into (6.37),

$$\begin{aligned}
\rho(k) &= \frac{1}{2\pi} + \cos k \int_{-\pi}^{\pi} \frac{1}{U'} R\left(\frac{\sin k - \sin k'}{U'}\right) \frac{dk'}{2\pi} \\
&= \frac{1}{2\pi} + \frac{\cos k}{\pi} \int_0^{\infty} \frac{J_0(\omega) \cos(\omega \sin k)}{1 + \exp(2U'\omega)} d\omega.
\end{aligned} \tag{6.45}$$

Substituting these into (6.42) we get:

$$e = -A - 4t \int_0^{\infty} \frac{J_0(\omega)J_1(\omega) d\omega}{\omega[1 + \exp(2U'\omega)]}. \tag{6.46}$$

6.2.5 Spinon excitation in the half-filled case

Consider the case of N_a even and $N_a/2$ odd. We assume that the external magnetic field h is zero. The excitation spectrum is represented analytically [78,91]. In this case the half-filled ground state is characterized by the integers:

$$\begin{aligned}
I_j &= \frac{N_a - 1}{2}, \frac{N_a - 3}{2}, ..., -\frac{N_a - 1}{2}, \\
J_\alpha &= \frac{N_a/2 - 1}{2}, \frac{N_a/2 - 3}{2}, ..., -\frac{N_a/2 - 1}{2}.
\end{aligned} \tag{6.47}$$

At first we consider the excitation where the number of electrons is the same and the number of down spins decreases from $N_a/2$ to $N_a/2 - 1$. Here we have

$$\begin{aligned}
I_j &= \frac{N_a}{2}, \frac{N_a - 2}{2}, ..., -\frac{N_a - 2}{2}, \\
J_\alpha &= \frac{N_a}{4}, \frac{N_a}{4} - 1, ..., \frac{N_a}{4} + 1 - r, \frac{N_a}{4} - 1 - r, ..., \frac{N_a}{4} + 1 - s, \\
&\quad \frac{N_a}{4} - 1 - s, ..., -\frac{N_a}{4}, \quad 0 \le r < s \le N_a/2.
\end{aligned} \tag{6.48}$$

There are two holes in the distribution of Λ. The situation is almost the same as with the calculation of the XXX model. The total momentum of this state is

$$K = \frac{2\pi}{N_a}(r + s) = 2\pi\left(\int_{\Lambda_r}^{\infty} \sigma(\Lambda)d\Lambda + \int_{\Lambda_s}^{\infty} \sigma(\Lambda)d\Lambda\right). \tag{6.49}$$

Using (6.31) and (6.32) we have the change of each quasi-momentum as follows:

$$\Delta k_j N_a = \pi + 2\tan^{-1}\frac{\sin k_j - \Lambda_{N_a/2}}{U'}$$

$$-2\pi\sum_{\alpha=1}^{N_a/2-1} a_1(\sin k_j - \Lambda_\alpha)(\cos k_j\Delta k_j - \Delta\Lambda_\alpha), \tag{6.50}$$

$$2\pi\sum_j(\Delta\Lambda_\alpha - \cos k_j\Delta k_j)a_1(\Lambda_\alpha - \sin k_j)$$

$$= \pi[\text{sign}(\Lambda_\alpha - \Lambda_r) + \text{sign}(\Lambda_\alpha - \Lambda_s)]$$

$$+2\pi\sum_{\beta=1}^{N_a/2-1} a_2(\Lambda_\alpha - \Lambda_\beta)(\Delta\Lambda_\alpha - \Delta\Lambda_\beta). \tag{6.51}$$

From (6.50) we have

$$2\pi\Delta k_j N_a\left(\frac{1}{2\pi} + \cos k_j\int a_1(\sin k_j - \Lambda)\sigma(\Lambda)d\Lambda\right)$$

$$= 2\pi N_a\int a_1(\sin k_j - \Lambda)\Delta\Lambda\sigma(\Lambda)d\Lambda. \tag{6.52}$$

The substitution of (6.37) yields

$$\rho(k_j)\Delta k_j = \int_{-\infty}^{\infty} a_1(\sin k_j - \Lambda)\Delta\Lambda\sigma(\Lambda)d\Lambda. \tag{6.53}$$

From (6.51) we have

$$\Delta\Lambda_\alpha\left\{\sum_j a_1(\Lambda_\alpha - \sin k_j) - \sum_\beta a_2(\Lambda_\alpha - \Lambda_\beta)\right\}$$

$$+\sum_\beta a_2(\Lambda_\alpha - \Lambda_\beta)\Delta\Lambda_\beta = \frac{1}{2}(\text{sign}(\Lambda_\alpha - \lambda_r) + \text{sign}(\Lambda_\alpha - \Lambda_s))$$

$$+\sum_j \cos k_j\Delta k_j a_1(\Lambda_\alpha - \sin k_j). \tag{6.54}$$

Using (6.38) and (6.53) we obtain

$$\Delta\Lambda_\alpha\sigma(\Lambda_\alpha) + \int a_2(\Lambda_\alpha - \Lambda')\sigma(\Lambda')\Delta\Lambda'd\Lambda'$$

$$= \frac{1}{2N_a}[\text{sign}(\Lambda_\alpha - \Lambda_r) + \text{sign}(\Lambda_\alpha - \Lambda_s)]. \tag{6.55}$$

Putting $N_a \Delta\Lambda\sigma(\Lambda) = J(\Lambda)$ we have an equation for $J(\Lambda)$:

$$J(\Lambda) + \int a_2(\Lambda - \Lambda')J(\Lambda')d\Lambda' = \frac{1}{2}[\text{sign}(\Lambda - \Lambda_r) + \text{sign}(\Lambda - \Lambda_s)]. \quad (6.56)$$

Thus the energy difference is

$$\Delta E = 2tN_a \int_{-\pi}^{\pi} \sin k\rho(k)\Delta k dk$$

$$= 2t \int_{-\pi}^{\pi} dk \sin k \int_{-\infty}^{\infty} d\Lambda a_1(\sin k - \Lambda)J(\Lambda)$$

$$= 2t \int_{-\pi}^{\pi} dk \sin k \int_{-\infty}^{\infty} d\Lambda \frac{1}{4U'} \text{sech}(\frac{\pi(\Lambda - \sin k)}{2U'})$$

$$\times \frac{1}{2}[\text{sign}(\Lambda - \Lambda_r) + \text{sign}(\Lambda - \Lambda_s)] = \epsilon(\Lambda_r) + \epsilon(\Lambda_s),$$

$$\epsilon(\Lambda) = \frac{t}{2U'} \int_{-\pi}^{\pi} dk \cos^2 k \ \text{sech}\left[\frac{\pi(\Lambda_r - \sin k)}{2U'}\right]. \quad (6.57)$$

Using (6.49) we have

$$K = q(\Lambda_r) + q(\Lambda_s), \quad q(\Lambda) = 2\pi \int_{\Lambda}^{\infty} dx \int_{-\pi}^{\pi} \frac{dk}{2\pi} \frac{1}{4U'} \text{sech}\left(\frac{\pi(x - \sin k)}{2U'}\right). \quad (6.58)$$

Then the excitation spectrum $\epsilon(q)$ is defined by the parameter Λ. The lower edge of the spinon spectrum is shown in Fig. (6.1). At $\Lambda \gg 1, U'$,

$$\epsilon = 4tI_1\left(\frac{\pi}{2U'}\right)\exp\left(-\frac{\Lambda\pi}{2U'}\right), \quad q = 2\pi I_0\left(\frac{\pi}{2U'}\right)\exp\left(-\frac{\Lambda\pi}{2U'}\right).$$

Thus the group velocity of this excitation is

$$v_s = 2t\frac{I_1(2t\pi/U)}{I_0(2t\pi/U)}. \quad (6.59)$$

The excitation spectrum in the sector $S_z = 1$ and $N = N_a$ is represented by two spinon excitations.

At $U = 0+$ we can put $(2U')^{-1}\text{sech}(\pi x/2U') = \delta(x)$. Then we have

$$K = \cos^{-1}\Lambda_r + \cos^{-1}\Lambda_s = q_1 + q_2,$$

and

$$\Delta E = 2t(\sqrt{1 - \Lambda_r^2} + \sqrt{1 - \Lambda_s^2}) = 2t|\sin q_1| + 2t|\sin q_2|.$$

At $U/t \gg 1$ the distribution of Λ becomes very broad on the real axis. On the other hand $\sin k$ is restricted to $[-1, 1]$. Then equation (6.57) becomes:

$$K = q_1 + q_2, \quad \Delta E = \frac{2\pi t^2}{U}[|\sin q_1| + |\sin q_2|]. \quad (6.60)$$

This mode corresponds to des Cloizeaux and Pearson's result as shown in

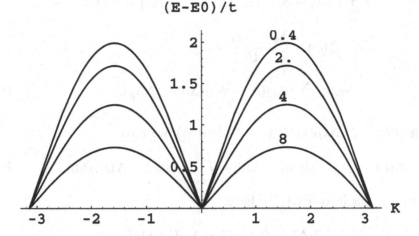

Fig. 6.1. Lower edge of the spinon excitation of the half-filled Hubbard model. We put $U/t = 0.4, 2.0, 4.0, 8.0$.

(3.45) if we put $-J = 4t^2/U$. In fact the half-filled Hubbard model at $U \gg t$ is represented by the antiferromagnetic Heisenberg model.

6.2.6 Energy gap of the charge excitation

Next we consider the case where $N = N_a - 1$ and $M = N_a/2 - 1$. The excitation spectrum for charge also has an analytical solution[70]. The half-filled ground state is characterized by (6.47). This excitation is

$$I_j = N'_a, N'_a - 1, ..., N'_a - r + 1, N'_a - r - 1, ..., -N'_a + 1,$$
$$J_\alpha = \frac{N'_a - 1}{2}, \frac{N'_a - 1}{2} - 1, ..., \frac{N'_a - 1}{2} - s + 1, \frac{N'_a - 1}{2} - s - 1, ..., -\frac{N'_a - 1}{2},$$

$$\tag{6.61}$$

where $N'_a \equiv N_a/2$. We have one hole in the I_j and one in the J_α. The former is called a holon and the latter is a spinon, which appeared in the previous subsection. The total momentum of this state is

$$K = \frac{2\pi}{N_a}\left(r + s - \frac{N_a - 2}{4}\right).$$

$$\tag{6.62}$$

The equations for the shift of the ks and Λs are

$$\Delta k_j N_a = \pi \, \text{sign}(k_j - k_r) + 2\tan^{-1}\frac{\sin k_j - \Lambda_{N_a/2}}{U'}$$

$$-2\pi \sum_{\alpha=1}^{N_a/2-1} a_1(\sin k_j - \Lambda_\alpha)(\cos k_j \Delta k_j - \Delta\Lambda_\alpha),$$

$$\tag{6.63}$$

$$2\pi \sum_j (\Delta\Lambda_\alpha - \cos k_j \Delta k_j)a_1(\Lambda_\alpha - \sin k_j) = \pi \, \mathrm{sign}(\Lambda_\alpha - \Lambda_r)$$

$$+2\tan^{-1}\frac{\Lambda_\alpha - \sin k_{N_a}}{U'}$$

$$+2\pi \sum_{\beta=1}^{N_a/2-1} a_2(\Lambda_\alpha - \Lambda_\beta)(\Delta\Lambda_\alpha - \Delta\Lambda_\beta). \tag{6.64}$$

Putting $D(k) = N_a\Delta k\rho(k), J(\Lambda) = N_a\Delta\Lambda\sigma(\Lambda)$, we have

$$D(k) = \frac{1}{2}\{1 + \mathrm{sign}(k - k_r)\} + \int_{-\infty}^{\infty} a_1(\sin k - \Lambda)J(\Lambda)\mathrm{d}\Lambda. \tag{6.65}$$

Substituting this into (6.64) we have:

$$J(\Lambda) + \int_{-\infty}^{\infty} a_2(\Lambda - \Lambda')J(\Lambda')\mathrm{d}\Lambda'$$

$$= \frac{1}{2}\mathrm{sign}(\Lambda - \Lambda_s) + \frac{1}{\pi}\tan^{-1}\frac{\Lambda - \sin k_r}{U'}. \tag{6.66}$$

The energy difference is

$$\Delta E = A + 2t\cos k_{N_a} + 2t\int_{-\pi}^{\pi}\sin k D(k)\mathrm{d}k$$

$$= A + 2t\cos k_r + 2t\int_{-\pi}^{\pi}\mathrm{d}k\sin k\int_{-\infty}^{\infty}\mathrm{d}\Lambda a_1(\sin k - \Lambda)J(\Lambda)$$

$$= \epsilon(\Lambda_s) + A + 2t\cos k_r + 2t\int_{-\pi}^{\pi}\cos^2 k\frac{1}{U'}R\left(\frac{\sin k_r - \sin k}{U'}\right)\mathrm{d}k$$

$$= A + 2t\left\{\cos k_r + 2\int_0^{\infty}\frac{J_1(\omega)\cos\omega(\sin k_r)}{\omega(e^{2U'\omega} + 1)}\mathrm{d}\omega\right\}. \tag{6.67}$$

Using (6.62) we have the momentum change,

$$\Delta K = q(\Lambda_s) + 2\pi\int_{k_r}^{\pi}\rho(k)\mathrm{d}k - \frac{\pi}{2}. \tag{6.68}$$

This holon excitation spectrum is shown in Fig. (6.2). The minimum point of ΔE is $k_r = \pm\pi, \Lambda_s = \pm\infty$. In the half-filled state the chemical potential should be taken as $A = U/2$. Thus the energy gap of a hole excitation in the half-filled state is

$$\Delta E = 2t\left\{U' - 1 + 2\int_0^{\infty}\frac{J_1(\omega)}{\omega(e^{2U'\omega} + 1)}\mathrm{d}\omega\right\} \tag{6.69}$$

at $\Delta K = \pm\frac{\pi}{2}$. This gap is shown in Fig. (6.3) as a function of U/t. This quantity is always positive at $U > 0$. To make a particle we also need the same energy. Then we can conclude that the half-filled state is always an insulator for positive U. It has been believed that the half-filled Hubbard model is metallic at $U < U_c$ and insulating at $U > U_c$ and that there is metal–insulator transition at some finite $U = U_c$. But Lieb and Wu[62] found that $U_c = 0$ for the 1D Hubbard model.

6.2.7 Susceptibility and magnetization curve of the half-filled case

We consider the case $Q = \pi$ of the coupled integral equations (6.37) and (6.38). Substituting (6.37) into (6.40) we have $n = 1$. The energy per site and equation for $\sigma(\Lambda)$ are

$$e = -A - h + \int_{-B}^{B} d\Lambda \sigma(\Lambda) \left[2h - 2t \int_{-\pi}^{\pi} \cos^2 k a_1(\Lambda - \sin k) dk \right],$$
(6.70)

$$\sigma(\Lambda) + \int_{-B}^{B} a_2(\Lambda - \Lambda') \sigma(\Lambda') d\Lambda' = \int_{-\pi}^{\pi} a_1(\Lambda - \sin k) \frac{dk}{2\pi}.$$
(6.71)

We change B to $B + \Delta B$. The change of energy is given as

$$\Delta e = 2\Delta B \sigma(B) \left[2h - 2t \int_{-\pi}^{\pi} \cos^2 k a_1(B - \sin k) dk \right]$$
$$+ \int_{-B}^{B} \Delta \sigma(\Lambda) \left[2h - 2t \int_{-\pi}^{\pi} \cos^2 k a_1(\Lambda - \sin k) dk \right] d\Lambda.$$
(6.72)

$\Delta \sigma(\Lambda)$ is the change in $\sigma(\Lambda)$ and satisfies

$$\Delta \sigma(\Lambda) + \int_{-B}^{B} a_2(\Lambda - \Lambda') \Delta \sigma(\Lambda') d\Lambda'$$
$$= -\Delta \sigma(B) \{ a_2(B - \Lambda) + a_2(-B - \Lambda) \}.$$
(6.73)

Then the change in the energy is written as follows,

$$\Delta e = 2\Delta B \sigma(B) \{ 2hF(B) - 2tG(B) \},$$
(6.74)

$$F(\Lambda) + \int_{-B}^{B} a_2(\Lambda - \Lambda') F(\Lambda') d\Lambda' = 1,$$
(6.75)

$$G(\Lambda) + \int_{-B}^{B} a_2(\Lambda - \Lambda') G(\Lambda') d\Lambda' = \int_{-\pi}^{\pi} \cos^2 k a_1(\Lambda - \sin k) dk.$$
(6.76)

Thus for a given B the corresponding magnetic field and magnetization are:

$$h = t \frac{G(B)}{F(B)}, \quad m = n_\uparrow - n_\downarrow = 1 - 2 \int_{-B}^{B} \sigma(\Lambda) d\Lambda.$$
(6.77)

At $B = 0$ we have $h = 2t[\sqrt{1 + U'^2} - U']$, $m = 1$. Above this magnetic field the system is completely magnetized. The magnetization curve is obtained

(a)

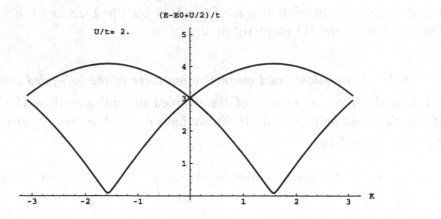

(b)

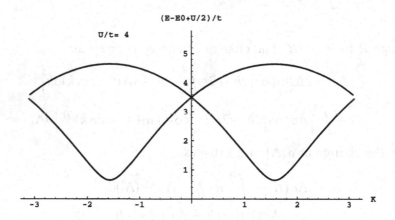

(c)

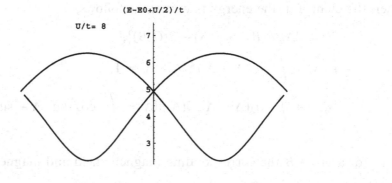

Fig. 6.2. The holon excitation of the half-filled Hubbard model at $U/t = 2.0, 4.0, 8.0$.

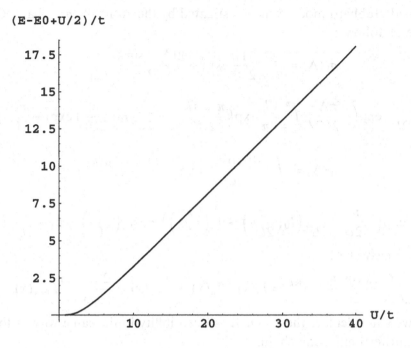

(E-E0+U/2)/t

Fig. 6.3. Energy gap for the charge excitation as a function of U.

by numerical calculations of these equations. At $B = \infty$ the magnetization is zero. To get the magnetic susceptibility for weak magnetic fields we must consider the behaviour of these equations at $B \gg 1, U'$.

Equation(6.71) is transformed as follows:

$$\sigma(\Lambda) - \int_B^\infty \frac{1}{U'}\Big\{R\Big(\frac{\Lambda - \Lambda'}{U'}\Big) + R\Big(\frac{\Lambda + \Lambda'}{U'}\Big)\Big\}\sigma(\Lambda')d\Lambda' = \sigma_0(\Lambda),$$

$$(6.78)$$

$$\sigma_0(\Lambda) \equiv \int_{-\pi}^\pi \frac{1}{4U'}\operatorname{sech}\Big(\frac{\pi(\Lambda - \sin k)}{2U'}\Big)\frac{dk}{2\pi}.$$

$$(6.79)$$

σ_0 is the solution at $B = \infty$. The energy (6.70) is as follows:

$$m = 2\int_B^\infty \sigma(\Lambda)d\Lambda, \tag{6.80}$$

$$e = e_0 - hm + 4t\int_B^\infty G_0(\Lambda)\sigma(\Lambda)d\Lambda, \tag{6.81}$$

$$G_0(\Lambda) = \int_{-\pi}^\pi \frac{dk\cos^2 k}{4U'}\operatorname{sech}\Big(\frac{\pi(\Lambda - \sin k)}{2U'}\Big). \tag{6.82}$$

$G_0(\Lambda)$ is the solution of (6.76) at $B = \infty$. The magnetization curve for the half-filled Hubbard model was investigated by the author[90]. $\sigma_0(\Lambda)$ and $G_0(\Lambda)$ behave as follows:

$$\sigma_0(\Lambda) = \int_{-\pi}^{\pi} \frac{1}{2U'} \exp\left(-\frac{\pi(\Lambda - \sin k)}{2U'}\right) \frac{dk}{2\pi}$$

$$= \frac{1}{2U'} \exp\left(-\frac{\pi\Lambda}{2U'}\right) \int_{-\pi}^{\pi} \frac{dk}{2\pi} \exp\left(\frac{\pi\cos k}{2U'}\right) = \frac{1}{2U'} I_0\left(\frac{\pi}{2U'}\right) \exp\left(-\frac{\pi\Lambda}{2U'}\right).$$

$$G_0(\Lambda) = \int_{-\pi}^{\pi} \frac{dk \cos^2 k}{2U'} \exp\left(-\frac{\pi(\Lambda - \sin k)}{2U'}\right)$$

$$= \exp\left(-\frac{\pi\Lambda}{2U'}\right) \frac{\pi}{2U'} \left(I_0\left(\frac{\pi}{2U'}\right) - I_2\left(\frac{\pi}{2U'}\right)\right) = 2I_1\left(\frac{\pi}{2U'}\right) \exp\left(-\frac{\pi\Lambda}{2U'}\right).$$

Here we have used

$$\int_{-\pi}^{\pi} \frac{dk \cos nk}{2\pi} e^{x\cos k} = I_n(x), \quad I_n(x) - I_{n+2}(x) = \frac{2(n+1)}{x} I_{n+1}(x).$$

Then we can calculate the magnetic susceptibility in the same way as for the XXX antiferromagnetic chain,

$$e - e_0 - 2h \int_B^{\infty} \sigma(\Lambda) d\Lambda + 8tI_1\left(\frac{\pi}{2U'}\right) \int_B^{\infty} \exp\left(\frac{-\pi\Lambda}{2U'}\right) \sigma(\Lambda) d\Lambda,$$

$$\sigma(\Lambda) - \int_B^{\infty} \frac{1}{U'} \left\{ R\left(\frac{\Lambda - \Lambda'}{U'}\right) + R\left(\frac{\Lambda + \Lambda'}{U'}\right) \right\} \sigma(\Lambda') d\Lambda'$$

$$= \frac{1}{2U'} I_0\left(\frac{\pi}{2U'}\right) \exp\left(\frac{-\pi\Lambda}{2U'}\right). \tag{6.83}$$

Putting $p(x) = 2U' \exp(\pi B/(2U'))\sigma(xU' + B)/I_0(\pi/(2U'))$ we have

$$e = e_0 - hI_0\left(\frac{\pi}{2U'}\right) \exp\left(\frac{-\pi B}{2U'}\right) \int_0^{\infty} p(x) dx$$

$$+ 8tI_0\left(\frac{\pi}{2U'}\right) I_1\left(\frac{\pi}{2U'}\right) \exp(-\pi B/U') \int_0^{\infty} e^{-\pi x/2} p(x) dx,$$

$$p(x) = \int_0^{\infty} \left\{ R(x - x') + R\left(x + x' + \frac{2B}{U'}\right) \right\} p(x') dx' + e^{-\pi x/2}. \tag{6.84}$$

Using the functions $a(x)$ and $b(x)$ defined in (3.60) we obtain the energy as function of B:

$$e = e_0 - 4h \exp\left(\frac{-\pi B}{2U'}\right) I_0\left(\frac{\pi}{2U'}\right) a\left(\frac{B}{U'}\right)$$

$$+ \frac{8t}{\pi} I_0\left(\frac{\pi}{2U'}\right) I_1\left(\frac{\pi}{2U'}\right) e^{-\pi B/U'} b\left(\frac{B}{U'}\right). \tag{6.85}$$

Then the magnetization and magnetic field are

$$h = \frac{4b_0 t}{\pi a_0} I_1\left(\frac{\pi}{2U'}\right) \exp\left(\frac{-\pi B}{2U'}\right)\left(1 + \frac{U'}{2\pi B}\right)^{-1},$$

$$m = 2a_0 I_0\left(\frac{\pi}{2U'}\right) \exp\left(\frac{-\pi B}{2U'}\right)\left(1 + \frac{U'}{2\pi B}\right). \tag{6.86}$$

a_0 and b_0 were given in (3.72). Thus m/h at a small magnetic field is given by

$$\chi \equiv \frac{m}{h} = \frac{1}{\pi t} \frac{I_0(\frac{\pi}{2U'})}{I_1(\frac{\pi}{2U'})}\left(1 + \frac{U'}{2\pi B}\right)^2. \tag{6.87}$$

These calculations were done by the author[90,91]. In the non-half-filled case one can calculate the susceptibility, although one must solve integral equations[83].

6.3 1/U' expansion

We analyze the analytic properties of the gap (6.69) and the ground state energy (6.46) in the half-filled case. The following expansion is very useful:

$$J_0(\omega)J_1(\omega) = \left\{\sum_{i=1}^{\infty} \frac{(-1)^i}{(i!)^2}\left(\frac{\omega}{2}\right)^{2i}\right\}\left\{\sum_{j=0}^{\infty} \frac{(-1)^j}{j!(j+1)!}\left(\frac{\omega}{2}\right)^{2j+1}\right\}$$

$$= \sum_{n=0}^{\infty} \frac{(-1)^n(2n+1)!}{[n!(n+1)!]^2}\left(\frac{\omega}{2}\right)^{2n+1}, \tag{6.88}$$

$$J_1(\omega) = \sum_{j=0}^{\infty} \frac{(-1)^j}{j!(j+1)!}\left(\frac{\omega}{2}\right)^{2j+1}. \tag{6.89}$$

Then we can show

$$\int_0^\infty \frac{J_0(\omega)J_1(\omega)}{\omega} e^{-2x\omega} d\omega = \sum_{n=0}^{\infty} \frac{(-1)^n(2n+1)!}{2^{2n+1}\{n!(n+1)!\}^2}$$

$$\times \int_0^\infty \omega^{2n} e^{-2x\omega} d\omega = \sum_{n=0}^{\infty} \frac{(-1)^n(2n+1)!!(2n-1)!!}{\{(2n+2)!!\}^2} \frac{1}{x^{2n+1}}$$

$$= x\left\{F\left(-\frac{1}{2}, \frac{1}{2}, 1; -\frac{1}{x^2}\right) - 1\right\}, \tag{6.90}$$

$$\int_0^\infty \frac{J_1(\omega)}{\omega} e^{-2x\omega} d\omega = \sum_{n=0}^{\infty} \frac{(-1)^n(2n-1)!!}{2^{2n+1}(n+1)!} \frac{1}{x^{2n+1}}$$

$$= x\left(\sqrt{1 + \frac{1}{x^2}} - 1\right). \tag{6.91}$$

Here $F(\alpha, \beta, \gamma; x)$ is the hypergeometric function. Using

$$1/(e^{2U'\omega} + 1) = e^{-2U'\omega} - e^{-4U'\omega} + -...,$$

we find

$$\Delta E = 2t\left[U' - 1 + 2\sum_{n=1}^{\infty}(-1)^{n-1}2nU'\left(\sqrt{1 + \frac{1}{(2nU')^2}} - 1\right)\right],$$

(6.92)

$$E/N_a = -A - 4t\sum_{n=1}^{\infty}(-1)^{n-1}nU'\left(F\left(-\frac{1}{2}, \frac{1}{2}; 1; -\frac{1}{(nU')^2}\right) - 1\right).$$

(6.93)

As functions of U', ΔE and E/N_a have branch points at $U' = \pm i/n$, $n = 1, 2, ...$. $U' = 0$ is the accumulation point of branch points. On the other hand expansion from $U' = \infty$ or $1/U'$ expansion gives

$$\Delta E = 2tU' - 2t + 4t\left[\frac{1}{2}\ln 2\ U'^{-1} - \left(\frac{1 \cdot 3}{2 \cdot 4}\right)\frac{\zeta(3)}{3}\left(1 - \frac{1}{2^2}\right)U'^{-3} + ...\right.$$

$$\left. + (-1)^{r-1}\left(\frac{(2r-1)!!}{(2r)!!}\right)\frac{\zeta(2r-1)}{2r-1}\left(1 - \frac{1}{2^{2r-2}}\right)U'^{-2r+1} + ...\right],$$

(6.94)

$$E/N_a = -A - 4t\left[\left(\frac{1}{2}\right)^2\ln 2\ U'^{-1} - \left(\frac{1 \cdot 3}{2 \cdot 4}\right)^2\frac{\zeta(3)}{3}\left(1 - \frac{1}{2^2}\right)U'^{-3} + ...\right.$$

$$\left. + (-1)^{r-1}\left(\frac{(2r-1)!!}{(2r)!!}\right)^2\frac{\zeta(2r-1)}{2r-1}\left(1 - \frac{1}{2^{2r-2}}\right)U'^{-2r+1} + ...\right].$$

(6.95)

These series converge at $|U'| > 1$. The expansion from $U' = 0$ is not convergent. So it is possible that we can get physically meaningful expansion from $U' = \infty$.

6.4 Perturbation expansion in the half-filled case

Let us consider the perturbation expansion of the general Hubbard Hamiltonian, putting

$$\mathcal{H}_0 = U\sum n_{i\uparrow}n_{i\downarrow} - A(n_{i\uparrow} + n_{i\downarrow}) - h(n_{i\uparrow} - n_{i\downarrow}),$$ (6.96)

$$\mathcal{H}_1 = -\sum_\sigma\sum_{i<j}t_{i,j}(c_{i\sigma}^\dagger c_{j\sigma} + c_{j\sigma}^\dagger c_{i\sigma}).$$ (6.97)

All sites are occupied by an electron with up or down spin. Then 2^{N_a} states degenerate at $t_{i,j} = h = 0$, $A = U/2$. The energy eigenvalue is $-UN_a/2$. Then in this subspace of the Hamiltonian a state is represented by $|\sigma_1, \sigma_2, ..., \sigma_{N_a}\rangle$. In the first order perturbation the matrix element $\langle\{\sigma\}|\mathcal{H}_1|\{\sigma'\}\rangle$ is always zero. In the second order perturbation one electron goes to the next site if the next site is occupied by an electron with opposite spin. This intermediate state has the higher energy $-UN_a/2 + U$. In the final state one electron returns to the vacant state and the unperturbed energy is again $-UN_a/2$. Then the energy of the ground states is described by the effective Hamiltonian:

$$\mathcal{H}_{\text{effective}} = -UN_a/2 + \sum_{i<j} \frac{t_{ij}t_{ji}}{U}(\sigma_i \cdot \sigma_j - 1). \tag{6.98}$$

Anderson showed that the XXX antiferromagnetic Hamiltonian can be derived from an itinerant electron model[3]. This is just the Heisenberg anti-ferromagnet with spin 1/2. Generally speaking the perturbation expansion of the L-fold degenerate ground state is given by $L \times L$ matrices. One can show that the odd-order elements are always zero. The fourth order term of the effective Hamiltonian is

$$U^{-3}\left[\sum_{i<j} t_{ij}^4(1 - \sigma_i \cdot \sigma_j) + \sum_{i<k} t_{ij}^2 t_{jk}^2(\sigma_i \cdot \sigma_k - 1) + \sum_{i<j<l, i<k, k\neq j,l}\right.$$
$$t_{ij}t_{jk}t_{kl}t_{li}(5(\sigma_j \cdot \sigma_k)(\sigma_i \cdot \sigma_l) + 5(\sigma_i \cdot \sigma_j)(\sigma_k \cdot \sigma_l)$$
$$-5(\sigma_j \cdot \sigma_k)(\sigma_i \cdot \sigma_l) - \sigma_i \cdot \sigma_j - \sigma_j \cdot \sigma_k - \sigma_k \cdot \sigma_l$$
$$\left. -\sigma_l \cdot \sigma_i - \sigma_i \cdot \sigma_k - \sigma_j \cdot \sigma_l + 1)\right]. \tag{6.99}$$

The third term gives the four spin interactions. This term cannot appear for the one-dimensional nearest neighbour Hubbard model because the square path is impossible on this lattice. This term was derived by the author[101]. On the two or three dimensional lattice this term plays an important role.

For the one-dimensional model the effective Hamiltonian becomes

$$\mathcal{H}_{\text{effective}} = -AN_a + \frac{t^2}{U}\sum_i (4\mathbf{S}_i \cdot \mathbf{S}_{i+1} - 1)$$
$$+ \frac{t^4}{U^3}\sum_i \left\{4(1 - 4\mathbf{S}_i \cdot \mathbf{S}_{i+1}) + (4\mathbf{S}_i \cdot \mathbf{S}_{i+2} - 1)\right\} + O\left(\frac{t^6}{U^5}\right). \tag{6.100}$$

Then the energy per site is given by

$$\frac{E}{N_a} = -A + \frac{t}{U'}\left(\langle\mathbf{S}_i \cdot \mathbf{S}_{i+1}\rangle_0 - \frac{1}{4}\right)$$
$$+ \frac{t}{U'^3}\left(-\frac{\langle\mathbf{S}_i \cdot \mathbf{S}_{i+1}\rangle_0}{4} + \frac{\langle\mathbf{S}_i \cdot \mathbf{S}_{i+2}\rangle_0}{16} + \frac{3}{64}\right) + O\left(\frac{t}{U'^5}\right), \tag{6.101}$$

where $\langle...\rangle_0$ is the correlation function of linear Heisenberg model. Comparing with (6.95) we have

$$\langle \mathbf{S}_i \cdot \mathbf{S}_{i+1} \rangle_0 = \frac{1}{4} - \ln 2 = -0.4431471806, \qquad (6.102)$$

$$\langle \mathbf{S}_i \cdot \mathbf{S}_{i+2} \rangle_0 = \frac{1}{4} - 4\ln 2 + \frac{9}{4}\zeta(3) = 0.1820393099. \qquad (6.103)$$

Thus we derive the second neighbour correlation function of XXX antiferromagnet from the ground state energy of the Hubbard model.

6.5 Asymptotic expansion from $U = 0$

As shown in the previous section the $U = 0$ point is a singular point and the usual expansion is not possible. But we need asymptotic behaviour of the energy gap and ground state energy in the half-filled case. For this purpose the following formula is very useful,

$$\sum_{n=1}^{\infty}(-1)^{n-1}(\sqrt{b^2 + (na)^2} - na) = -\frac{a}{4} + \frac{b}{2} + \int_{b/a}^{\infty} \frac{\sqrt{(ay)^2 - b^2}}{\sinh(\pi y)}dy,$$

$$a, b > 0. \qquad (6.104)$$

The l.h.s. is given by the following contour integral

$$\int_C \frac{1}{2i\sin\pi z}\{\sqrt{b^2 + a^2 z^2} - az\}dz. \qquad (6.105)$$

The contour is taken so that it surrounds the positive real z axis. We deform this path to the imaginary axis. See Fig. (6.4). We should note that the integrand has a pole at $z = 0$ and branch cuts at $z = iy$, $|y| \geq b/a$. Thus we get (6.104). Substituting this into (6.92) we have

$$\Delta E = 4t\int_{1/2U'}^{\infty} \frac{\sqrt{y^2 - (2U')^{-2}}}{\sinh\pi y}dy. \qquad (6.106)$$

Then at $U' \ll 1$ this behaves as

$$\Delta E \simeq 8t\exp(-\pi/(2U'))\int_0^{\infty} e^{-\pi t}\sqrt{t(t + U'^{-1})}dt \simeq \frac{4t}{\pi\sqrt{U'}}e^{-\pi/(2U')}. \qquad (6.107)$$

Then the energy gap is exponentially small for small U'.

To calculate the ground state energy we should note the following identity,

$$F(-\frac{1}{2}, \frac{1}{2}; 1; -x^2) = \frac{2}{\pi}\int_0^{\pi/2}\sqrt{1 + x^2\sin^2\phi}\,d\phi. \qquad (6.108)$$

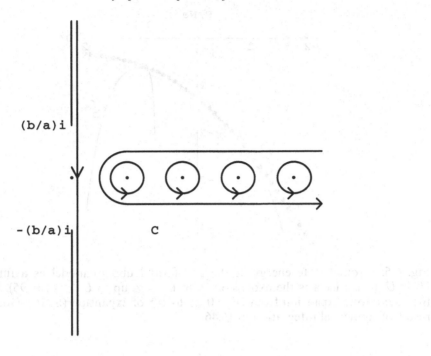

(b/a)i

−(b/a)i

C

Fig. 6.4. Contour integral on the complex plane. The path is deformed to the imaginary axis.

Then equation (6.93) is transformed as

$$(E/N_a + A)/(4t) = -\frac{2}{\pi}\int_0^{\pi/2}\sum_{n=1}^{\infty}(-1)^{n-1}(\sqrt{(nU')^2 + \sin^2\phi} - nU')d\phi. \quad (6.109)$$

Substituting (6.104) we have

$$= \frac{2}{\pi}\int_0^{\pi/2}d\phi\left(\frac{U'}{4} - \frac{\sin\phi}{2} - \int_{\sin\phi/U'}^{\infty}\frac{\sqrt{y^2 - (\sin\phi/U')^2}}{\sinh(\pi y)}dy\right). \quad (6.110)$$

If we change the order of integration of this double integral we have

$$= \frac{U'}{4} - \frac{1}{\pi} - \int_0^{\infty}\frac{G(U'y)}{\sinh(\pi y)}dy, \quad (6.111)$$

where

$$G(x) = \begin{cases} \dfrac{2}{\pi}\displaystyle\int_0^x\sqrt{\dfrac{x^2 - t^2}{1 - t^2}}dt, & \text{at } 0 \le x \le 1, \\[4mm] \dfrac{2}{\pi}\displaystyle\int_0^1\sqrt{\dfrac{x^2 - t^2}{1 - t^2}}dt, & \text{at } x \ge 1. \end{cases} \quad (6.112)$$

E/Na

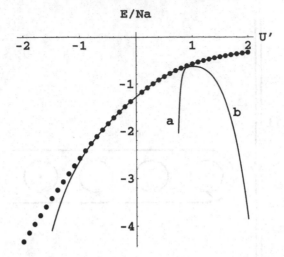

Fig. 6.5. Ground state energy of the half-filled Hubbard model as a function of $U' \equiv U/(4t)$. Line a is the expansion from $U' = \infty$ up to U'^{-21} of (6.95). Line b is the asymptotic expansion from $U' = 0$ up to U'^{6} of expansion (6.116). Dots are the result of numerical integration of (6.46).

This is written in complete elliptic integrals

$$
G(x) = \begin{cases} \dfrac{2}{\pi}\big\{E(x) - (1 - x^2)K(x)\big\}, & \text{at } 0 \le x \le 1, \\[4mm] \dfrac{2}{\pi}E(x^{-1})/x, & \text{at } x \ge 1. \end{cases} \tag{6.113}
$$

At $|x| < 1$, $G(x)$ is expanded as

$$
G(x) = \sum_{n=1}^{\infty}\Big(\frac{(2n-1)!!}{2^n n!}\Big)^2 \frac{2n}{(2n-1)^2}x^{2n}. \tag{6.114}
$$

We can show that

$$
\int_0^{\infty}\frac{x^{2n}}{\sinh(\pi x)}\,\mathrm{d}x = 2\int_0^{\infty}x^{2n}(e^{-\pi x} + e^{-3\pi x} + e^{-5\pi x} + ...)\,\mathrm{d}x
$$
$$
= \frac{2(2n)!}{\pi^{2n+1}}(1 + 3^{-(2n+1)} + 5^{-(2n+1)} + ...)
$$
$$
= \frac{2(2n)!}{\pi^{2n+1}}(1 - 2^{-(2n+1)})\zeta(2n + 1). \tag{6.115}
$$

Then we obtain an asymptotic expansion from $U' = 0$,

$$
(E/N_a + A)/(4t) = -\frac{1}{\pi} + \frac{U'}{4} - \frac{7\zeta(3)}{4\pi^3}U'^2 - \frac{93\zeta(5)}{32\pi^5}U'^4 - ...
$$

$$-\left(\frac{((2n-3)!!)^3(2n-1)(2^{2n+1}-1)}{(2n-2)!!2^{2n}}\right)\frac{\zeta(2n+1)}{\pi^{2n+1}}U'^{2n} - \dots$$

$$(6.116)$$

This is an asymptotic expansion and the convergence radius is zero. This was obtained by Economou and Poulopoulos[22]. Metzner and Vollhardt[68] calculated analytically the second-order term of the perturbation expansion of the ground state energy and certified that it coincides with the U'^2 term of equation (6.116).

Part two
Finite temperature integral equations for un-nested systems

7
Repulsive delta-function bosons

7.1 Uniqueness of the solution

In chapter 2 we consider the problem of bosons interacting via a repulsive delta-function potential with periodic boundary conditions. The ground state energy is obtained from a Fredholm-type integral equation. For this model Yang and Yang[123] found a non-linear integral equation which gives the free energy at a given temperature. This theory is important because it consists of the basic theory for thermodynamics of other solvable models.

The quasi-momenta k_js are determined by equation (2.30) for a set of quantum numbers I_j. The equation is

$$k_j L = 2\pi I_j - \sum_{l=1}^{N} 2\tan^{-1}\frac{k_j - k_l}{c}, \quad j = 1, ..., N, \tag{7.1}$$

where the I_js are integers (half-odd integers) for odd (even) N. This is equivalent to the following equations:

$$\frac{\partial B(k_1, k_2, ..., k_N)}{\partial k_j} = 0,$$

$$B(k_1, k_2, ..., k_N) \equiv \frac{L}{2}\sum k_j^2 - 2\pi I_j k_j + \sum_{j<l}\theta_1(k_j - k_l),$$

$$\theta_1(x) = x\tan^{-1}(x/c) - \frac{c}{2}\ln(1 + (x/c)^2). \tag{7.2}$$

The extremal point of the function B is the solution of (7.1) and gives an eigenstate of the Hamiltonian. Let us consider the following $N \times N$ matrix:

$$B_{jl} = \frac{\partial^2 B}{\partial k_j \partial k_l} = \delta_{jl}\left(L + \sum_m \frac{2c}{c^2 + (k_j - k_m)^2}\right) - \frac{2c}{c^2 + (k_j - k_l)^2}. \tag{7.3}$$

This matrix is always positive definite because

$$\sum_{lj} u_l B_{lj} u_j = L \sum u_l^2 + \sum_{l<j} \frac{2c}{c^2 + (k_j - k_l)^2} (u_j - u_l)^2 \geq 0, \qquad (7.4)$$

for an arbitrary real vector $\{u_j\}$. The function B is a concave function in N-dimensional space. Thus solutions of equation (7.1) for given set of $\{I_j\}$ are unique.

7.2 Holes of quasi-momenta and their distribution function

In the limit of $c \to \infty$ this set of eigenstates is complete. Let us consider the following function:

$$h(k) = k + \frac{2}{L} \sum_j \tan^{-1} \frac{k - k_j}{c}. \qquad (7.5)$$

The position of a hole is defined by

$$h(k_h) = \frac{2\pi}{L} \times \text{unoccupied (half-odd) integer.}$$

We denote the distribution function of holes as $\rho^h(k)$ and that of particles as $\rho(k)$. In the thermodynamic limit we have

$$h(k) = 2\pi \int^k \rho(t) + \rho^h(t) dt. \qquad (7.6)$$

Differentiating this yields

$$2\pi(\rho(k) + \rho^h(k)) = 1 + 2 \int_{-\infty}^{\infty} \frac{c\rho(k')dk'}{c^2 + (k - k')^2}. \qquad (7.7)$$

From equation (2.25) we have

$$e = \int_{-\infty}^{\infty} k^2 \rho(k) dk, \quad n = \int_{-\infty}^{\infty} \rho(k) dk. \qquad (7.8)$$

The entropy of the distribution between k and $k + dk$ is the logarithm of the number of orderings of $L\rho(k)dk$ particles and $L\rho^h(k)dk$ holes

$$\ln \frac{[L(\rho(k) + \rho^h(k))dk]!}{[L\rho(k)]![L\rho^h(k)]!} = Ldk[(\rho(k) + \rho^h(k)) \ln(\rho(k) + \rho^h(k))$$

$$-\rho(k) \ln \rho(k) - \rho^h(k) \ln \rho^h(k)].$$

Here we use the Stirling formula $\ln(n!) \simeq n(\ln n - 1)$. Thus the entropy per unit length is

$$s = \int_{-\infty}^{\infty} (\rho(k) + \rho^h(k)) \ln(\rho(k) + \rho^h(k)) - \rho(k) \ln \rho(k) - \rho^h(k) \ln \rho^h(k) dk. \qquad (7.9)$$

7.3 Thermodynamic equilibrium

The free energy per unit length $f = e - Ts$ must be minimized under the condition that n is constant. f and n are functionals of $\rho(k)$ and $\rho^h(k)$. Thus we should determine $\rho(k)$ and $\rho^h(k)$ such as to minimize $f - An$. Next we look for a parameter A such that n takes on its required value, in the variational calculus A is called a Lagrange multiplier. At the minimum point the variation of $f - An$ must be zero for any infinitesimal variation of functions

$$0 = \delta \int_{-\infty}^{\infty} (k^2 - A)\rho(k) - T\{(\rho(k) + \rho^h(k)) \ln(\rho(k) + \rho^h(k))$$
$$-\rho(k) \ln \rho(k) - \rho^h(k) \ln \rho^h(k)\} dk$$
$$= \int \left\{ k^2 - A - T \ln\left(\frac{\rho(k) + \rho^h(k)}{\rho(k)}\right) \right\} \delta\rho(k)$$
$$-T \ln\left(\frac{\rho(k) + \rho^h(k)}{\rho^h(k)}\right) \delta\rho^h(k) dk. \tag{7.10}$$

$\delta\rho(k)$ and $\delta\rho^h(k)$ are not independent. From (7.7) we have

$$\delta\rho^h(k) = -\delta\rho(k) + \frac{1}{\pi} \int_{-\infty}^{\infty} \frac{c\delta\rho(k')dk'}{c^2 + (k - k')^2}.$$

Substituting this into (7.10) one obtains

$$0 = \int_{-\infty}^{\infty} \delta\rho(k) \left\{ k^2 - A - T \ln\left(\frac{\rho^h(k)}{\rho(k)}\right) \right.$$
$$\left. -\frac{T}{\pi} \int_{-\infty}^{\infty} dq \frac{c}{c^2 + (k - q)^2} \ln\left(1 + \frac{\rho(q)}{\rho^h(q)}\right) \right\} dk. \tag{7.11}$$

This equation must hold for any arbitrary infinitesimal change of the function $\rho(k)$. Thus

$$T \ln\left(\frac{\rho^h(k)}{\rho(k)}\right) = k^2 - A - \frac{T}{\pi} \int_{-\infty}^{\infty} \frac{c}{c^2 + (k - q)^2} \ln\left(1 + \frac{\rho(q)}{\rho^h(q)}\right) dq, \tag{7.12}$$

must be satisfied at thermodynamic equilibrium. If we put $\epsilon(k) = T \ln(\rho^h(k)/\rho(k))$, this becomes

$$\epsilon(k) = k^2 - A - \frac{T}{\pi} \int_{-\infty}^{\infty} \frac{c}{c^2 + (k - q)^2} \ln\left(1 + \exp\left(-\frac{\epsilon(q)}{T}\right)\right) dq. \tag{7.13}$$

$\epsilon(k)$ can be determined by iteration. This function has physical meaning as the excitation energy for an elementary excitation, as will be shown in the next section. From equation (7.7) we have

$$2\pi\rho(k)\left(1 + \exp\left(\frac{\epsilon(k)}{T}\right)\right) = 1 + 2c\int_{-\infty}^{\infty}\frac{\rho(q)dq}{c^2 + (k - q)^2}. \tag{7.14}$$

From this equation we can determine $\rho(k)$.

If equation (7.13) is differentiated with respect to the chemical potential A, one obtains a linear integral equation

$$\frac{\partial\epsilon(k, A)}{\partial A} = -1 + \int dq\frac{c}{c^2 + (k - q)^2}\frac{1}{1 + \exp(\frac{\epsilon(k)}{T})}\frac{\partial\epsilon(q, A)}{\partial A}. \tag{7.15}$$

Comparing this with (7.14) we find

$$\frac{\partial\epsilon(k, A)}{\partial A} = -2\pi\rho(k)\left(1 + \exp\left(\frac{\epsilon(k)}{T}\right)\right) = -2\pi(\rho(k) + \rho^h(k)). \tag{7.16}$$

The entropy density (7.9) is

$$s = \int[(\rho + \rho^h)\ln(1 + \exp(-\epsilon(k)/T)) + \rho(k)\epsilon(k)/T]dk.$$

The free energy density is

$$f = \int(k^2 - \epsilon(k))\rho(k) - T(\rho(k) + \rho^h(k))\ln\left(1 + \exp\left(-\frac{\epsilon(k)}{T}\right)\right)dk. \tag{7.17}$$

Substituting (7.13) into (7.17) one obtains

$$f = \int\left[A + \frac{T}{\pi}\int\frac{dqc}{c^2 + (k - q)^2}\ln\left(1 + \exp\left(-\frac{\epsilon(q)}{T}\right)\right)\right]\rho(k)$$
$$-T(\rho(k) + \rho^h(k))\ln\left(1 + \exp\left(-\frac{\epsilon(k)}{T}\right)\right)dk$$
$$= \int A\rho(k)dk - T\int \ln\left(1 + \exp\left(-\frac{\epsilon(k)}{T}\right)\right)$$
$$\times\left[\rho(k) + \rho^h(k) - \frac{1}{\pi}\int\frac{c}{c^2 + (k - q)^2}\rho(q)dq\right]dk. \tag{7.18}$$

Substituting (7.7) we have a very simple expression for the free energy density:

$$f = An - T\int\frac{dk}{2\pi}\ln\left(1 + \exp\left(-\frac{\epsilon(k)}{T}\right)\right). \tag{7.19}$$

The thermodynamic potential density $g = f - An$ is:

$$g(T, A) = -T\int\ln\left(1 + \exp\left(-\frac{\epsilon(k)}{T}\right)\right)\frac{dk}{2\pi}. \tag{7.20}$$

For a given temperature T and chemical potential A we can calculate $\epsilon(k)$

through the non-linear integral equation (7.13) and obtain the thermodynamic potential g from (7.20). All thermodynamic quantities are derived from g through the thermodynamic relations

$$n = -\frac{\partial g}{\partial A}, \quad s = -\frac{\partial g}{\partial T}, \quad e = g + An + Ts, \ldots \quad (7.21)$$

The pressure p is $-g$.

7.4 Elementary excitations

We consider the change of energy and momentum when one particle is moved from thermodynamic equilibrium. Assume that the l-th particle is removed and one particle is added between the m-th and $m + 1$-th particles,

$$\{I_1 > I_2 > I_3 > \ldots > I_N\} \to$$
$$\{I_1 > I_2 > \ldots > I_m > I_l' > I_{m+1} > \ldots > I_{l-1} > I_{l+1} > \ldots > I_N\}$$

The Bethe-ansatz equations for the original state are

$$Lk_j = 2\pi I_j - \sum_{n=1}^{N} 2\tan^{-1}\frac{k_j - k_n}{c}, \quad j = 1, \ldots, N. \quad (7.22)$$

For the excited state they are

$$Lk_j' = 2\pi I_j - 2\tan^{-1}\frac{k_j' - k_l'}{c} - \sum_{n\neq m} 2\tan^{-1}\frac{k_j' - k_n'}{c}, \quad j \neq l$$

$$\quad (7.23)$$

$$Lk_l' = 2\pi I_l' - \sum_{n\neq m} 2\tan^{-1}\frac{k_l' - k_n'}{c}. \quad (7.24)$$

The change of momentum and energy are

$$\Delta K = k_l' - k_l + \sum_{n\neq l}(k_n' - k_n),$$

$$\Delta E = k_l'^2 - k_l^2 + \sum_{n\neq l}(k_n'^2 - k_n^2). \quad (7.25)$$

Subtracting (7.22) from (7.23) we have

$$\Delta k_j L + \sum \frac{2c}{c^2 + (k_j - k_l)^2}(\Delta k_j - \Delta k_l)$$

$$= 2\tan^{-1}\frac{k_j - k_l}{c} - 2\tan^{-1}\frac{k_j - k_l'}{c}. \quad (7.26)$$

Then equation for the back-flow $J(k) = \rho(k)\Delta kL$ is

$$\left(1 + \frac{\rho^h(k)}{\rho(k)}\right)J(k;k_l',k_l) - \int \frac{c}{\pi(c^2 + (k - k')^2)}J(k';k_l',k_l)dk'$$

$$= \frac{1}{\pi}\left(\tan^{-1}\frac{k - k_l}{c} - \tan\frac{k - k_l'}{c}\right). \tag{7.27}$$

Energy and momentum changes are

$$\Delta K(k_l',k_l) = k_l' - k_l + \int J(k;k_l',k_l)dk,$$

$$\Delta E(k_l',k_l) = k_l'^2 - k_l^2 + \int 2kJ(k;k_l',k_l)dk. \tag{7.28}$$

It is clear that $\Delta K(k_l,k_l) = \Delta E(k_l',k_l) = 0$. The differential of these with respect to k_l' is

$$\frac{\partial \Delta K}{\partial k_l'} = 1 + \int u(k;k_l')dk,$$

$$\frac{\partial \Delta E}{\partial k_l'} = 2k_l' + \int 2ku(k;k_l')dk, \tag{7.29}$$

where

$$u(k;k_l') \equiv \frac{\partial}{\partial k_l'}J(k;k_l',k_l),$$

$$(1 + \eta(k))u(k;k_l') - \int a(k - k')u(k';k_l')dk' = a(k - k_l'). \tag{7.30}$$

Using this integral equation we find that $u(k;k_l')$ is given by the infinite series:

$$u(k;k_l') = \frac{1}{1 + \eta(k)}\left[a(k - k_l') + \int dk_1 \frac{a(k - k_1)}{1 + \eta(k_1)}a(k_1 - k_l')\right.$$

$$\left. + \int\int dk_1 dk_2 \frac{a(k - k_1)}{1 + \eta(k_1)}\frac{a(k_1 - k_2)}{1 + \eta(k_2)}a(k_2 - k_l') + ...\right]. \tag{7.31}$$

If we substitute this into (7.29), we get an expression for $\frac{\partial \Delta K}{\partial k_l'}$ and $\frac{\partial \Delta E}{\partial k_l'}$ in terms of an infinite series:

$$\frac{\partial \Delta K}{\partial k_l'} = 1 + \int dk_1 a(k_l' - k_1)\frac{1}{1 + \eta(k_1)}$$

$$+ \int\int dk_1 dk_2 a(k_l' - k_1)\frac{1}{1 + \eta(k_1)}a(k_1 - k_2)\frac{1}{1 + \eta(k_2)} + ...,$$

$$\frac{\partial \Delta E}{\partial k_l'} = 2k_l' + \int dk_1 a(k_l' - k_1)\frac{1}{1 + \eta(k_1)}2k_1$$

$$+ \int\int dk_1 dk_2 a(k_l' - k_1)\frac{1}{1 + \eta(k_1)}a(k_1 - k_2)\frac{1}{1 + \eta(k_2)}2k_2 + ...$$

Substituting this into (7.29) we get an expression for $\frac{\partial \Delta K}{\partial k'_l}$ and $\frac{\partial \Delta E}{\partial k'_l}$:

$$\frac{\partial \Delta K(k'_l, k_l)}{\partial k'_l} = 1 + \int a(k'_l - k_1) \frac{1}{1 + \eta(k_1)} \frac{\partial \Delta K(k_1, k_l)}{\partial k_1} dk_1,$$

$$\frac{\partial \Delta E(k'_l, k_l)}{\partial k'_l} = 2k'_l + \int a(k'_l - k_1) \frac{1}{1 + \eta(k_1)} \frac{\partial \Delta E(k_1, k_l)}{\partial k_1} dk_1. \tag{7.32}$$

On the other hand we have equations for $2\pi(\rho(k) + \rho^h(k))$ and $\partial \epsilon(k)/\partial k$ from (7.7) and (7.13):

$$2\pi(\rho(k) + \rho^h(k)) = 1 + \int a(k - k') \frac{2\pi(\rho(k') + \rho^h(k'))}{1 + \eta(k')} dk',$$

$$\frac{\partial \epsilon(k)}{\partial k} = 2k + \int a(k - k') \frac{1}{1 + \eta(k')} \frac{\partial \epsilon(k')}{\partial k'} dk'. \tag{7.33}$$

The solution of non-singular linear integral equation is unique. Thus

$$\frac{\partial \Delta K(k'_l, k_l)}{\partial k'_l} = 2\pi(\rho(k'_l) + \rho^h(k'_l)),$$

$$\frac{\partial \Delta E(k'_l, k_l)}{\partial k'_l} = \frac{\partial \epsilon(k'_l)}{\partial k'_l}. \tag{7.34}$$

Integrating these one obtains

$$\Delta K(k'_l, k_l) = 2\pi \int_{k_l}^{k'_l} \rho(k) + \rho^h(k) dk,$$

$$\Delta E(k'_l, k_l) = \epsilon(k'_l) - \epsilon(k_l). \tag{7.35}$$

Thus $\epsilon(k) = T \ln(\rho^h(k)/\rho(k))$ has a physical meaning as the energy of an elementary excitation. In the limit of $T \to 0$ this excitation coincides with the excitation calculated in §2.1.7.

7.5 Some special limits

7.5.1 The $c = \infty$ limit

In the thermodynamic limit the system is equivalent to ideal spinless fermions. In this limit $c/(c^2 + (k - q)^2)$ in the integrand of (7.13) is zero. Thus $\epsilon(k) = k^2 - A$. From equation (7.14) we have

$$\rho(k) = \frac{1}{2\pi} \frac{1}{\exp[(k^2 - A)/T] + 1}. \tag{7.36}$$

From (7.20),

$$g = -\frac{1}{2\pi} \int dk \ln\left(1 + \exp\left(-\frac{k^2 - A}{T}\right)\right). \tag{7.37}$$

This is equivalent to (2.8).

7.5.2 The $c = 0+$ limit

In this limit the integration kernel $c/(c^2 + (k - q)^2)$ can be replaced by $\pi\delta(k - q)$. Equation (7.13) becomes $\epsilon(k) = k^2 - A - T \ln(1 + \exp(-\epsilon(k)/T))$. We obtain

$$\epsilon(k) = T \ln(\exp((k^2 - A)/T) - 1). \tag{7.38}$$

The Gibbs free energy per site g, pressure p, $\rho(k)$ and $\rho^h(k)$ are

$$g = -p = T \int \ln\left(1 - \exp\left(-\frac{k^2 - A}{T}\right)\right)\frac{\mathrm{d}k}{2\pi}, \quad \rho^h(k) = \frac{1}{2\pi},$$

$$\rho(k) = \frac{1}{2\pi}\frac{1}{\exp((k^2 - A)/T) - 1}. \tag{7.39}$$

These results coincide with those for ideal bosons in §2.1.1.

7.5.3 The $T = 0+$ limit

Generally speaking $\epsilon(k)$ is a monotonically increasing function of k^2. At $T = 0+$, assume that $\epsilon(q_0) = 0$. Thus

$$\rho(k) = 0 \quad \text{for} \quad k^2 > q_0^2, \quad \rho^h(k) = 0 \quad \text{for} \quad k^2 < q_0^2.$$

Equations (7.13) and (7.7) at $k < q_0$ are

$$\epsilon(k) = k^2 - A + \frac{c}{\pi}\int_{-q_0}^{q_0}\frac{\epsilon(q)\mathrm{d}q}{c^2 + (k - q)^2},$$

$$2\pi\rho(k) = 1 + 2c\int_{-q_0}^{q_0}\frac{\rho(q)\mathrm{d}q}{c^2 + (k - q)^2}. \tag{7.40}$$

The last equation is equivalent to the Lieb–Liniger equation (2.32).

The above theory was introduced by Yang and Yang[123]. They solved the equation numerically[124]. Very surprisingly it seems that this simple non-linear equation gives the exact free energy in the thermodynamic limit of repulsive bosons. The next problem is to find the thermodynamic Bethe-ansatz equation for other soluble models. As the next simplest case we treat the $S = 1/2$ XXX chain.

8

Thermodynamics of the XXX chain

8.1 String solution of an infinite XXX chain

The author obtained the thermodynamic Bethe-ansatz equations for the XXX model[93]. We consider the wave function in equation (3.18) assuming that N is infinity. Particle coordinates n_j move from $-\infty$ to ∞. Assume that $\Im k_n \leq \Im k_{n-1} \leq ... \leq \Im k_1$. The wave function is written as follows

$$a(n_1, n_2, ..., n_M) = (z_1 z_2 .. z_M)^{n_1} \sum_P A(P) \prod_{j=2}^{M} (\prod_{l=j}^{M} z_{Pl})^{n_{j+1} - n_j},$$

$$A(P) = \epsilon(P) \prod_{j<l}(x_{Pj} - x_{Pl} - 2i), \quad z_j = e^{ik_j} = \left(\frac{x_j + i}{x_j - i}\right). \tag{8.1}$$

In the infinite system, a satisfies the following boundary conditions:

$$| \lim_{n_1,...,n_r \to -\infty} a(n_1, n_2, ..., n_M)| < \infty,$$

$$| \lim_{n_{M-r+1},...,n_M \to \infty} a(n_1, n_2, ..., n_M)| < \infty. \tag{8.2}$$

From this condition we find that $|z_1 z_2 ... z_M| = 1$ and that $A(P) = 0$ if one of $| \prod_{l=j}^{M} z_{Pl}|$ is greater than one. If $\Im k_n < \Im k_{n-1} < ... < \Im k_1$, we have

$$A(I) \neq 0, \quad A(P \neq I) = 0, \quad |z_1 z_2 ... z_n| = 1. \tag{8.3}$$

These conditions are satisfied only if

$$x_j = \alpha + (n + 1 - 2j)i, \quad j = 1, 2, ..., n. \tag{8.4}$$

From $|z_1 z_2 ... z_M| = |\frac{\alpha + ni}{\alpha - ni}| = 1$, we find that α must be real. The condition

$$(\prod_{l=j+1}^{M} z_l) = |\frac{\alpha + (n - 2j)i}{\alpha - ni}| \leq 1, \quad j = 1, 2, ..., n - 1,$$

is automatically satisfied for general n. Thus strings with arbitrary length are possible for the XXX Heisenberg chain. Moreover the following type of string is impossible:

$$x_1 = \alpha + (2+\beta)i, \ x_2 = \alpha + \beta i, \ x_3 = \alpha - \beta i, \ x_4 = \alpha - (2+\beta)i, \qquad (8.5)$$

where α, β are real and $\beta \neq 1$.

8.2 String hypothesis for a long XXX chain

Now we return to the problem of a length N chain with periodic boundary condition. We assume that all rapidities x_j belong to bound states with $n = 1, 2, \dots$. For bound states of n xs the real parts of all the xs are the same and the imaginary parts are $(n-1)i, (n-3)i, \dots, -(n-1)i$ within the accuracy of $O(\exp(-\delta N))$. This assumption seems to be too strong and there are some counter examples in some extreme cases. This is a very controversial point of the thermodynamic Bethe-ansatz equations for soluble models, except for the repulsive boson case, which has no string solutions. But equations obtained using the string hypothesis seem to give the correct free energy and other thermodynamic quantities.

Consider the case where M_n bound states of n xs exist. We designate the xs as

$$x_\alpha^{n,j}, \quad \alpha = 1, 2, \dots, M_n,$$
$$x_\alpha^{n,j} = x_\alpha^n + i(n+1-2j) + \text{deviation}. \qquad (8.6)$$

From (3.17),

$$e^N(x_\alpha^{n,j}) = \prod_{(m,\beta) \neq (n,\alpha)} e\left(\frac{x_\alpha^{n,j} - x_\beta^m}{m-1}\right) e\left(\frac{x_\alpha^{n,j} - x_\beta^m}{m+1}\right)$$
$$\times \prod_{j' \neq j} e\left(\frac{x_\alpha^{n,j} - x_\alpha^{n,j'}}{2}\right), \qquad j = 1, 2, \dots, n. \qquad (8.7)$$

Here $e(x) \equiv (x+i)/(x-i)$. The last product is delicate because the numerator or denominator may become very small. If we take the product of these n equations, these delicate terms are cancelled and we have

$$e^N(x_\alpha^n/n) = \prod_{j=1}^{n} e^N(x_\alpha^{n,j}) = \prod_{(m,\beta) \neq (n,\alpha)} E_{nm}(x_\alpha^n - x_\beta^m), \qquad (8.8)$$

where

$$E_{nm}(x) \equiv$$

$$
\begin{cases}
e\left(\dfrac{x}{|n-m|}\right) e^2\left(\dfrac{x}{|n-m|+2}\right) ... e^2\left(\dfrac{x}{n+m-2}\right) e\left(\dfrac{x}{n+m}\right) \\
\qquad\qquad\qquad \text{for} \quad n \neq m, \\
e^2\left(\dfrac{x}{2}\right) e^2\left(\dfrac{x}{4}\right) ... e^2\left(\dfrac{x}{2n-2}\right) e\left(\dfrac{x}{2n}\right) \quad \text{for} \quad n = m.
\end{cases}
\tag{8.9}
$$

The logarithm of these equations gives

$$
N\theta(x_\alpha^n/n) = 2\pi I_\alpha^n + \sum_{(m,\beta)\neq(n,\alpha)} \Theta_{nm}(x_\alpha^n - x_\beta^m),
\tag{8.10}
$$

where

$$
\theta(x) \equiv 2\tan^{-1}(x),
\tag{8.11}
$$

and

$$\Theta_{nm}(x) \equiv$$

$$
\begin{cases}
\theta\left(\dfrac{x}{|n-m|}\right) + 2\theta\left(\dfrac{x}{|n-m|+2}\right) + ... + 2\theta\left(\dfrac{x}{n+m-2}\right) + \theta\left(\dfrac{x}{n+m}\right) \\
\qquad\qquad\qquad \text{for} \quad n \neq m, \\
2\theta\left(\dfrac{x}{2}\right) + 2\theta\left(\dfrac{x}{4}\right) + ... + 2\theta\left(\dfrac{x}{2n-2}\right) + \theta\left(\dfrac{x}{2n}\right) \quad \text{for} \quad n = m.
\end{cases}
\tag{8.12}
$$

I_α^n is an integer (half-odd integer) if $N - M_n$ is odd (even) and should satisfy:

$$
|I_\alpha^n| \leq \frac{1}{2}\left(N - 1 - \sum_{m=1}^{\infty} t_{nm} M_m\right), \quad t_{nm} \equiv 2\min(n,m) - \delta_{nm}.
\tag{8.13}
$$

We can prove the number of sets $\{I_\alpha^n\}$ is $C_M^N - C_{M-1}^N$ under the condition $M = \sum_{n=1}^{\infty} nM_n$. For details see Appendix B.

The energy of this state is given by:

$$
E(\{I_\alpha^n\}) = N\left(-h - \frac{J}{4}\right) + \sum_{n,\alpha}\left(\frac{2Jn}{(x_\alpha^n)^2 + n^2} + 2hn\right).
\tag{8.14}
$$

We can construct wave functions through (3.3), (3.5) and (3.12) at $S = S_z = N/2 - M$. Wave functions at $S_z = S - 1, S - 2, ..., -S$ are obtained by applying the operator S_{total}^-. The energies for these cases are $E - 2h, E - 4h, ..., E - 2(N - 2M)h$. Then the total number of states which are generated by the string assumption and descending operator at $S_z = N/2 - M$ is C_M^N. Therefore the total number of states is 2^N. This coincides with the true total number of states. It is expected that all eigenfunctions constructed in the above mentioned way should be a complete set. The partition function of this system is written as follows:

$$\mathscr{L} = \sum_{M=0}^{[N/2]} \frac{1 - \exp(-2(N+1-2M)h/T)}{1 - \exp(-2h/T)} \sum_{\{I_\alpha^n\}} \exp[-T^{-1}E(\{I_\alpha^n\})]. \quad (8.15)$$

The free energy is given by $G = -T \ln \mathscr{L}$. We define functions $h_n(x)$ by

$$h_n(x) \equiv \theta_n(x) - N^{-1} \sum_{(m,\alpha)} \Theta_{nm}(x - x_\alpha^m). \quad (8.16)$$

We can define holes in an n-string sea by the solution of

$$2\pi J_\beta^n/N = h_n(x_\beta^n), \quad (8.17)$$

where J_β^n are omitted integers or half-odd integers.

8.3 Thermodynamic Bethe-ansatz equations for the XXX chain

In the thermodynamic limit, we define distribution functions of n-strings and holes of n-strings as $\rho_n(x)$ and $\rho_n^h(x)$. The numbers of strings and holes between x and $x + dx$ are $\rho_n(x)Ndx$ and $\rho_n^h(x)Ndx$. Thus we have

$$2\pi \int^x \rho_n(t) + \rho_n^h(t)dt = \theta_n(x) - \sum_{m=1}^{\infty} \int_{-\infty}^{\infty} \Theta_{nm}(x - y)\rho_m(y)dy. \quad (8.18)$$

Differentiating with respect to x, we obtain the integral relation

$$a_n(x) = \rho_n(x) + \rho_n^h(x) + \sum_{m=1}^{\infty} T_{nm} * \rho_m(x). \quad (8.19)$$

Here $T_{nm}(x)$ is a function defined by

$$T_{nm}(x) \equiv \begin{cases} a_{|n-m|}(x) + 2a_{|n-m|+2}(x) + 2a_{|n-m|+4}(x) + \cdots \\ \\ \quad +2a_{n+m-2}(x) + a_{n+m}(x) \qquad \text{for } n \neq m, \\ \\ 2a_2(x) + 2a_4(x) + \cdots + 2a_{2n-2}(x) + a_{2n}(x) \quad \text{for } n = m. \end{cases} \quad (8.20)$$

$a_n(x)$ is a function defined by

$$a_n(x) \equiv \frac{1}{\pi} \frac{n}{x^2 + n^2}, \quad a_0(x) \equiv \delta(x). \quad (8.21)$$

We denote the convolution $\int_{-\infty}^{\infty} a(x - y)b(y)dy$ as $a * b(x)$ for two arbitrary functions $a(x)$ and $b(x)$.

The energy per site is

$$e = -\left(\frac{J}{4}+h\right) + \sum_{n=1}^{\infty} \int_{-\infty}^{\infty} g_n(x)\rho_n(x)dx,$$

$$g_n(x) \equiv 2\pi J a_n(x) + 2nh. \tag{8.22}$$

The total entropy per site is:

$$s = \sum_{n=1}^{\infty} \int_{-\infty}^{\infty} \rho_n(x)\ln\left(1 + \frac{\rho_n^h(x)}{\rho_n(x)}\right) + \rho_n^h(x)\ln\left(1 + \frac{\rho_n(x)}{\rho_n^h(x)}\right)dx. \tag{8.23}$$

$e - Ts$ should be minimized at thermodynamic equilibrium. Consider the functional variation of the free energy with respect to $\rho_n(x)$ and $\rho_n^h(x)$

$$0 = \delta e - T\delta s = \sum_{n=1}^{\infty} \int dx \left[g_n(x) - T\ln\left(1 + \frac{\rho_n^h(x)}{\rho_n(x)}\right)\right]\delta\rho_n(x)$$

$$- T\ln\left(1 + \frac{\rho_n(x)}{\rho_n^h(x)}\right)\delta\rho_n^h(x). \tag{8.24}$$

From equation (8.19) we have

$$\delta\rho_n^h(x) = -\delta\rho_n(x) - \sum_{m=1}^{\infty} T_{nm} * \delta\rho_m(x). \tag{8.25}$$

Substituting these into (8.24) yields

$$0 = T\sum_{n=1}^{\infty} \int \left\{\frac{g_n(x)}{T} - \ln\eta_n(x) + \sum_{m=1}^{\infty} T_{nm} * \ln(1 + \eta_m^{-1}(x))\right\}\delta\rho_n(x)dx, \tag{8.26}$$

where $\eta_n(x) \equiv \rho_n^h(x)/\rho_n(x)$. Thus we have integral equations for an infinite number of unknown $\eta_n(x)$,

$$\ln\eta_n(x) = \frac{g_n(x)}{T} + \sum_{m=1}^{\infty} T_{nm} * \ln(1 + \eta_m^{-1}(x)). \tag{8.27}$$

The free energy per site becomes as follows:

$$f = e - Ts = -\left(\frac{J}{4}+h\right) + \sum_{n=1}^{\infty} \int g_n\rho_n - T[\rho_n\ln(1 + \eta_n) + \rho_n^h\ln(1 + \eta_n^{-1})]dx.$$

We eliminate ρ_n^h using (8.19):

$$f = -\left(\frac{J}{4}+h\right) - T\sum_{n=1}^{\infty} \int \ln(1 + \eta_n^{-1})a_n(x)$$

$$+ \rho_n\left[\ln\eta_n - \frac{g_n}{T} - \sum_{m=1}^{\infty} T_{nm} * \ln(1 + \eta_m^{-1})\right]dx. \tag{8.28}$$

Using (8.27) one sees that the inside of the brackets on the r.h.s. is zero. So we have

$$f = -\left(\frac{J}{4} + h\right) - T \sum_{n=1}^{\infty} \int a_n(x) \ln(1 + \eta_n^{-1}(x)) dx. \tag{8.29}$$

From the $n = 1$ case of (8.27) we have

$$\ln(1 + \eta_1) = \frac{2\pi J a_1(x) + 2h}{T} + \sum_{l=1}^{\infty} (a_{l-1} + a_{l+1}) * \ln(1 + \eta_l^{-1}). \tag{8.30}$$

Operating $\int dx s(x)$ on this equation yields

$$\int dx s(x) \ln(1 + \eta_1) = \frac{2\pi J}{T} \int s(x) a_1(x) dx + \frac{h}{T}$$

$$+ \sum_{l=1}^{\infty} \int a_l(x) \ln(1 + \eta_l^{-1}(x)) dx.$$

Then equation (8.29) is transformed as follows:

$$f = J\left(\ln 2 - \frac{1}{4}\right) - T \int s(x) \ln(1 + \eta_1(x)) dx. \tag{8.31}$$

Solutions η_n of (8.27) are functions of x, J, T and h. Differentiating (8.27) with respect to J yields

$$\frac{2\pi J a_n(x)}{T} = \frac{1}{\eta_n} \frac{\partial \eta_n}{\partial J} + \sum_m T_{nm} * \frac{1}{(1 + \eta_m)\eta_m} \frac{\partial \eta_m}{\partial J}. \tag{8.32}$$

Comparing this with (8.19) we have

$$\rho_n = \frac{T}{2\pi} \frac{1}{(1 + \eta_n)\eta_n} \frac{\partial \eta_n}{\partial J}, \quad \rho_n + \rho_n^h = \frac{T}{2\pi} \frac{\partial \ln \eta_n}{\partial J}. \tag{8.33}$$

By the definition (8.20) we have:

$$a_1 * (T_{n-1,m} + T_{n+1,m}) - (a_0 + a_2) * T_{n,m} = (\delta_{n-1,m} + \delta_{n+1,m}) a_1. \tag{8.34}$$

Using equations (8.27) and (8.34) yields

$$(a_0 + a_2) * \ln \eta_1(x) = \frac{2\pi J a_1(x)}{T} + a_1 * \ln(1 + \eta_2(x)), \tag{8.35}$$

$$(a_0 + a_2) * \ln \eta_n(x) = a_1 * \ln(1 + \eta_{n-1}(x))(1 + \eta_{n+1}(x)),$$

$$n = 2, 3, \ldots. \tag{8.36}$$

Equations (8.35) and (8.36) are not sufficient to determine all the $\eta_n(x)$, as they do not contain h. Taking the $n = 1$ case of (8.27),

$$\ln \eta_1 = \frac{2\pi J a_1(x) + 2h}{T} + a_2 * \ln(1 + \eta_1^{-1}) + \sum_{j=2}^{\infty} (a_{j-1} + a_{j+1}) * \ln(1 + \eta_j^{-1}). \tag{8.37}$$

Substituting (8.35) and (8.36) we can eliminate $\eta_j, j < n$ for a given integer n,

$$\frac{2h}{T} = a_n * \ln \eta_{n+1} - a_{n+1} * \ln(1 + \eta_n) - a_{n+2} * \ln(1 + \eta_{n+1}^{-1})$$

$$- \sum_{l=n+2}^{\infty} (a_{l-1} + a_{l+1}) * \ln(1 + \eta_l^{-1}). \tag{8.38}$$

Thus we have

$$\ln \eta_{n+1} = \frac{2h}{T} + a_1 * \ln \eta_n + a_2 * \ln(1 + \eta_{n+1}^{-1})$$

$$+ \sum_{l=n+2}^{\infty} (a_{l-n-1} + a_{l-n+1}) * \ln(1 + \eta_l^{-1}). \tag{8.39}$$

For large n, $\ln(1 + \eta_n^{-1}) = o(n^{-2})$ and therefore:

$$\lim_{n\to\infty} \ln \eta_{n+1} - a_1 * \ln \eta_n = \frac{2h}{T}, \tag{8.40}$$

or

$$\lim_{n\to\infty} \frac{\ln \eta_n}{n} = \frac{2h}{T}. \tag{8.41}$$

Thus the following equations determine η_n,

$$\ln \eta_1(x) = \frac{2\pi J}{T} s(x) + s * \ln(1 + \eta_2(x)), \tag{8.42}$$

$$\ln \eta_n(x) = s * \ln(1 + \eta_{n-1}(x))(1 + \eta_{n+1}(x)), \tag{8.43}$$

$$\lim_{n\to\infty} \frac{\ln \eta_n}{n} = \frac{2h}{T}. \tag{8.44}$$

where

$$s(x) = \frac{1}{4}\mathrm{sech}\left(\frac{\pi x}{2}\right), \quad s * f(x) = \int s(x - y)f(y)\mathrm{d}y. \tag{8.45}$$

8.4 Some special cases and expansions

8.4.1 The $J/T \to 0$ case

In the limit $J/T \to 0$ and $h/T \geq 0$ we can expect that $\eta_n(x)$ is independent of x, because there are no x dependent terms in equations (8.42), (8.43), (8.44) and (8.45). As $\int \mathrm{d}x s(x) = 1/2$, equations (8.42), (8.43), (8.44) and (8.45) become

$$\eta_n^2 = (1 + \eta_{n-1})(1 + \eta_{n+1}), \tag{8.46}$$

$$\eta_1^2 = 1 + \eta_2, \lim_{n\to\infty} \ln \eta_n/n = 2h/T. \tag{8.47}$$

Equation (8.46) is a difference equation of second order. It is similar to a differential equation of second order and contains two arbitrary parameters. The general solution of this equation is:

$$\eta_n = \left(\frac{az^n - a^{-1}z^{-n}}{z - z^{-1}}\right)^2 - 1. \tag{8.48}$$

Parameters a and z are determined by (8.47) and we have $a = z, z = \exp(h/T)$ and

$$\eta_n = \left(\frac{\sinh[(n+1)h/T]}{\sinh[h/T]}\right)^2 - 1 \quad \text{for} \quad h > 0$$

$$\eta_n = (n+1)^2 - 1 \quad \text{for} \quad h = 0. \tag{8.49}$$

Substituting this into (8.31) we obtain the free energy, magnetization and entropy:

$$f = -T \ln[2 \cosh h/T], \quad m = 2s_z = -\partial f/\partial h = \tanh h/T,$$

$$s = -\partial f/\partial T = \ln[2 \cosh(h/T)] - (h/T)\tanh(h/T). \tag{8.50}$$

At $h = 0$ the entropy per site is $\ln 2$. This corresponds to the fact that the number of states per site is two.

8.4.2 High-temperature expansion or small J expansion

For the XXX chain system we can perform the high-temperature expansion of the free energy density from the definition,

$$f/T = -N^{-1} \ln \mathrm{Tr} \exp(-\mathcal{H}/T). \tag{8.51}$$

This is expanded as a power series of $1/T$. Assume that $\mathcal{H} = \mathcal{H}_0 + \mathcal{H}_1$ and $\mathcal{H}_0$ and $\mathcal{H}_1$ commute with each other. Then the exponential of $\mathcal{H}$ can be expanded as follows,

$$\exp(-\mathcal{H}/T) = \exp(-\mathcal{H}_0/T)\left(1 - T^{-1}\frac{\mathcal{H}_1}{1!} + T^{-2}\frac{\mathcal{H}_1^2}{2!} - + \dots\right). \tag{8.52}$$

Thus,

$$f/T = -N^{-1} \ln \mathrm{Tr} \exp(-\mathcal{H}_0/T) + \frac{\langle \mathcal{H}_1 \rangle}{NT} - \frac{\langle \mathcal{H}_1^2 \rangle - \langle \mathcal{H}_1 \rangle^2}{2!NT^2}$$

$$+ \frac{\langle \mathcal{H}_1^3 \rangle - 3\langle \mathcal{H}_1^2 \rangle\langle \mathcal{H}_1 \rangle + 2\langle \mathcal{H}_1 \rangle^3}{3!NT^3} - + \dots, \tag{8.53}$$

where

$$\langle X \rangle \equiv \frac{\mathrm{Tr} \exp(-\mathcal{H}_0/T)X}{\mathrm{Tr} \exp(-\mathcal{H}_0/T)}. \tag{8.54}$$

If we set

$$\mathscr{H}_0 = -2h \sum_{l=1}^{N} S_l^z, \quad \mathscr{H}_1 = -J \sum_{l=1}^{N} S_l^x S_{l+1}^x + S_l^y S_{l+1}^y + S_l^z S_{l+1}^z, \quad (8.55)$$

for the Hamiltonian (3.1), we obtain the J/T expansion of the free energy at fixed h/T:

$$f/T = -\ln(2 \cosh h/T) - \frac{J}{4T} \tanh^2(h/T)$$

$$-\frac{J^2}{32T^2}(3 + 2 \tanh^2(h/T) - 3 \tanh^4(h/T)) + O((J/T)^3). \quad (8.56)$$

The calculation to higher orders can be done by the use of a linked cluster expansion. Higher order terms are polynomials of $\tanh(h/T)$.

Apparently the expression for the free energy in (8.50) coincides with the first term of the above expansion. Writing $\ln(\eta_n + 1)$ as the expansion

$$\ln(\eta_n(x) + 1) = \ln\left(\frac{1}{\alpha_n - 1}\right) + \sum_{l=1}^{\infty} f_n^{(l)}\left(\frac{J}{T}\right)^l,$$

$$\alpha_n \equiv \frac{\sinh^2(h(n+1)/T)}{\sinh(hn/T) \sinh(h(n+2)/T)}, \quad (8.57)$$

we obtain an expansion of $\ln \eta_n(x)$:

$$\ln \eta_n(x) = \ln \frac{\alpha_n}{\alpha_n - 1} + \frac{J}{T} \alpha_n f_n^{(1)}$$

$$+ \left(\frac{J}{T}\right)^2 \left(\alpha_n f_n^{(2)} + (\alpha_n - \alpha_n^2)\frac{(f_n^{(1)})^2}{2}\right) + O\left(\left(\frac{J}{T}\right)^3\right). \quad (8.58)$$

Substituting these expansions into (8.42), (8.43), (8.44) and (8.45) and taking first order terms in J/T, linear integral equations for $f_n^{(1)}(x)$ are obtained,

$$\alpha_1 f_1^{(1)}(x) - s * f_2^{(1)}(x) = 2\pi s(x), \quad (8.59)$$

$$\alpha_n f_n^{(1)}(x) - s * (f_{n-1}^{(1)}(x) + f_{n+1}^{(1)}(x)) = 0, \quad (8.60)$$

$$\lim_{n \to \infty} \frac{\alpha_n f_n(x)}{n} = 0. \quad (8.61)$$

The r.h.s. of these equations are the inhomogeneous terms of the integral equations. The Fourier transform of these equations is

$$(e^{|\omega|} + e^{-|\omega|})\alpha_n \tilde{f}_n^{(1)}(\omega) = \tilde{f}_{n-1}^{(1)}(\omega) + \tilde{f}_{n+1}^{(1)}(\omega). \quad (8.62)$$

The general solution of this difference equation is

$$\tilde{f}_n^{(1)}(\omega) =$$
$$A(\omega)\left[\frac{\sinh((n+2)h/T)}{\sinh((n+1)h/T)}e^{-n|\omega|} - \frac{\sinh(nh/T)}{\sinh((n+1)h/T)}e^{-(n+2)|\omega|}\right]$$
$$+B(\omega)\left[\frac{\sinh((n+2)h/T)}{\sinh((n+1)h/T)}e^{n|\omega|} - \frac{\sinh(nh/T)}{\sinh((n+1)h/T)}e^{(n+2)|\omega|}\right].$$

$$(8.63)$$

From the boundary conditions we have

$$A(\omega) = \frac{\pi}{\cosh h/T}, \quad B(\omega) = 0. \tag{8.64}$$

Thus,

$$\tilde{f}_1^{(1)}(\omega) = \frac{\pi}{\cosh(h/T)}\left[\frac{\sinh 3h/T}{\sinh 2h/T}e^{-|\omega|} - \frac{\sinh h/T}{\sinh 2h/T}e^{-3|\omega|}\right],$$
$$f_1^{(1)}(x) = \frac{\pi}{\cosh(h/T)}\left[\frac{\sinh 3h/T}{\sinh 2h/T}a_1(x) - \frac{\sinh h/T}{\sinh 2h/T}a_3(x)\right]. \tag{8.65}$$

Substituting this into (8.31) we obtain the second term of the J/T expansion (8.56). The higher order terms can be calculated by solving the linear integral equations for $f_n^{(2)}, f_n^{(3)}, \ldots$. The equations are similar to (8.59), (8.60) and (8.61) except for the inhomogeneous terms, which are given by lower order $f_n(l)$.

8.4.3 The low-temperature limit

At low temperature $\ln \eta_n$ diverges as $1/T$. So we should define the following functions:

$$\epsilon_n(x) = T \ln \eta_n(x). \tag{8.66}$$

The integral equations become

$$\epsilon_1(x) = 2\pi J s(x) + s * T \ln\left(1 + \exp\left(\frac{\epsilon_2(x)}{T}\right)\right),$$
$$\epsilon_n(x) = s * T \ln\left(1 + \exp\left(\frac{\epsilon_{n-1}(x)}{T}\right)\right)\left(1 + \exp\left(\frac{\epsilon_{n+1}(x)}{T}\right)\right),$$
$$\lim_{n\to\infty} \frac{\epsilon_n(x)}{n} = 2h. \tag{8.67}$$

The free energy expression becomes

$$f = -\left(\frac{J}{4} + h\right) - T\sum_{n=1}^{\infty}\int a_n(x)\ln(1 + \exp(-\epsilon_n(x)/T))dx$$

$$= J\left(\ln 2 - \frac{1}{4}\right) - T \int s(x) \ln(1 + \exp(\epsilon_1(x)/T))dx. \tag{8.68}$$

The $T = 0$ limit of these equations is

$$\epsilon_1(x) = 2\pi J s(x) + s * \epsilon_2^+(x),$$

$$\epsilon_n(x) = s * (\epsilon_{n-1}^+(x) + \epsilon_{n+1}^+(x)),$$

$$\lim_{n\to\infty} \frac{\epsilon_n(x)}{n} = 2h, \tag{8.69}$$

$$f = -\left(\frac{J}{4} + h\right) + \sum_{n=1}^{\infty} \int a_n(x)\epsilon_n^-(x)dx$$

$$= J\left(\ln 2 - \frac{1}{4}\right) - \int s(x)\epsilon_1^+(x)dx,$$

$$\epsilon_n^+(x) \equiv \begin{cases} \epsilon_n(x), & \text{for } \epsilon_n(x) \geq 0 \\ 0, & \text{for } \epsilon_n(x) < 0. \end{cases}$$

$$\epsilon_n^-(x) \equiv \begin{cases} 0, & \text{for } \epsilon_n(x) \geq 0 \\ \epsilon_n(x), & \text{for } \epsilon_n(x) < 0. \end{cases} \tag{8.70}$$

In the ferromagnetic case $J > 0$ we have

$$\epsilon_n(x) = \epsilon_n^+(x) = 2\pi J a_n(x) + 2hn, \quad n = 1, 2, ..., \tag{8.71}$$

and therefore $f = -(\frac{J}{4} + h)$. This is the ground state energy of the ferromagnetic case.

In the antiferromagnetic case $J < 0$ we have

$$\epsilon_n(x) = \epsilon_n^+(x) = a_{n-1} * \epsilon_1^+(x) + 2(n-1)h, \quad n = 2, 3, ... \tag{8.72}$$

The equation which determines ϵ_1 is

$$\epsilon_1(x) = -2\pi|J|s(x) + h + \int_{|y|>B} R(x-y)\epsilon_1(y)dy, \quad \epsilon_1(\pm B) = 0. \tag{8.73}$$

In the limit of $h \to 0$, B becomes infinite. We have $\epsilon_1(x) = -2\pi|J|s(x)$ and $f = J(\ln 2 - \frac{1}{4})$. These results coincide with those of chapter 3.

8.4.4 The fugacity expansion

In the case of very large h the free energy can be expanded as a series in $z = \exp(-h/T)$. z is called the fugacity. From equations (8.22) and (8.27) we have expansions of η_n^{-1} as follows:

$$\eta_n^{-1} = z^{2n} \exp\left(-\frac{2\pi J}{T} a_n(x)\right) \exp\left[-T_{nm} * \left(\eta_m^{-1} - \frac{1}{2}\eta_m^{-2} + -...\right)\right]. \tag{8.74}$$

The expansion of η_1^{-1} and η_2^{-1} up to z^4 is

$$\eta_1^{-1} = z^2 \exp\left(-\frac{2\pi J}{T} a_1(x)\right)$$

$$\times \left(1 - z^2 \int a_2(x-y)\exp\left(-\frac{2\pi J}{T} a_1(y)\right) dy\right) + O(z^6),$$

$$\eta_2^{-1} = z^4 \exp\left(-\frac{2\pi J}{T} a_2(x)\right) + O(z^6). \tag{8.75}$$

As η_n^{-1} becomes small, (8.29) is more convenient than (8.31),

$$f = -\frac{J}{4} - h - T \sum_{n=1}^{\infty} \int dx a_n(x)\left(\eta_n^{-1} - \frac{1}{2}\eta_n^{-2} + -...\right).$$

Substituting (8.75) we obtain

$$f = -\frac{J}{4} - h - z^2 T \int a_1(x)\exp\left(-\frac{2\pi J}{T} a_1(x)\right) dx$$

$$-z^4 T\left\{\int a_2(x)\exp\left(-\frac{2\pi J}{T} a_2(x)\right) - \frac{1}{2}a_1(x)\exp\left(-\frac{4\pi J}{T} a_1(x)\right) dx\right.$$

$$\left. - \int dx \int dy a_1(x)\exp\left(-\frac{2\pi J}{T}(a_1(x) + a_1(y))\right) a_2(x-y)\right\} + O(z^6). \tag{8.76}$$

Putting $x = \tan u$, $y = \tan v$, we have

$$\left(f + \frac{J}{4} + h\right)/T$$

$$= z^2 e^{-K} I_0(K) + z^4\left\{-\frac{1}{2}e^{-2K}I_0(2K) + e^{-K/2}I_0(K/2)\right.$$

$$\left. -\frac{1}{\pi^2}\int_0^\pi \int_0^\pi \frac{e^{2K(1-\cos\omega_1\cos\omega_2)}(1-\cos\omega_1\cos\omega_2)}{1 - 2\cos\omega_1\cos\omega_2 + \cos^2\omega_1} d\omega_1 d\omega_2\right\}$$

$$+O(z^6), \tag{8.77}$$

where $K \equiv J/T$, $\omega_1 = \pi + u + v$, $\omega_2 = u - v$ and $I_0(x)$ is a modified Bessel function. This result is the same as that of Katsura[48].

   The strings are stable in the chain of infinite length with the condition that there are only a few down-spins. Some counter examples of the string assumption are found in some special limits. Nevertheless the thermodynamic Bethe-ansatz equation seems to give the exact free energy even in the case when the density of down-spins is comparable to the density of up-spins. This non-linear integral equation contains an infinite number of unknown functions. To solve this equation one needs to do numerical calculations by computer. For the conventional analysis we can use the high-temperature expansion method or the exact diagonalization method. For the investigation of the low-temperature thermodynamics of solvable models this method is the only way. For the spin $1/2$ ferromagnetic XXX chain, the susceptibility diverges as $T^{-\gamma}$ and the specific heat behaves as T^{α}. By numerical calculation using thermodynamic Bethe-ansatz equations, it was established that $\gamma = 2$ and $\alpha = 1/2$[109, 81].

9

Thermodynamics of the XXZ model

9.1 Thermodynamic equations for the XXZ model for $\Delta > 1$

Gaudin derived a set of thermodynamic Bethe-ansatz equations for $\Delta > 1$[30]. The wave function for M down-spins in an infinite lattice becomes

$$f(n_1, n_2, ..., n_M) = (z_1 z_2 ... z_M)^{n_1} \sum_P A(P) \prod_{j=2}^{M} (\prod_{l=j}^{M} z_{Pl})^{n_{j+1}-n_j},$$

$$z_j = e^{ik_j} = \left(\frac{\sin \frac{\phi}{2}(x_j + i)}{\sin \frac{\phi}{2}(x_j - i)} \right). \tag{9.1}$$

This corresponds to (8.1). From the normalizability condition of the wave function we have

$$A(I) \neq 0, A(P \neq I) = 0, \quad |z_1 z_2 ... z_M| = 1$$

$$| \prod_{l=j+1}^{M} z_l | \leq 1, \quad j = 1, 2, 3, ..., M-1. \tag{9.2}$$

These conditions are satisfied only if

$$x_j = \alpha + (M + 1 - 2j)i, \quad j = 1, 2, ..., M, \quad -Q < \alpha \leq -Q. \tag{9.3}$$

We can show that

$$| \prod_{l=j}^{M} z_l | = \left| \frac{\sin \frac{\phi}{2}(\alpha + i(M - 2j))}{\sin \frac{\phi}{2}(\alpha - iM)} \right|$$

$$= \sqrt{\frac{\cosh \phi(M - 2j) - \cos \phi \alpha}{\cosh \phi n - \cos \phi \alpha}} \leq 1, \quad 1 \leq j \leq M-1,$$

for arbitrary M. Thus a string with arbitrary length is possible for the XXZ chain for $|\Delta| > 1$. In the case $|\Delta| < 1$ the string condition is more complicated than for the case $|\Delta| \geq 1$. From (4.11) we have the following equation corresponding to (8.8):

$$e_n^N(x_\alpha^n) = \prod_{j=1}^{n} e^N(x_\alpha^{n,j}) = \prod_{(m,\beta)\neq(n,\alpha)} E_{nm}(x_\alpha^n - x_\beta^m). \tag{9.4}$$

Here

$$e_n(x) = \frac{\sin\frac{\phi}{2}(x+in)}{\sin\frac{\phi}{2}(x-in)}, \tag{9.5}$$

$$E_{nm}(x) \equiv \begin{cases} e_{|n-m|}(x)e_{|n-m|+2}^2(x)e_{|n-m|+4}^2(x)...e_{n+m-2}^2(x)e_{n+m}(x) \\ \qquad\qquad\text{for}\quad n \neq m, \\ \\ e_2^2(x)e_4^2(x)...e_{2n-2}^2(x)e_{2n}(x) \quad \text{for}\quad n = m. \end{cases} \tag{9.6}$$

x_α^n is the real part of the α-th string for strings of length n. The logarithm of (9.4) is

$$N\theta_n(x_\alpha^n) = 2\pi I_\alpha^n + \sum_{(m,\beta)\neq(n,\alpha)} \Theta_{nm}(x_\alpha^n - x_\beta^m), \tag{9.7}$$

where

$$\theta_n(x) = 2\tan^{-1}\left(\frac{\tan\frac{x\phi}{2}}{\tanh\frac{n\phi}{2}}\right) + 2\pi\left[\frac{\phi x + \pi}{2\pi}\right],$$

and

$$\Theta_{nm}(x) \equiv \begin{cases} \theta_{|n-m|}(x) + 2\theta_{|n-m|+2}(x) + ... + 2\theta_{n+m-2}(x) + \theta_{n+m}(x) \\ \qquad\qquad\text{for}\quad n \neq m, \\ \\ 2\theta_2(x) + 2\theta_4(x) + ... + 2\theta_{2n-2}(x) + \theta_{2n}(x) \quad \text{for}\quad n = m. \end{cases} \tag{9.8}$$

The function $\theta_n(x)$ is a quasi-periodic function which satisfies

$$\theta_n(x + 2jQ) = \theta_n(x) + 2\pi j, \quad j = \text{integer}.$$

We consider the energy of the general eigenstates which is given by the set of quantum numbers $\{I_\alpha^n\}$,

$$E(\{I_\alpha^n\}) = N\left(-h - \frac{J\Delta}{4}\right) + \sum_{n,\alpha}\left(\frac{2\pi J\sinh\phi}{\phi}a_n(x_\alpha^n) + 2hn\right), \tag{9.9}$$

where

$$a_n(x) = \frac{1}{2\pi}\frac{\phi\sinh n\phi}{\cosh n\phi - \cos\phi x}. \tag{9.10}$$

The partition function of the XXZ model is as follows:

$$\mathscr{Z} = \sum_{M=0}^{[N/2]} \left[1 + (1 - \delta_{N,2M}) \left(\exp - \frac{(N-2M)h}{T} \right) \right] \sum_{\{I_\alpha^n\}} \exp[-T^{-1} E(\{I_\alpha^n\})].$$

(9.11)

Corresponding to (8.16) we define the following functions:

$$h_n(x) \equiv \theta_n(x) - N^{-1} \sum_{(m,\alpha)} \Theta_{nm}(x - x_\alpha^m).$$

(9.12)

Using this function we can determine the position of holes for n-strings. We set distribution functions of particles and holes of n-strings as $\rho_n(x)$ and $\rho_n^h(x)$. By the equation (9.7) we have conditions for these two kinds of functions,

$$\mathbf{a}_n(x) = \rho_n(x) + \rho_n^h(x) + \sum_m \mathbf{T}_{nm} * \rho_m(x).$$

(9.13)

Here $\mathbf{a}_n$ was defined in (9.10) and

$$\mathbf{T}_{nm}(x) \equiv \begin{cases} \mathbf{a}_{|n-m|}(x) + 2\mathbf{a}_{|n-m|+2}(x) + 2\mathbf{a}_{|n-m|+4}(x) + \dots \\ \quad + 2\mathbf{a}_{n+m-2}(x) + \mathbf{a}_{n+m}(x) \qquad \text{for} \quad n \neq m, \\ \\ 2\mathbf{a}_2(x) + 2\mathbf{a}_4(x) + \dots + 2\mathbf{a}_{2n-2}(x) + \mathbf{a}_{2n}(x) \quad \text{for} \quad n = m. \end{cases}$$

(9.14)

Here the meaning of a convolution of two periodic functions $\mathbf{a}$ and $\mathbf{b}$ with periodicity $2Q$ is redefined:

$$\mathbf{a} * \mathbf{b}(x) \equiv \int_{-Q}^{Q} \mathbf{a}(x - y)\mathbf{b}(y)\mathrm{d}y.$$

(9.15)

The energy per site is

$$e = -\left(\frac{J\Delta}{4} + h \right) + \sum_{n=1}^{\infty} \int_{-Q}^{Q} g_n(x)\rho_n(x)\mathrm{d}x,$$

$$g_n(x) \equiv \frac{2\pi J \sinh \phi}{\phi} a_n(x) + 2nh.$$

(9.16)

The entropy per site s is

$$s = \sum_{n=1}^{\infty} \int_{-Q}^{Q} \rho_n(x) \ln\left(1 + \frac{\rho_n^h(x)}{\rho_n(x)} \right) + \rho_n^h(x) \ln\left(1 + \frac{\rho_n(x)}{\rho_n^h(x)} \right) \mathrm{d}x.$$

(9.17)

The condition of minimizing the free energy $e - Ts$ yields equations for $\eta_n(x) \equiv \rho_n^h(x)/\rho_n(x)$,

$$\ln \eta_n(x) = \frac{g_n(x)}{T} + \sum_{m=1}^{\infty} \mathbf{T}_{nm} * \ln(1 + \eta_m^{-1}(x)).$$

(9.18)

This set of equations is equivalent to the following:

$$\ln \eta_1(x) = \frac{2\pi J \sinh \phi}{T\phi} s(x) + s * \ln(1 + \eta_2(x)), \qquad (9.19)$$

$$\ln \eta_n(x) = s * \ln(1 + \eta_{n-1}(x))(1 + \eta_{n+1}(x)), \qquad (9.20)$$

$$\lim_{n\to\infty} \frac{\ln \eta_n}{n} = \frac{2h}{T}, \qquad (9.21)$$

where

$$s(x) = \frac{1}{4} \sum_{n=-\infty}^{\infty} \text{sech}\left(\frac{\pi(x - 2nQ)}{2}\right). \qquad (9.22)$$

The free energy per site is

$$f = -\left(\frac{J\Delta}{4} + h\right) - T \sum_{n=1}^{\infty} \int_{-Q}^{Q} a_n(x) \ln(1 + \eta_n^{-1}(x)) dx. \qquad (9.23)$$

Corresponding to (8.31) we have another expression for the free energy:

$$f = J\left(\frac{2\pi \sinh \phi}{\phi} \int_{-Q}^{Q} a_1(x)s(x)dx - \frac{\Delta}{4}\right) - T \int_{-Q}^{Q} s(x) \ln(1 + \eta_1(x)) dx. \qquad (9.24)$$

Equations (8.42)–(8.44) and (9.19)–(9.21) have almost the same structure. These equations are called the Gaudin–Takahashi equations[30, 93].

9.2 Theory for the $|\Delta| < 1$ XXZ model

9.2.1 String solution of the infinite XXZ model with $|\Delta| < 1$

The shapes of strings for $|\Delta| < 1$ is quite different from those for $|\Delta| \geq 1$. Takahashi and Suzuki[108] proposed a condition on the strings and constructed thermodynamic integral equations. Later Hida[37] and Fowler and Zotos[27] derived these conditions from the normalizability condition of the string wave function for an infinite chain. For $|\Delta| < 1$ there are two kinds of strings, one of which has its centre on the real axis and the other is centred on the $p_0 i$ axis,

$$x_j = \alpha + (n + 1 - 2j)i, \qquad (9.25)$$

$$x_j = \alpha + (n + 1 - 2j)i + p_0 i. \qquad (9.26)$$

We designate that the string of the former type has parity $v = 1$ and that the latter has parity $v = -1$. Applying the normalizability condition to the form (9.25) yields

$$1 > \left| \prod_{l=j+1}^{n} z_l \right| = \left| \frac{\sinh \frac{\gamma}{2}(\alpha + i(n - 2j))}{\sinh \frac{\gamma}{2}(\alpha - in)} \right| = \sqrt{\frac{\cosh \gamma\alpha - \cos \gamma(n - 2j)}{\cosh \gamma\alpha - \cos \gamma n}}.$$

Thus $\cos \gamma n < \cos \gamma(n - 2j)$ for $j = 1, 2, 3, ..., n - 1$. For (9.26) we have

$$1 > \left| \prod_{l=j+1}^{n} z_l \right| = \left| \frac{\cosh \frac{\gamma}{2}(\alpha + i(n - 2j))}{\cosh \frac{\gamma}{2}(\alpha - in)} \right| = \sqrt{\frac{\cosh \gamma\alpha + \cos \gamma(n - 2j)}{\cosh \gamma\alpha + \cos \gamma n}},$$

and therefore $\cos \gamma n > \cos \gamma(n - 2j)$ for $j = 1, 2, 3, ..., n - 1$. Then from the normalizability condition we get

$$0 < \cos\left(\frac{\pi}{2}(1 - v) + (n - 2j)\gamma\right) - \cos\left(\frac{\pi}{2}(1 - v) + n\gamma\right)$$
$$= 2v \sin((n - j)\gamma) \sin(j\gamma), \quad \text{for} \quad j = 1, 2, ..., n - 1. \tag{9.27}$$

This is equivalent to

$$\exp(\pi i([(n - j)\gamma/\pi] + [j\gamma/\pi])) = v, \quad \text{for} \quad j = 1, 2, ..., n - 1, \tag{9.28}$$

$$\frac{j\gamma}{\pi} \neq \left[\frac{j\gamma}{\pi}\right], \quad \text{for} \quad j = 1, 2, ..., n - 1, \tag{9.29}$$

where $[x]$ denotes the maximum integer less than or equal to x (Gauss' symbol). For rational $p_0 = \pi/\gamma$, (9.29) is a strong condition. If $p_0 = n_1/n_2$ and n_1 and n_2 are co-prime, a string greater than n_1 cannot satisfy at least one of (9.29). Thus $n \geq n_1 + 1$ strings are forbidden. Moreover for a $n = n_1$ string, the momentum is always 0 or π. So this string also has no meaning in thermodynamics. Next we seek the number n and parity v which satisfies (9.28) within $n < n_1$. Equation (9.28) is equivalent to

$$\left[\frac{(n - j)\gamma}{\pi}\right] + \left[\frac{j\gamma}{\pi}\right] \equiv \left[\frac{(n - j - 1)\gamma}{\pi}\right] + \left[\frac{(j + 1)\gamma}{\pi}\right] \quad (\text{Mod } 2),$$

$$j = 1, 2, ..., n - 2,$$

$$\left[\frac{(n - 1)\gamma}{\pi}\right] \equiv \frac{1 - v}{2} \quad (\text{Mod } 2).$$

As $[(n - j)\gamma/\pi] - [(n - j - 1)\gamma/\pi]$ is 0 or 1 and $[j\gamma/\pi] - [(j + 1)\gamma/\pi]$ is 0 or -1, we obtain

$$\left[\frac{(n - j)\gamma}{\pi}\right] + \left[\frac{j\gamma}{\pi}\right] = \left[\frac{(n - j - 1)\gamma}{\pi}\right] + \left[\frac{(j + 1)\gamma}{\pi}\right], \quad j = 1, 2, ..., n - 2.$$

These are strong restrictions on the parity v and the length n. The above conditions are equivalent to the following conditions which were given in reference[108]. The length n of a string should satisfy

$$2 \sum_{j=1}^{n-1} [j\gamma/\pi] = (n - 1)[(n - 1)\gamma/\pi], \tag{9.30}$$

$$v \sin\{(n - 1)\gamma\} \geq 0. \tag{9.31}$$

This condition was first introduced by the author[108] under the assumption that these strings form a complete half-filled state. Later Hida[37] and Fowler and Zotos[27] showed that conditions (9.30) and (9.31) can be re-derived from the normalizability condition of the wave function for $N \to \infty$ and finite M. For a given value of Δ (or γ) we can determine the series of n which satisfies the condition. This series becomes finite if γ/π is a rational number, and in this case the number of unknown functions becomes finite. We consider the $\gamma = \pi/v$ case. For $n = 1$, both $v = 1$ and -1 are possible. For $n = 2, 3, ..., v-1$ only $v = +1$ states are possible. These excitations have the following energy and momentum:

$$E = -2J\frac{\sin\gamma\,\sin(n\gamma)}{v\cosh(\alpha\gamma) - \cos n\gamma} + 2nh, \tag{9.32}$$

$$K = -i\ln\left[-\frac{\sinh\frac{1}{2}(\alpha\gamma + i(1-v)\pi/2 + in\gamma)}{\sinh\frac{1}{2}(\alpha\gamma + i(1-v)\pi/2 - in\gamma)}\right]. \tag{9.33}$$

The energy and momentum have the following relation:

$$E = -J\sin\gamma\frac{\cos n\gamma - \cos K}{\sin n\gamma} + 2nh. \tag{9.34}$$

The momentum is restricted to the region

$$|K| < \pi - \left(n\gamma - \pi\left[\frac{n\gamma}{\pi}\right]\right) \quad \text{for} \quad v = 1,$$

$$\pi \geq |K| > \pi - \left(n\gamma - \pi\left[\frac{n\gamma}{\pi}\right]\right) \quad \text{for} \quad v = -1.$$

Then for $n = v$ the energy and momentum are always zero. Only one state is obtained from this string solution. So we exclude this $n = v$ state from the thermodynamics of this case. So v string states $(1, +), (2, +), ..., (v-1, +), (1, -)$ play important roles. Especially at $\Delta = 0$, $\gamma = \pi/2$, $v = 2$ we have only the string states $(1, +), (1, -)$. These are single states at momentum $|K| < \pi/2$ and $|K| > \pi/2$.

Next we consider the $\gamma = \pi/(v_1 + 1/v_2)$ case. The following set of strings satisfy the condition (9.30) and (9.31),

$$(1, +), (2, +), ..., (v_1 - 1, +), (1, -), (1 + v_1, +), (1 + 2v_1, -),$$

$$..., (1 + (v_2 - 1)v_1, (-1)^{v_2-1}), (v_1, (-1)^{v_2}).$$

Then $v_1 + v_2$ strings are necessary to describe the thermodynamics of this case.

For a general rational number between 0 and 1, we can express it by a

Table 9.1. *Length n_j, parity v_j and q_j of strings for some rational values of γ/π*

j	$\frac{1}{5}$			$\frac{3}{16} =$		$\frac{1}{5+\frac{1}{3}}$	$\frac{13}{69} =$		$\frac{1}{5+\frac{1}{3+\frac{1}{4}}}$
1	1	+	4	1	+	13/3	1	+	56/13
2	2	+	3	2	+	10/3	2	+	43/13
3	3	+	2	3	+	7/3	3	+	30/13
4	4	+	1	4	+	4/3	4	+	17/13
5	1	−	−1	1	−	−3/3	1	−	−13/13
6				6	+	−2/3	6	+	−9/13
7				11	−	−1/3	11	−	−5/13
8				5	+	1/3	5	+	4/13
9							21	−	3/13
10							37	+	2/13
11							53	−	1/13
12							16	−	−1/13

continued fraction with length l:

$$\frac{\gamma}{\pi} = \frac{1|}{|v_1|} + \frac{1|}{|v_2|} + ... + \frac{1|}{|v_l|}, \quad v_1, v_2, ..., v_{l-1} \geq 1, \quad v_l \geq 2. \tag{9.35}$$

We define the following series of numbers $y_{-1}, y_0, y_1, ..., y_l$ and $m_0, m_1, ..., m_l$ as

$$y_{-1} = 0, \quad y_0 = 1, \quad y_1 = v_1 \quad \text{and} \quad y_i = y_{i-2} + v_i y_{i-1},$$

$$m_0 = 0, \quad m_i = \sum_{k=1}^{i} v_k. \tag{9.36}$$

The general rule to determine the parity v and length n is as follows:

$$n_j = y_{i-1} + (j - m_i)y_i, \quad v_j = (-1)^{[(n_j-1)/p_0]} \quad \text{for} \quad m_i < j < m_{i+1},$$

$$n_{m_l} = y_{l-1}, \quad v_{m_l} = (-1)^l. \tag{9.37}$$

The number of strings is m_l. We give examples for some rational numbers in Tables (9.1) and (9.2). We put x_j^α as the real part of strings with parity v_j and length n_j. α takes values from 1 to M_j. We find the following relations for these series of numbers:

$$n_j = \frac{1}{2}[(1 - 2\delta_{m_i,j})n_{j-1} + n_{j+1}], \quad \text{for} \quad m_i \leq j \leq m_{i+1} - 2,$$

$$n_j = (1 - 2\delta_{m_{i-1},j})n_{j-1} + n_{j+1}, \quad \text{for} \quad j = m_i - 1, \quad i < l,$$

$$n_0 = 0, \quad n_{m_l} + n_{m_l-1} = y_l. \tag{9.38}$$

(a) $\gamma = \pi/5$

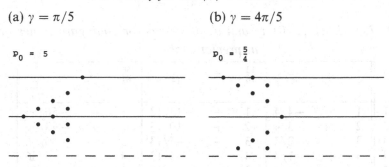

(b) $\gamma = 4\pi/5$

(c) $\gamma = 3\pi/16$

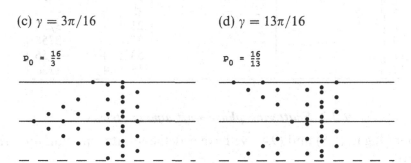

(d) $\gamma = 13\pi/16$

(e) $\gamma = 13\pi/69$

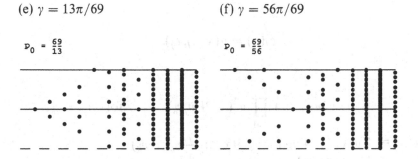

(f) $\gamma = 56\pi/69$

Fig. 9.1. Strings on the complex plane for given values of $p_0 = \pi/\gamma$. If we change γ to $\pi - \gamma$, the strings are almost the same. But imaginary parts of γx_j shift by π.

Table 9.2. *Length n_j, parity v_j and q_j of strings for conjugate values of γ/π in previous table*

j	$\frac{4}{5}$	$=$	$\frac{1}{1+\frac{1}{4}}$	$\frac{13}{16}$	$=$	$\frac{1}{1+\frac{1}{4+\frac{1}{3}}}$	$\frac{56}{69}$	$=$	$\frac{1}{1+\frac{1}{4+\frac{1}{3+\frac{1}{4}}}}$
1	1	$-$	$-4/4$	1	$-$	$-13/13$	1	$-$	$-56/56$
2	2	$+$	$-3/4$	2	$+$	$-10/13$	2	$+$	$-43/56$
3	3	$-$	$-2/4$	3	$-$	$-7/13$	3	$-$	$-30/56$
4	4	$+$	$-1/4$	4	$+$	$-4/13$	4	$+$	$-17/56$
5	1	$+$	$1/4$	1	$+$	$3/13$	1	$+$	$13/56$
6				6	$+$	$2/13$	6	$+$	$9/56$
7				11	$+$	$1/13$	11	$+$	$5/56$
8				5	$-$	$-1/13$	5	$-$	$-4/56$
9							21	$+$	$-3/56$
10							37	$-$	$-2/56$
11							53	$+$	$-1/56$
12							16	$+$	$1/56$

9.2.2 Scattering phase shift among strings

Corresponding to (8.8) and (9.4) we have a Bethe-ansatz equation for strings as follows:

$$\{e_j(x_\alpha^j)\}^N = -\prod_{k=1}^{m_l}\prod_{\beta=1}^{M_k} E_{j,k}(x_\alpha^j - x_\beta^k), \tag{9.39}$$

where

$$e_j(x) = g(x; n_j, v_j), \tag{9.40}$$

$$E_{j,k}(x) = \begin{cases} g(x; 2n_j, v_j v_k)\displaystyle\prod_{l=1}^{n_j-1} g^2(x; 2l, v_j v_k) & \text{for } n_j = n_k, \\[2mm] g(x; (n_j + n_k), v_j v_k)g(x; |n_j - n_k|, v_j v_k) & \\ \quad \times \displaystyle\prod_{l=1}^{\min(n_j,n_k)-1} g^2(x; |n_j - n_k| + 2l, v_j v_k) & \text{for } n_j \neq n_k, \end{cases} \tag{9.41}$$

$$g(x; n, +) = \frac{\sinh\frac{\gamma}{2}(x + in)}{\sinh\frac{\gamma}{2}(x - in)}, \quad g(x; n, -) = -\frac{\cosh\frac{\gamma}{2}(x + in)}{\cosh\frac{\gamma}{2}(x - in)}. \tag{9.42}$$

The logarithm of (9.39) yields

$$N\theta_j(x_\alpha^j) = 2\pi I_\alpha^j + \sum_{k=1}^{m_l}\sum_{\beta=1}^{M_k} \Theta_{jk}(x_\alpha^j - x_\beta^k), \quad \alpha = 1, 2, ..., M_j, \tag{9.43}$$

where

$$\theta_j(x) = f(x; n_j, v_j), \quad \Theta_{jk}(x) = f(x; |n_j - n_k|, v_j v_k) +$$

$$f(x; n_j + n_k, v_j v_k) + 2 \sum_{i=1}^{min(n_j, n_k)-1} f(x; |n_j - n_k| + 2i, v_j v_k), \tag{9.44}$$

and

$$f(x; n, v) = \begin{cases} 0 & \text{for } n\gamma/\pi = \text{integer}, \\ 2v \tan^{-1}\{(\cot(n\gamma/2))^v \tanh(\gamma x/2)\} & \text{otherwise}. \end{cases} \tag{9.45}$$

The quantity I_α^j is an integer (or half-odd integer) for M_j odd (or even), which is located in the region

$$|I_\alpha^j| < \frac{1}{2\pi} \left| N\theta_j(\infty) - \sum_{k=1}^{m_l} M_k \Theta_{jk}(\infty) \right|. \tag{9.46}$$

The function $f(x; n_j, v_j)$ is a monotonically increasing function for $m_{2i} \le j < m_{2i+1}$ and a monotonically decreasing function for $m_{2i-1} \le j < m_{2i}$.

9.2.3 Bethe-ansatz equation for the XXZ model with $|\Delta| < 1$

Following Yang and Yang we define particles and holes of strings. We obtain an integral relation between the distribution functions ρ_j and ρ_j^h of particles and holes of strings in the thermodynamic limit,

$$a_j(x) = \text{sign}(q_j)(\rho_j(x) + \rho_j^h(x)) + \sum_{k=1}^{m_l} T_{j,k} * \rho_k(x). \tag{9.47}$$

Here

$$a_j(x) \equiv (2\pi)^{-1} \frac{\mathrm{d}}{\mathrm{d}x} \theta_j(x), \quad T_{j,k}(x) \equiv (2\pi)^{-1} \frac{\mathrm{d}}{\mathrm{d}x} \Theta_{j,k}(x). \tag{9.48}$$

The symbol $a * b$ denotes the convolution of $a(x)$ and $b(x)$ as follows:

$$a * b(x) = \int_{-\infty}^{\infty} a(x - y)b(y)\mathrm{d}y. \tag{9.49}$$

The functions $a_j(x)$ and their Fourier transforms $\tilde{a}_j(\omega)$ are written as

$$a_j(x) = \frac{1}{2\pi} \frac{\gamma \sin \gamma q_j}{\cosh \gamma x + \cos \gamma q_j}, \quad \tilde{a}_j(\omega) = \frac{\sinh q_j \omega}{\sinh p_0 \omega}, \tag{9.50}$$

$$q_j \equiv (-1)^i(p_i - (j - m_i)p_{i+1}), \quad \text{for } m_i \le j < m_{i+1}, \tag{9.51}$$

where the p_i, $i = 0, 1, ..., l$ are defined by:

$$p_0 = \pi/\gamma, \quad p_1 = 1, \quad v_i = [p_{i-1}/p_i], \quad p_i = p_{i-2} - p_{i-1}v_{i-1}. \tag{9.52}$$

For the series q_j we find the following relations:

$$q_j = \frac{1}{2}[(1 - 2\delta_{m_i,j})q_{j-1} + q_{j+1}], \quad \text{for } m_i \leq j \leq m_{i+1} - 2,$$
$$q_j = (1 - 2\delta_{m_{i-1},j})q_{j-1} + q_{j+1}, \quad \text{for } j = m_i - 1, \ i < l,$$
$$q_0 = p_0, \quad n_{m_l} + n_{m_l-1} = 0. \tag{9.53}$$

The Fourier transform of $T_{j,k}(x)$ is given by

$$\tilde{T}_{j,k}(\omega) = \tilde{T}_{k,j}(\omega)$$
$$= 2\text{sign}(q_j)\coth(p_{i+1}\omega)\frac{\sinh((p_0 - |q_j|)\omega)\sinh(q_k\omega)}{\sinh(p_0\omega)}$$
$$+\delta_{j,m_l-1}\delta_{k,m_l} - \delta_{j,k}, \quad \text{for } j \leq k, \ m_i < j \leq m_{i+1}. \tag{9.54}$$

At $j = 1$ we have

$$\tilde{T}_{1,k}(\omega) = \text{sign}(q_1)2\cosh\omega a_k(\omega) - \delta_{1,k}. + \delta_{2,k}\delta_{2,m_l}. \tag{9.55}$$

The last term appear only at $\Delta = 0, \gamma = \pi/2, p_0 = 2$. At $j = m_l$ we have

$$\tilde{T}_{m_l,k}(\omega) = -\tilde{T}_{m_l-1,k}(\omega),$$
$$\tilde{T}_{m_l,m_l}(\omega) = -\tilde{T}_{m_l-1,m_l}(\omega) = \frac{\sinh((p_0 - 2p_l)\omega)}{\sinh(p_0\omega)}. \tag{9.56}$$

The energy and entropy per site are given by

$$e = -\left(\frac{J\Delta}{4} + h\right) + \sum_{j=1}^{m_l}\int_{-\infty}^{\infty} g_j(x)\rho_j(x)dx,$$
$$g_j(x) \equiv -\frac{2\pi J \sin\gamma}{\gamma}a_j(x) + 2n_jh, \tag{9.57}$$

and

$$s = \sum_{j=1}^{m_l}\int_{-\infty}^{\infty} \rho_j \ln\left(1 + \frac{\rho_j^h}{\rho_j}\right) + \rho_j^h \ln\left(1 + \frac{\rho_j}{\rho_j^h}\right)dx. \tag{9.58}$$

To minimize the free energy density $e - Ts$ with respect to ρ_j, we have

$$\delta(e - Ts) = \sum_j \int g_j(x)\delta\rho_j(x) - T\left\{\delta\rho_j \ln\left(1 + \frac{\rho_j^h}{\rho_j}\right) + \delta\rho_j^h \ln\left(1 + \frac{\rho_j}{\rho_j^h}\right)\right\}dx.$$

The variation of (9.47) gives

$$\delta\rho_j^h = -\delta\rho_j - \text{sign}(q_j)\sum T_{jk} * \delta\rho_k.$$

Thus we get

$$\delta(e - Ts) = T \sum_j \int dx \delta \rho_j(x)$$

$$\times \left\{ \frac{g_j(x)}{T} - \ln\left(\frac{\rho_j^h}{\rho_j}\right) + \sum_k \text{sign}(q_k) T_{j,k} * \ln\left(1 + \frac{\rho_k}{\rho_k^h}\right) \right\}.$$

At the thermodynamic equilibrium one obtains the following non-linear equations determining $\eta_j(x) \equiv \rho_j^h(x)/\rho_j(x)$:

$$\ln \eta_j(x) = g_j(x)/T + \sum_{k=1}^{m_l} \text{sign}(q_k) T_{k,j} * \ln(1 + \eta_k^{-1}(x)), \quad j = 1, ..., m_l. \quad (9.59)$$

The free energy is given as follows:

$$f = e - Ts = -\left(\frac{J\Delta}{4} + h\right) + \sum_{j=1}^{m_l} \int_{-\infty}^{\infty}$$

$$\rho_j(x)[g_j(x) - T \ln \eta_j(x)] - T(\rho_j + \rho_j^h) \ln(1 + \eta_j^{-1}) dx. \quad (9.60)$$

If we substitute (9.59) into the first bracket and (9.47) into the second, the $T_{j,k}$ terms are cancelled and we get

$$f = -\left(\frac{J\Delta}{4} + h\right) - T \sum_{j=1}^{m_l} \text{sign}(q_j) \int a_j(x) \ln(1 + \eta_j^{-1}(x)) dx. \quad (9.61)$$

If one uses the $j = 1$ case of equation (9.59) and (9.55) one obtains

$$f = -\frac{J\Delta}{4} - \text{sign}(q_1) \frac{2\pi J \sin \gamma}{\gamma} \int a_1(x) s_1(x) dx$$

$$-T \int s_1(x)[\ln(1 + \eta_1(x)) + \delta_{2,m_l} \ln(1 + \eta_2^{-1}(x))] dx. \quad (9.62)$$

The second term in the bracket is non-zero only at $\Delta = 0$. From equations (9.50)–(9.52) we get the following relations:

$$a_j - s_i * ((1 - 2\delta_{m_{i-1},j})a_{j-1} + a_{j+1}) = 0$$
$$\text{for } m_{i-1} \leq j \leq m_i - 2,$$
$$a_{m_i-1} - (1 - 2\delta_{m_{i-1},m_i-1})s_i * a_{m_i-2} - d_i * a_{m_i-1} - s_{i+1} * a_{m_i} = 0$$
$$\text{for } i < l,$$
$$a_{m_l-1}(x) = -a_{m_l}(x) = s_l * a_{m_l-2}, \quad (9.63)$$

where

$$a_0(x) = \delta(x),$$

$$s_i(x) \equiv \int_{-\infty}^{\infty} \frac{d\omega}{4\pi} \frac{e^{i\omega x}}{\cosh(p_i\omega)} = \frac{1}{4p_i}\operatorname{sech}\frac{\pi x}{2p_i},$$

$$d_i(x) \equiv \int_{-\infty}^{\infty} \frac{d\omega}{4\pi} \frac{e^{i\omega x}\cosh((p_i - p_{i+1})\omega)}{\cosh(p_i\omega)\cosh(p_{i+1}\omega)}. \tag{9.64}$$

Using (9.54) one can show the following relations:

$$T_{j,k} - s_i * ((1 - 2\delta_{m_{i-1},j})T_{j-1,k} + T_{j+1,k})$$
$$= (-1)^{i+1}(\delta_{j-1,k} + \delta_{j+1,k})s_i$$
$$\text{for } m_{i-1} \leq j \leq m_i - 2,$$
$$T_{m_i-1,k} - (1 - 2\delta_{m_{i-1},m_i-1})s_i * T_{m_i-2,k} - d_i * T_{m_i-1,k}$$
$$-s_{i+1} * T_{m_i,k} = (-1)^{i+1}(\delta_{m_i-2,k}s_i + \delta_{m_i-1,k}d_i - \delta_{m_i,k}s_{i+1})$$
$$\text{for } i = 1, 2, ..., l - 1,$$
$$T_{m_l-1,k} = -T_{m_l,k} = s_l * T_{m_l-2,k} + \operatorname{sign}(q_k)\delta_{m_l-2,k}s_l, \tag{9.65}$$

with $T_{0,k} = 0$. Using (9.63) and (9.65) one can rewrite (9.47) as follows:

$$\rho_j + \rho_j^h = s_i * (\rho_{j-1}^h + \rho_{j+1}^h) \text{ for } m_{i-1} \leq j \leq m_i - 2,$$
$$\rho_{m_i-1} + \rho_{m_i-1}^h = s_i * \rho_{m_i-2}^h + d_i * \rho_{m_i-1}^h - s_{i+1} * \rho_{m_i}^h,$$
$$\rho_{m_l-1} + \rho_{m_l-1}^h = \rho_{m_l} + \rho_{m_l}^h = s_l * \rho_{m_l-1}^h. \tag{9.66}$$

with $\rho_0^h = \delta(x)$. Equations (9.59) are rewritten as

$$\ln(1 + \eta_0) = -\frac{2\pi J \sin\gamma}{\gamma T}\delta(x),$$
$$\ln\eta_j = (1 - 2\delta_{m_{i-1},j})s_i * \ln(1 + \eta_{j-1}) + s_i * \ln(1 + \eta_{j+1}),$$
$$\text{for } m_{i-1} \leq j \leq m_i - 2, j \neq m_l - 2$$
$$\ln\eta_{m_i-1} = (1 - 2\delta_{m_{i-1},m_i-1})s_i * \ln(1 + \eta_{m_i-2})$$
$$+d_i * \ln(1 + \eta_{m_i-1}) + s_{i+1} * \ln(1 + \eta_{m_i}), \text{ for } i < l$$
$$\ln\eta_{m_l-2} = (1 - 2\delta_{m_{l-1},m_l-2})s_l * \ln(1 + \eta_{m_l-3})$$
$$+s_l * \ln((1 + \eta_{m_l-1})(1 + \eta_{m_l}^{-1})),$$
$$\ln\eta_{m_l-1} - y_l h/T = y_l h/T - \ln\eta_{m_l} = s_l * \ln(1 + \eta_{m_l-2}). \tag{9.67}$$

Then if we write $\ln\kappa(x) = \ln\eta_{m_l-1} - y_l h/T$ we obtain integral equations with

$m_l - 1$ unknown functions

$$\ln(1 + \eta_0) = -\frac{2\pi J \sin \gamma}{\gamma T}\delta(x),$$

$$\ln \eta_j = (1 - 2\delta_{m_{i-1},j})s_i * \ln(1 + \eta_{j-1}) + s_i * \ln(1 + \eta_{j+1})$$

$$\text{for } m_{i-1} \le j \le m_i - 2, j \ne m_l - 2,$$

$$\ln \eta_{m_i-1} = (1 - 2\delta_{m_{i-1},m_i-1})s_i * \ln(1 + \eta_{m_i-2})$$

$$+ d_i * \ln(1 + \eta_{m_i-1}) + s_{i+1} * \ln(1 + \eta_{m_i}) \text{ for } i < l,$$

$$\ln \eta_{m_l-2} = (1 - 2\delta_{m_{l-1},m_l-2})s_l * \ln(1 + \eta_{m_l-3})$$

$$+ s_l * \ln(1 + 2\cosh(y_l h/T)\kappa + \kappa^2),$$

$$\ln \kappa(x) = s_l * \ln(1 + \eta_{m_l-2}). \tag{9.68}$$

9.3 Some special limits

9.3.1 The $T \to \infty$ or $J \to 0$ limit

In equations (9.68), $\ln(1+\eta_0)$ becomes zero and the $\eta_j(x)$ are all independent of x. Thus we get the following difference equation:

$$\eta_j^2 = (1 + \eta_{j-1})^{1-2\delta_{m_{i-1},j}}(1 + \eta_{j+1})$$

$$\text{for } m_{i-1} \le j \le m_i - 2, j \ne m_l - 2,$$

$$\eta_{m_i-1}^2 = (1 + \eta_{m_i-2})^{1-2\delta_{m_{i-1},m_i-1}}(1 + \eta_{m_i-1})\ln(1 + \eta_{m_i})$$

$$\text{for } i < l,$$

$$\eta_{m_l-2}^2 = (1 + \eta_{m_l-3})^{1-2\delta_{m_{l-1},m_l-2}}(1 + 2\cosh(y_l h/T)\kappa + \kappa^2),$$

$$\kappa^2 = (1 + \eta_{m_l-2}). \tag{9.69}$$

The solution of this set of equations is

$$\eta_j = \left(\frac{\sinh(n_j + y_{i-1})h/T}{\sinh(y_{i-1}h/T)}\right)^2 - 1$$

$$\text{for } m_{i-1} < j \le m_i, \quad j \le m_l - 2,$$

$$\kappa = \frac{\sinh(n_{m_l-2} + y_{l-1})h/T}{\sinh(y_{l-1}h/T)}. \tag{9.70}$$

For $j = 1$ we have $\eta_1 = (2\cosh h/T)^2 - 1$. Substituting this into (9.62) we find the free energy,

$$f/T = -\ln(2\cosh h/T). \tag{9.71}$$

At $h = 0$ this gives that the entropy per site is $\ln 2$, as it should be.

9.3.2 The case $J > 0$, $-1 < \Delta < 0$ and $T \to 0$

We define $\epsilon_j(x) = T \ln \eta_j(x)$ and $\epsilon_j^+(x) = T \ln(1 + \eta_j(x))$. One can show that ϵ_j, $j \neq 1$ is always positive. The equation (9.59) gives

$$\epsilon_1(x) = -\frac{2\pi J \sin \gamma}{\gamma} a(x, 1) + 2h - \int a(x - y, 2)\epsilon_1^-(y)dy. \tag{9.72}$$

If $\epsilon_1(x) < 0$ at $|x| < B$ and $\epsilon_1(x) > 0$ at $|x| > B$, then one obtains a linear integral equation for $\rho_1(x)$,

$$\rho_1(x) + \int_{-B}^{B} a(x - y, 2)\rho_1(y)dy = a(x, 1). \tag{9.73}$$

This is equivalent to the Fredholm type equation obtained in (4.40).

9.3.3 The case $J > 0$, $0 < \Delta < 1$ and $T \to 0$

In this case $v_1 = 1$ because $\pi/2 < \gamma < \pi$ and $1 < p_0 < 2$. We should note that the length 1 and parity + string changed from 1 to $v_2 + 1$. One can show that ϵ_j, $j \neq v_2 + 1$ is always positive. The equation (9.59) gives

$$\epsilon_{v_2+1}(x) = -\frac{2\pi J \sin \gamma}{\gamma} a(x, 1) + 2h - \int a(x - y, 2)\epsilon_{v_2+1}^-(y)dy. \tag{9.74}$$

If $\epsilon(x)_{v_2+1} < 0$ at $|x| < B$ and $\epsilon_{v_2+1}(x) > 0$ at $|x| > B$, then one obtains a linear integral equation for $\rho_{v_2+1}(x)$,

$$\rho_{v_2+1}(x) + \int_{-B}^{B} a(x - y, 2)\rho_{v_2+1}(y)dy = a(x, 1). \tag{9.75}$$

This is equivalent to the Fredholm type equation obtained in (4.40).

9.3.4 The $\Delta = 0$ case

Here we have $\gamma = \pi/2$, $v_1 = 2$, $l = 1$ and $m = 2$. Then from the first and last equations of (9.67) we have

$$\ln \eta_1(x) = 2h + 4Js_1(x)/T, \quad \ln \eta_2(x) = 2h - 4Js_1(x)/T. \tag{9.76}$$

Substituting these into (9.61) or (9.62) we get the free energy, which coincides with (1.19).

10

Thermodynamics of the XYZ model

10.1 Bethe-ansatz equation for the XYZ model

The low temperature behaviour of the XYZ model using thermodynamic Bethe-ansatz equations was investigated by the author[96,97]. In the Bethe-ansatz equation (4.15), the function $\sinh(\gamma x/2)$ is merely replaced by the elliptic theta functions. We should consider equation (5.54),

$$\left(\frac{H_l(i\zeta(x_l + i))}{H_l(i\zeta(x_l - i))}\right)^N = -e^{-2\pi i v'/p_0} \prod_{j=1}^{N/2} \frac{H_l(i\zeta(x_l - x_j + 2i))}{H_l(i\zeta(x_l - x_j - 2i))},$$

$$\sum_l^{N/2} x_l = Qv' + ip_0 v, \quad Q = K(l')/\zeta, \quad p_0 = K(l)/\zeta. \tag{10.1}$$

In the limit $l \to 0$ it becomes

$$\left(\frac{\sin(i\zeta(x_l + i))}{\sin(i\zeta(x_l - i))}\right)^N = -e^{-2\pi i v'/p_0} \prod_{j=1}^{N/2} \frac{\sin(i\zeta(x_l - x_j + 2i))}{\sin(i\zeta(x_l - x_j - 2i))}. \tag{10.2}$$

This is an equivalent equation to (4.15), if we assume that the x_ls are finite and $v' = 0$. ζ becomes $\gamma/2$ and K_l becomes $\pi/2$. So it is natural that we assume the same type of strings which can be determined by $p_0 = K_l/\zeta$,

$$x_j = \alpha + (n + 1 - 2j)i,$$
$$x_j = \alpha + (n + 1 - 2j)i + p_0 i, \quad -Q < \alpha \le Q. \tag{10.3}$$

The solution becomes doubly periodic. So we consider the distribution of solutions for $-Q < \Re x \le Q$ and $-p_0 < \Im x \le p_0$. It is expected that the same kind of strings appear in the case of the XXZ model at $\pi/\gamma = p_0$. We

145

can determine the n_js and q_js as (9.38) and (9.51),

$$\{e_j(x_\alpha^j)\}^N = -\exp(-2\pi i v'/p_0) \prod_{k=1}^{m_l} \prod_{\beta=1}^{M_k} E_{j,k}(x_\alpha^j - x_\beta^k),$$

$$v' = \frac{1}{Q} \sum_{j=1}^{m_l} \sum_{\alpha=1}^{M_j} n_j x_\alpha^j, \tag{10.4}$$

where

$$e_j(x) = g(x; n_j, v_j), \tag{10.5}$$

$$g(x; n, +) = \frac{H_l(i\zeta(x+in))}{H_l(i\zeta(x-in))},$$

$$g(x; n, -) = -\frac{H_l(K_l + i\zeta(x+in))}{H_l(K_l + i\zeta(x-in))}, \tag{10.6}$$

$$E_{j,k}(x) = \begin{cases} g(x; 2n_j, v_j v_k) \displaystyle\prod_{l=1}^{n_j-1} g^2(x; 2l, v_j v_k) & \text{for } n_j = n_k, \\[2ex] g(x; (n_j+n_k), v_j v_k) g(x; |n_j-n_k|, v_j v_k) \\[1ex] \times \displaystyle\prod_{l=1}^{\min(n_j,n_k)-1} g^2(x; |n_j-n_k|+2l, v_j v_k) & \text{for } n_j \neq n_k. \end{cases} \tag{10.7}$$

Taking the logarithm of (10.4) we have

$$N\theta_j(x_\alpha^j) = 2\pi I_\alpha^j - 2\pi v'/p_0 + \sum_{k=1}^{m_l} \sum_{\beta=1}^{M_k} \Theta_{j,k}(x_\alpha^j - x_\beta^k), \quad \alpha = 1, 2, ..., M_j, \tag{10.8}$$

where

$$\theta_j(x) = \mathbf{f}(x; n_j, v_j), \quad \Theta_{jk}(x) = \mathbf{f}(x; |n_j-n_k|, v_j v_k) +$$

$$\mathbf{f}(x; n_j+n_k, v_j v_k) + 2 \sum_{i=1}^{\min(n_j,n_k)-1} \mathbf{f}(x; |n_j-n_k|+2i, v_j v_k), \tag{10.9}$$

and $\mathbf{f}(x, n, v)$ is defined by

$$\mathbf{f}(x, n, v) = f(x, n, v) + \sum_{l=1}^{\infty} f(x - 2lQ, n, v) + f(x + 2lQ, n, v).$$

$f(x, n, v)$ was defined in (9.45). An eigenstate should be identified by the set of quantum numbers I_j^α. From (5.74) the energy must be

$$E = -NJ_zR - \frac{J_z\pi \text{sn}2\zeta}{\zeta} \sum_{l=1}^{N/2} \mathbf{a}(x_l, 1)$$

$$= -NJ_zR - \frac{J_z\pi \text{sn}2\zeta}{\zeta} \sum_{j=1}^{m_l} \sum_{\alpha=1}^{M_j} \mathbf{a}_j(x_\alpha^j),$$

$$R \equiv \frac{1}{4}\left[1 - \frac{\pi \text{sn}2\zeta}{\zeta}(\mathbf{a}(0,1) + \mathbf{a}(Q,1))\right],$$

$$\mathbf{a}_j(x) \equiv \frac{1}{2Q}\left[\frac{q_j}{p_0} + 2\sum_{l=1}^{\infty} \frac{\sinh(q_j\pi l/Q)}{\sinh(p_0\pi l/Q)}\cos(\pi jx/Q)\right], \qquad (10.10)$$

where q_j was defined in (9.51). The number of zeros must be $N/2$,

$$N/2 = \sum_{j=1}^{m_l} n_j M_j. \qquad (10.11)$$

Thus the energy per site is given by

$$e = -J_zR - \frac{J_z\pi \text{sn}2\zeta}{\zeta} \sum_{j=1}^{m_l} \int_{-Q}^{Q} \mathbf{a}_j(x)\rho_j(x)dx. \qquad (10.12)$$

The entropy per site is

$$s = \sum_{j=1}^{m_l} \int_{-Q}^{Q} \rho_j \ln\left(1 + \frac{\rho_j^h}{\rho_j}\right) + \rho_j^h \ln\left(1 + \frac{\rho_j}{\rho_j^h}\right)dx. \qquad (10.13)$$

From (10.9) we have the relation between $\rho_j(x)$ and $\rho_j^h(x)$:

$$\mathbf{a}_j(x) = \text{sign}(q_j)(\rho_j(x) + \rho_j^h(x)) + \sum_{k=1}^{m_l} \mathbf{T}_{j,k} * \rho_k(x). \qquad (10.14)$$

Moreover from (10.11),

$$\frac{1}{2} = m \equiv \sum_{j=1}^{m_l} n_j \int_{-Q}^{Q} \rho_j(x)dx. \qquad (10.15)$$

Next we need a Lagrange multiplier to guarantee the condition (10.15). One should minimize $e - Ts + 2hm$ under conditions (10.14), and after that the multiplier h should be chosen so that (10.15) is satisfied. Just in the same way as (9.59) was derived, we get the following integral equations for $\eta_j(x) = \rho_j^h(x)/\rho_j(x)$,

$$\ln \eta_j(x) = \mathbf{g}_j(x)/T + \sum_{k=1}^{m_l} \text{sign}(q_k)\mathbf{T}_{k,j} * \ln(1 + \eta_k^{-1}(x)), \quad j = 1, ..., m_l. \quad (10.16)$$

Here *, $g_j(x)$ and $T_{j,k}$ are

$$f * g(x) = \int_{-Q}^{Q} f(x - y)g(y)dy,$$

$$g_j(x) \equiv -\frac{J_z \pi \text{sn} 2\zeta}{\zeta} a_j(x) + 2n_j h,$$

$$T_{j,k}(x) = \frac{1}{2Q} \sum_{n=-\infty}^{\infty} e^{i\pi nx/Q} \tilde{T}_{j,k}\left(\frac{n}{2Q}\right) = \sum_{n} T_{j,k}(x - 2nQ).$$

The quantity $g \equiv e - Ts + 2hm$ is given as follows:

$$g(J_z, T, h) = -J_z R$$

$$+ \sum_{j=1}^{m_l} \int_{-Q}^{Q} \rho_j(x)[g_j(x) - T \ln \eta_j(x)] - T[\rho_j + \rho_j^h] \ln(1 + \eta_j^{-1})dx$$

$$= -J_z R - T \sum_{j=1}^{m_l} \text{sign}(q_j) \int_{-Q}^{Q} a_j(x) \ln(1 + \eta_j^{-1}(x))dx. \qquad (10.17)$$

Corresponding to (9.62) this is

$$g = -J_z R + h - \text{sign}(q_1)\frac{\pi J_z \text{sn} 2\zeta}{\zeta} \int_{-Q}^{Q} a_1(x)s_1(x)dx$$

$$-T \int_{-Q}^{Q} s_1(x)[\ln(1 + \eta_1(x)) + \delta_{2,m_l} \ln(1 + \eta_2^{-1}(x))]dx. \qquad (10.18)$$

Then m should be determined by

$$m = \frac{1}{2}\frac{\partial g}{\partial h} = \frac{1}{2} - \frac{1}{2} \int_{-Q}^{Q} s_1(x)(1 + \eta_1(x))^{-1}\frac{\partial \eta_1(x)}{\partial h}dx. \qquad (10.19)$$

The equation (10.16) is also equivalent to the following block tridiagonal equations:

$$\ln(1 + \eta_0) = -\frac{\pi J_z \text{sn} 2\zeta}{\zeta T}\delta(x),$$

$$\ln \eta_j = (1 - 2\delta_{m_{i-1},j})s_i * \ln(1 + \eta_{j-1}) + s_i * \ln(1 + \eta_{j+1}),$$

$$\text{for} \quad m_{i-1} \leq j \leq m_i - 2, j \neq m_l - 2$$

$$\ln \eta_{m_i-1} = (1 - 2\delta_{m_{i-1},m_i-1})s_i * \ln(1 + \eta_{m_i-2})$$

$$+ d_i * \ln(1 + \eta_{m_i-1}) + s_{i+1} * \ln(1 + \eta_{m_i}), \quad \text{for} \quad i < l$$

$$\ln \eta_{m_l-2} = (1 - 2\delta_{m_{l-1},m_l-2})s_l * \ln(1 + \eta_{m_l-3})$$

$$+ s_l * \ln(1 + 2\cosh(y_l h/T)\kappa + \kappa^2),$$

$$\ln \kappa = s_l * \ln(1 + \eta_{m_l-2}). \qquad (10.20)$$

In this equation the parameter h appears only in the $\cosh(y_l h/T)$ term.

So $\eta_1(x, h)$ is an even function of h and $\frac{\partial \eta_1(x)}{\partial h}|_{h=0} = 0$. Using (10.19) we find $m = 1/2$ and the condition (10.15) is satisfied at $h = 0$. Thus equation (10.20) becomes

$$\ln(1 + \eta_0) = -\frac{\pi J_z \text{sn}2\zeta}{\zeta T} \delta(x),$$

$$\ln \eta_j = (1 - 2\delta_{m_{i-1},j}) s_i * \ln(1 + \eta_{j-1}) + s_i * \ln(1 + \eta_{j+1})$$
$$\text{for } m_{i-1} \leq j \leq m_i - 2, j \neq m_l - 2,$$

$$\ln \eta_{m_i-1} = (1 - 2\delta_{m_{i-1},m_i-1}) s_i * \ln(1 + \eta_{m_i-2})$$
$$+ d_i * \ln(1 + \eta_{m_i-1}) + s_{i+1} * \ln(1 + \eta_{m_i}) \text{ for } i < l,$$

$$\ln \eta_{m_l-2} = (1 - 2\delta_{m_{l-1},m_l-2}) s_l * \ln(1 + \eta_{m_l-3})$$
$$+ 2s_l * \ln(1 + \kappa),$$

$$\ln \kappa = s_l * \ln(1 + \eta_{m_l-2}). \tag{10.21}$$

Corresponding to (9.62) the free energy is

$$f = -J_z R - \text{sign}(q_1) \frac{\pi J_z \text{sn}2\zeta}{\zeta} \int_{-Q}^{Q} a_1(x) s_1(x) dx$$

$$- T \int_{-Q}^{Q} s_1(x) [\ln(1 + \eta_1(x)) + \delta_{2,m_l} \ln(1 + \eta_2^{-1}(x))] dx. \tag{10.22}$$

We can calculate the free energy of the XYZ model in a zero external field.

10.2 Some special limits

10.2.1 The $T \to \infty$ or $J \to 0$ limit

In equations (10.21), $\ln(1+\eta_0)$ becomes zero and the $\eta_j(x)$ are all independent of x. Thus we get the following difference equation:

$$\eta_j^2 = (1 + \eta_{j-1})^{1-2\delta_{m_{i-1},j}} (1 + \eta_{j+1})$$
$$\text{for } m_{i-1} \leq j \leq m_i - 2, j \neq m_l - 2,$$

$$\eta_{m_i-1}^2 = (1 + \eta_{m_i-2})^{1-2\delta_{m_{i-1},m_i-1}} (1 + \eta_{m_i-1}) \ln(1 + \eta_{m_i})$$
$$\text{for } i < l,$$

$$\eta_{m_l-2}^2 = (1 + \eta_{m_l-3})^{1-2\delta_{m_{l-1},m_l-2}} (1 + 2\kappa + \kappa^2),$$

$$\kappa^2 = (1 + \eta_{m_l-2}). \tag{10.23}$$

The solution of this set of equations is

$$\eta_j = \left(\frac{n_j + y_{i-1}}{y_{i-1}}\right)^2 - 1$$
$$\text{for} \quad m_{i-1} < j \le m_i, \quad j \le m_l - 2,$$
$$\kappa = \frac{n_{m_l-2} + y_l}{y_l}. \tag{10.24}$$

We have $\eta_1 = 3$ for $p_0 \ne 2$ and $\kappa = \eta_1 = \eta_2 = 1$ for $p_0 = 2$. Substituting this into (10.22) we find the free energy,

$$f/T = -\ln 2. \tag{10.25}$$

This gives that the entropy per site is $\ln 2$, as it should be.

10.2.2 The $J_z > 0$, $J_x < 0$ and $T \to 0$ limit

We define $\epsilon_j(x) = T \ln \eta_j(x)$ and $\epsilon_j^+(x) = T \ln(1 + \eta_j(x))$. One can show that ϵ_j, $j \ne 1$ is always positive. The equation (10.20) gives

$$\epsilon_1(x) = -\frac{\pi J_z \operatorname{sn}2\zeta}{\zeta} \mathbf{a}(x, 1) - \int_{-Q}^{Q} \mathbf{a}(x - y, 2)\epsilon_1(y)dy. \tag{10.26}$$

Solving this we have

$$\epsilon_1(x) = -\frac{\pi J_z \operatorname{sn}2\zeta}{\zeta} s_1(x), \quad \epsilon_j(x) = 0, \quad j > 1. \tag{10.27}$$

Substituting this into (10.22) we have

$$f = -J_z R - \frac{\pi J_z \operatorname{sn}2\zeta}{\zeta} \int_{-Q}^{Q} \mathbf{a}_1(x)s_1(x)dx. \tag{10.28}$$

This is just the ground state energy given in (5.80).

10.2.3 The $J_z > 0$, $J_x > 0$ and $T \to 0$ limit

Here we have $2 > p_0 > 1$ and $v_1 = 1$. One can show that ϵ_j, $j \ne v_2 + 1$ is always positive. The equation (10.20) gives

$$\epsilon_{v_2+1}(x) = -\frac{\pi J_z \operatorname{sn}2\zeta}{\zeta} \mathbf{a}(x, 1) - \int_{-Q}^{Q} \mathbf{a}(x - y, 2)\epsilon_{v_2+1}(y)dy. \tag{10.29}$$

Solving this we have

$$\epsilon_{v_2+1}(x) = -\frac{\pi J_z \operatorname{sn}2\zeta}{\zeta} s_2(x), \quad \epsilon_j(x) = 0, \quad j > v_2 + 1. \tag{10.30}$$

10.2.4 The $J_x = 0$ case

Here we have $\gamma = \pi/2$, $v_1 = 2$, $l = 1$ and $m = 2$. Then from the first and last equations of (9.67) we have

$$\ln \eta_1(x) = 4J\mathbf{s}_1(x)/T, \quad \ln \eta_2(x) = -4J\mathbf{s}_1(x)/T. \qquad (10.31)$$

Substituting these into (9.61) or (9.62) we get the free energy, which coincides with (1.19) at $h = 0$.

11

Low-temperature thermodynamics

11.1 The XXZ model

11.1.1 The XXZ model at $2h > J(1 - \Delta)$, $\Delta < 1$

Johnson and McCoy investigated the low-temperature thermodynamics at $\Delta > 1$ using the thermodynamic Bethe-ansatz equations[45]. This paper did not treat the thermodynamics of the XXZ model at $|\Delta| < 1$ or the XYZ model. In this case the ground state is a completely ordered state. The elementary excitations are bound states of down-spins. One down-spin state has an energy momentum relation as follows:

$$\epsilon_1(K) = 2h + J(\Delta - \cos K).$$

The lowest energy state is $K = 0$. Then in this case the system has the energy gap $E_g = 2h - J(1 - \Delta)$ and the low-temperature specific heat behaves as

$$T^{-3/2} e^{-E_g/T}.$$

The magnetization $m = -\partial f / \partial h$ behaves as

$$1 - O(T^{-1/2} e^{-E_g/T}).$$

The system is almost completely magnetized.

11.1.2 The case $J(1 - \Delta)/2 > h > 0$

At $\Delta < -1$ and $h < h_c$ the system has an energy gap and magnetization begins at $h = h_c$. Here h_c is obtained by (4.35). Thus we can give the following phase diagram for the low-temperature behaviour of the XXZ model in magnetic field. The shaded region is the gapless phase, and

152

generally speaking the low-temperature specific heat is proportional to the temperature. The solution of the thermodynamic Bethe-ansatz equation is as follows:

$$\epsilon_1(x) = -As(x) + s * \epsilon_2^+(x),$$

$$\epsilon_n(x) = a_{n-1} * \epsilon_1^-(x) + 2(n-1)h.$$

Only the $n = 1$ solution is relevant to the low-temperature thermodynamics. The other excitation has a finite energy gap and its contribution to the thermodynamics is exponentially small. Inside the shaded region, the specific heat is proportional to the temperature T. The coefficient is inversely proportional to the velocity of excitations. On the boundary of the gapless and gapfull regions the velocity becomes zero and the specific heat is proportional to $T^{1/2}$. This is because the dispersion of excitations is $\epsilon \simeq K^2$.

11.2 Roger's dilogarithm and specific heat at $h = 0$

11.2.1 Specific heat of the XXX antiferromagnet

Consider the entropy density at low temperature. As $\rho_n(x) = \rho_n(-x)$, we can write the entropy density as follows:

$$s = 2 \sum_{n=1}^{\infty} \int_0^{\infty} \rho_n(x) \ln(1 + \eta_n(x)) + \rho_n^h(x) \ln(1 + \eta_n^{-1}(x)) dx. \tag{11.1}$$

From equations (8.42), (8.43) and (8.44) we find that the integral equation for $\eta_n(x)$ at low temperature is

$$\ln \eta_1(x) = -\frac{\pi J}{T} e^{-\pi x/2} + s * \ln(1 + \eta_2(x)),$$
$$\ln \eta_n(x) = s * \ln(1 + \eta_{n-1}(x))(1 + \eta_{n+1}(x)),$$
$$\lim_{n \to \infty} \frac{\ln \eta_n}{n} = 0. \tag{11.2}$$

For $x \to \infty$, $\eta_n = (n+1)^2 - 1$, and for $x \simeq 0$, $\eta_n \simeq n^2 - 1$. $\rho_n(x)$ is given by the differential of $\eta_n(x)$ with respect to J,

$$\rho_n(x) = \frac{T}{2\pi} \frac{1}{(1 + \eta_n(x))\eta_n(x)} \frac{\partial \eta_n}{\partial J}.$$

Differentiating equation (11.2) with respect to J, we have the linear integral equations,

$$\frac{1}{\eta_1}\frac{\partial \eta_1}{\partial J} = -\frac{\pi}{T}e^{-\pi x/2} + s * \frac{1}{1+\eta_2}\frac{\partial \eta_2}{\partial J},$$

$$\frac{1}{\eta_j}\frac{\partial \eta_j}{\partial J} = s * \left\{ \frac{1}{1+\eta_{j-1}}\frac{\partial \eta_{j-1}}{\partial J} + \frac{1}{1+\eta_{j+1}}\frac{\partial \eta_{j+1}}{\partial J} \right\},$$

$$\lim_{n \to \infty} \frac{1}{n\eta_n}\frac{\partial \eta_n}{\partial J} = 0. \tag{11.3}$$

On the other hand, differentiation with respect to x yields similar integral equations,

$$\frac{1}{\eta_1}\frac{\partial \eta_1}{\partial x} = \frac{\pi^2 J}{2T}e^{-\pi x/2} + s * \frac{1}{1+\eta_2}\frac{\partial \eta_2}{\partial x},$$

$$\frac{1}{\eta_j}\frac{\partial \eta_j}{\partial x} = s * \left\{ \frac{1}{1+\eta_{j-1}}\frac{\partial \eta_{j-1}}{\partial x} + \frac{1}{1+\eta_{j+1}}\frac{\partial \eta_{j+1}}{\partial x} \right\},$$

$$\lim_{n \to \infty} \frac{1}{n\eta_n}\frac{\partial \eta_n}{\partial x} = 0. \tag{11.4}$$

Thus we find

$$\frac{\partial \eta_n}{\partial J} = -\frac{2}{\pi J}\frac{\partial \eta_n}{\partial x}.$$

Then equation (11.1) becomes:

$$s = \frac{2T}{\pi^2 J}\sum_{n=1}^{\infty}\int_{n^2-1}^{(n+1)^2-1}\frac{1}{y(1+y)}\ln(1+y) + \frac{1}{1+y}\ln\left(1+\frac{1}{y}\right)dy$$

$$= \frac{2T}{\pi^2 J}\int_0^{\infty}\frac{1}{y(1+y)}\ln(1+y) + \frac{1}{1+y}\ln\left(1+\frac{1}{y}\right)dy. \tag{11.5}$$

Putting $u = 1/(1+y)$ we have

$$s = \frac{2T}{\pi^2 J}\int_0^1 \frac{1}{u}\ln\frac{1}{1-u} + \frac{1}{1-u}\ln\frac{1}{u}du$$

$$= \frac{4T}{\pi^2 J}\int_0^1 \frac{1}{u}\ln\frac{1}{1-u}du$$

$$= \frac{4T}{\pi^2 J}\left(1 + \frac{1}{2^2} + \frac{1}{3^2} + \frac{1}{4^2} + ...\right) = \frac{2T}{3J}. \tag{11.6}$$

Thus the low-temperature specific heat per site is given by

$$C = T\frac{\partial s}{\partial T} = \frac{2T}{3J}.$$

We find that the low-temperature specific heat is proportional to the temperature.

11.3 The ferromagnetic chain and modified spin-wave theory

11.3.1 Numerical analysis of the thermodynamic Bethe-ansatz equation

The low-temperature behaviour of the XYZ model was analyzed by means of the thermodynamic Bethe-ansatz equations. The XXZ case is gapless for $|\Delta| \leq 1$ and the low-temperature specific heat is proportional to the temperature[96],

$$\lim_{T \to 0} \frac{C}{T} = \frac{2\gamma}{3J \sin \gamma}, \quad \gamma \equiv \cos^{-1}(-\Delta), \quad 0 \leq \gamma \leq \pi. \tag{11.7}$$

For $\Delta < -1$ and $\Delta > 1$ the system has a gap and the specific heat behaves as $\exp(-\alpha/T)$. The coefficient diverges as $\Delta \to 1$, namely at the ferromagnetic XXX point. Yang and Yang[119–122] calculated the longitudinal magnetic susceptibility using the Bethe-ansatz method,

$$\lim_{T \to 0} \chi = \frac{4\gamma}{J(\pi - \gamma)\pi \sin \gamma}. \tag{11.8}$$

This also diverges in the limit $\Delta \to 1$.

The next problems are the low-temperature behaviours of the magnetic susceptibility and specific heat of the ferromagnetic XXX model. In the analytical calculation of the Bethe-ansatz equations, the simplest XXX case remains unknown. This problem was investigated by many authors. Let us assume that $C \sim T^{-\alpha}$ and $\chi \sim T^{-\gamma}$. Baker *et al.*[4] estimated $\gamma \simeq 1.6$ using the series expansions from high temperature. Bonner and Fisher[16] estimated that $\alpha \simeq -0.4$ and $\gamma \simeq 1.8$. Quantum Monte Carlo calculation yielded $\gamma \simeq 1.3 - 1.75$[21,63]. But the numerical calculation of the thermodynamic Bethe-ansatz equation gave the final answer that $\alpha = -0.5$ and $\gamma = 2.0$ [81,82,109,116]. From the data at $T \geq 0.002J$, the following low-temperature expansions were obtained[109],

$$f = J\left\{-1.042\left(\frac{T}{J}\right)^{3/2} + 1.00\left(\frac{T}{J}\right)^2 - 0.9\left(\frac{T}{J}\right)^{5/2} + O\left(\frac{T}{J}\right)^3\right\},$$

$$\chi = J^{-1}\left\{0.1667\left(\frac{J}{T}\right)^2 + 0.581\left(\frac{J}{T}\right)^{3/2} + 0.68\left(\frac{J}{T}\right)^1 + O\left(\frac{J}{T}\right)^{1/2}\right\}. \tag{11.9}$$

11.3.2 Spin-wave calculation of the 1D ferromagnetic chain

For a long time it was believed that the spin-wave theory is not applicable for one- or two-dimensional Heisenberg models where there is no long range order at finite temperatures. But the first term of the free energy in equation (11.9) is the same as that of the conventional spin-wave theory, although the second term is completely different. This means that the spin-wave theory is valid in some sense. In the conventional spin-wave theory the chemical potential of the spin-wave is proportional to the external magnetic field. In zero magnetic field there appear infrared divergences in the calculation of physical quantities. If we assume that the chemical potential should be determined by the condition of zero magnetization, we can suppress this divergence and the calculation of physical quantities becomes possible. Using the mean field type approximation, the author found the following expansion of the free energy[102, 103]:

$$
f = T\Big[-\frac{\zeta(\tfrac{3}{2})}{(2\pi)^{1/2}}\Big(\frac{T}{2SJ}\Big)^{1/2} + \frac{T}{4S^2 J}
$$
$$
+\Big(\frac{1}{2S^2}\frac{\zeta(\tfrac{1}{2})}{(2\pi)^{1/2}} - \frac{1}{8}\frac{\zeta(\tfrac{5}{2})}{(2\pi)}\Big)\Big(\frac{T}{2SJ}\Big)^{3/2} + O(T^2)\Big],
$$
$$
\chi = \frac{8}{3}S^4 J T^{-2}\Big[1 - \frac{3}{S}\frac{\zeta(\tfrac{1}{2})}{(2\pi)^{1/2}}\Big(\frac{T}{2SJ}\Big)^{1/2}
$$
$$
+\frac{3}{S^2}\frac{\zeta^2(\tfrac{1}{2})}{2\pi}\Big(\frac{T}{2SJ}\Big) + O(T^{3/2})\Big]. \tag{11.10}
$$

At $S = 1/2$ these become

$$
f = T\Big[-1.0421869\Big(\frac{T}{J}\Big)^{1/2} + (T/J) - 1.2320919\Big(\frac{T}{J}\Big)^{3/2} + O(T^2)\Big],
$$
$$
\chi = \frac{J}{T^2}\Big[\frac{1}{6} + 0.5825974\Big(\frac{T}{J}\Big)^{1/2} + 0.6788396\Big(\frac{T}{J}\Big) + O(T^{3/2})\Big].
$$
$$
\tag{11.11}
$$

In the third term of the free energy the spin-wave result is apparently different from thermodynamic Bethe-ansatz calculations, equation (11.9). This means that the spin-wave theory works to give correct results for the first few terms of the low-temperature expansion. Yamada[115] gave the low-temperature expansion of the correlation length by the Bethe-ansatz method for the quantum transfer-matrix. The low-temperature expansion also coincides with the spin-wave result. The thermodynamic Bethe-ansatz method was used for the analysis of the quasi one-dimensional Heisenberg ferromagnet[110].

11.4 The antiferromagnetic XXX model

At $\Delta = 1-$, $\gamma = 0+$, and we have $\chi = 4/(J\pi^2)$. The low-temperature behaviour of the susceptibility shows a strong logarithmic singularity. The

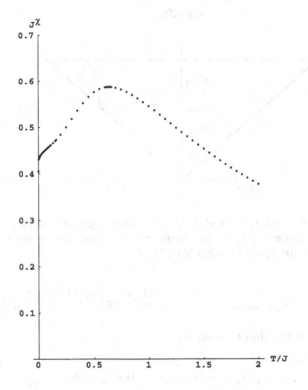

Fig. 11.1. Magnetic susceptibility of the antiferromagnetic Heisenberg chain as a function of temperature. There is a logarithmic anomaly at low temperature.

susceptibility has an infinite slope. This point has been found by the Bethe-ansatz calculation of the quantum transfer matrix[23]. Recently this singularity has been experimentally verified[89, 69]. Such a logarithmic anomaly is found also in the antiferromagnetic ladder lattice with three legs using the quantum Monte Carlo calculations[28].

11.5 The XYZ model at $|J_x| \le J_y < J_z$

In the case $J_y < J_z$ the ground state of the Hamiltonian (5.41) has a gap. Then the free energy at low temperature is written as follows:

$$f(T) - f(0) = T^{-1/2} \exp(-\Delta_f/T). \tag{11.12}$$

It was pointed out that there are two gaps for the free energy at $J_x > 0$[96],

$$\Delta_{\text{spinon}} = \frac{J_z \operatorname{sn}(2\zeta, l) K_u' u'}{2\zeta} \tag{11.13}$$

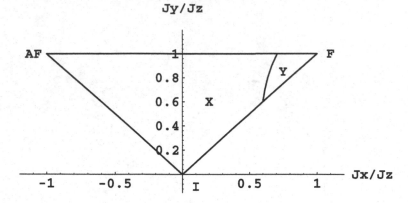

Fig. 11.2. The plane of J_x/J_z and J_y/J_z and the energy gap. In region X the spinon gap given by equation (11.13) is dominant in free energy. In region Y the free energy has a spin-wave gap given by equation (11.14).

$$\Delta_{\text{spin-wave}} = \frac{J_z \text{sn}(2\zeta, l) K_u' u'}{\zeta} \text{sn}\left(\frac{K_u'(K_l - \zeta)}{\zeta}, u'\right), \qquad (11.14)$$

where l, ζ and u are determined by

$$l = \sqrt{\frac{J_z^2 - J_y^2}{J_z^2 - J_x^2}}, \quad J_x/J_z = -\text{cn}(2\zeta, l), \quad 0 < \zeta < K_l, \quad \frac{K_u}{K_u'} = \frac{K_{l'}}{\zeta}. \qquad (11.15)$$

We call these the spinon gap and the spin-wave gap. These are the minimal energies of elementary excitations. For $J_x \leq 0$ we have a spinon gap (5.92) only. As shown in Fig. 11.2, the spinon gap is lower than the spin-wave gap in a wide region X. But near the point F(1,1) there is a region Y where the spin-wave gap which was defined by (5.102) becomes lower. So the free energy at low temperature at $J_x > 0$ is represented by these two gaps, as was shown in Ref. (xcvi). On the line $J_x = J_y$ (line FI in Fig. 11.2) we have

$$\Delta_{\text{spinon}} = \sqrt{J_z^2 - J_y^2}/2, \quad \Delta_{\text{spin-wave}} = J_z - J_y.$$

The crossover of the gap in free energy occurs at $J_y/J_z = 0.6$. On the other hand, at $J_z = J_y$ (line AF-F in Fig. 11.2) we have $l = 0, l' = 1, K_{l'} = \infty$ and $g' = 0$. So the gaps become zero. The free energy behaves as $f(T) = f(0) - cT^2$ except at the point F. At the point F $(J_x = J_y = J_z)$ we have $f(T) = f(0) - cT^{1.5}$.

Part three

Finite temperature integral equations for nested systems

12

$S = 1/2$ fermions with repulsive potential in the continuum

12.1 Derivation of the thermodynamic equations

Lai[55, 56] and Takahashi[94] derived the thermodynamic Bethe-ansatz equation for delta-function fermions. Recapping the equations (2.97) and (2.98),

$$\exp(ik_j L) = \prod_{\alpha=1}^{M} \left(\frac{k_j - \Lambda_\alpha + ic'}{k_j - \Lambda_\alpha - ic'} \right), \quad j = 1, ..., N, \tag{12.1}$$

$$\prod_{j=1}^{N} \left(\frac{\Lambda_\alpha - k_j + ic'}{\Lambda_\alpha - k_j - ic'} \right) = - \prod_{\beta=1}^{M} \left(\frac{\Lambda_\alpha - \Lambda_\beta + ic}{\Lambda_\alpha - \Lambda_\beta - ic} \right),$$

$$\alpha = 1, ..., M. \tag{12.2}$$

If we have one solution of this equation we can construct an eigenfunction

$$\Psi(x_1 s_1, x_2 s_2, ..., x_N s_N)$$

of the Hamiltonian (2.1)

$$\mathscr{H} = - \sum_{i=1}^{n} \frac{\partial^2}{\partial x_i^2} + 2c \sum_{i<j} \delta(x_i - x_j),$$

with $S = S_z = \frac{N}{2} - M$. We can generate $N - 2M + 1$ states by the application of the spin descending operator S^-. The energy eigenvalue is $\sum_{j=1}^{N} k_j^2$.
Conjecture 1. If a set of solutions $(k_1, k_2, ..., k_N; \Lambda_1, \Lambda_2, ..., \Lambda_M)$ of (12.1) and (12.2) contains a complex k (or Λ), its complex conjugate $\bar{k}$ (or $\bar{\Lambda}$) is in the set of ks or (Λs).

This means that the distributions of ks and Λs are symmetric with respect to the real axis. We can prove that for a symmetric state, the ks must be real for the repulsive case ($c > 0$). If $\Im k_j > 0$ the absolute value of the l.h.s. of (12.1) is less than unity, but the absolute value of the r.h.s. is greater than unity. Then $\Im k_j > 0$ must be excluded. If $\Im k_j < 0$ the absolute value of

159

the l.h.s. is greater than unity, but the absolute value of the r.h.s. becomes less than unity. Then we have $\Im k_j = 0$ for $c > 0$. For the $c < 0$ case we have strings with pairs of ks. In the thermodynamic Bethe-ansatz equations the treatment of the ks at $c > 0$ becomes very simple. Next we consider the distribution of Λs on the complex plane. The equation (12.2) resembles the equation for the XXX model if we read $2\Lambda_j/c$ as x_j in (3.17). In the limit $c \to \infty$, the k_js remain finite and the two equations become equivalent. It is very natural to think that we have string states as in §8.2. We write the Λs as $\Lambda_\alpha^{n,j}$,

$$\Lambda_\alpha^{n,j} = \Lambda_\alpha^n + (n + 1 - 2j)c'i + \text{deviation}, \quad j = 1, 2, ..., n. \tag{12.3}$$

Assume that there are M_n strings. The equation (12.1) becomes

$$\exp(ik_j L) = \prod_{n=1}^{\infty} \prod_{\alpha=1}^{M_n} e\left(\frac{k_j - \Lambda_\alpha^n}{nc'}\right), \quad j = 1, 2, ..., N. \tag{12.4}$$

We consider the product

$$\prod_{j=1}^{N} e\left(\frac{\Lambda_\alpha^n - k_j}{nc'}\right). \tag{12.5}$$

By (12.3) this is transformed as

$$\prod_{j=1}^{N} \prod_{l=1}^{n} e\left(\frac{\Lambda_\alpha^{n,j} - k_j}{c'}\right). \tag{12.6}$$

Substituting (12.2) we have

$$(12.6) = \prod_{l=1}^{n}\left\{-\prod_{(m,\beta)} \prod_{h=1}^{m} e\left(\frac{\Lambda_\alpha^{n,l} - \Lambda_\beta^{m,h}}{c}\right)\right\}$$

$$= \prod_{l=1}^{n}\left\{\prod_{(m,\beta)\neq(n,\alpha)} \prod_{h=1}^{m} e\left(\frac{\Lambda_\alpha^{n,l} - \Lambda_\beta^{m,h}}{c}\right)\right\}. \tag{12.7}$$

Using (12.3) we have

$$\prod_{j=1}^{N} e\left(\frac{\Lambda_\alpha^n - k_j}{nc'}\right) = -\prod_{m=1}^{\infty} \prod_{\beta=1}^{M_m} E_{nm}\left(\frac{\Lambda_\alpha^n - \Lambda_\beta^m}{c'}\right), \quad \alpha = 1, 2, ..., M_n, \quad n \geq 1, \tag{12.8}$$

where $E_{nm}(x)$ is defined in (8.9). The logarithm of (12.4) and (12.8) is

$$k_j L = 2\pi I_j - \sum_{n=1}^{\infty} \sum_{\alpha=1}^{M_n} \theta\left(\frac{k_j - \Lambda_\alpha^n}{nc'}\right), \quad j = 1, 2, ..., N, \tag{12.9}$$

$$\sum_{j=1}^{N} \theta\left(\frac{\Lambda_\alpha^n - k_j}{nc'}\right) = 2\pi J_\alpha^n + \sum_{m=1}^{\infty} \sum_{\beta=1}^{M_m} \Theta_{nm}\left(\frac{\Lambda_\alpha^n - \Lambda_\beta^m}{c'}\right),$$

$$\alpha = 1, 2, ..., M_n, \quad n \geq 1, \tag{12.10}$$

Here $\theta(x) \equiv 2\tan^{-1} x$, $-\pi < \theta \leq \pi$ and the function $\Theta_{nm}(x)$ was defined in (8.12). I_j are distinct integers (half-odd integers) for even (odd) $M_1 + M_2 + ...$,

$$I_j \equiv M_1 + M_2 + ... \pmod{1}. \tag{12.11}$$

The J_α^n are distinct integers (half-odd integers) for odd (even) $N - M_n$, which satisfy

$$J_\alpha^n \equiv N - M_n + \frac{1}{2} \pmod{1},$$

$$|J_\alpha^n| \leq \frac{1}{2}(N - 1 - \sum t_{nm} M_m), \quad t_{nm} \equiv 2\min(n, m) - \delta_{nm}. \tag{12.12}$$

Given a set of integers $\{I_j, J_\alpha^n\}$ which satisfies (12.11) and (12.12), we can determine a set of k_j and Λ_α^n through equations (12.9) and (12.10). For a set of integers $\{I_j, J_\alpha^n\}$, there is a set of omitted integers which satisfy (12.11) and (12.12) but is not contained in $\{I_j, J_\alpha^n\}$. We define holes of k and holes of Λ^n as solutions of

$$Lh(k) = 2\pi \times (\text{omitted } I),$$

$$Lj_n(\Lambda^n) = 2\pi \times (\text{omitted } J^n),$$

where

$$h(k) \equiv k + \frac{1}{L} \sum_{n=1}^{\infty} \sum_{\alpha=1}^{M_n} \theta\left(\frac{k - \Lambda_\alpha^n}{nc'}\right), \tag{12.13}$$

$$j_n(\Lambda) \equiv \frac{1}{L} \sum_{j=1}^{N} \theta\left(\frac{\Lambda - k_j}{nc'}\right) - \frac{1}{L} \sum_{m=1}^{\infty} \sum_{\beta=1}^{M_m} \Theta_{nm}\left(\frac{\Lambda - \Lambda_\beta^m}{c'}\right). \tag{12.14}$$

We consider a very large system. We put the distribution functions of ks and Λ^ns as $\rho(k)$ and $\sigma_n(k)$, and those of holes as $\rho^h(k)$ and $\sigma_n^h(k)$. By the definition of holes we have:

$$\frac{d}{dk}h(k) = 2\pi(\rho(k) + \rho^h(k)), \tag{12.15}$$

$$\frac{d}{dk}j_n(k) = 2\pi(\sigma_n(k) + \sigma_n^h(k)). \tag{12.16}$$

Equations (12.13) and (12.14) are written as follows:

$$h(k) = k + \sum_{n=1}^{\infty} \int \theta\left(\frac{k-k'}{nc'}\right)\sigma_n(k')dk', \tag{12.17}$$

$$j_n(k) = \int \theta\left(\frac{k-k'}{nc'}\right)\rho(k')dk' - \int \Theta_{nm}\left(\frac{k-k'}{c'}\right)\sigma_m(k')dk'. \tag{12.18}$$

Hereafter we denote $\int_{-\infty}^{\infty} dk$ as $\int dk$. Substituting (12.17) and (12.18) into equations (12.15) and (12.16) we have

$$\frac{1}{2\pi} = \rho(k) + \rho^h(k) - \sum_{n=1}^{\infty} a_n * \sigma_n(k), \tag{12.19}$$

$$a_n * \rho(k) = \sigma_n(k) + \sigma_n^h(k) + \sum_{m=1}^{\infty} T_{nm} * \sigma_m(k), \tag{12.20}$$

where a_n is a function defined by

$$a_n(k) \equiv \frac{1}{\pi}\frac{n|c'|}{(nc')^2 + k^2}, \quad a_0(k) \equiv \delta(k), \tag{12.21}$$

and

$$T_{nm}(k) \equiv \begin{cases} a_{|n-m|}(k) + 2a_{|n-m|+2}(k) + 2a_{|n-m|+4}(k) + \dots \\ \quad +2a_{n+m-2}(k) + a_{n+m}(k) \quad \text{for} \quad n \neq m, \\ \\ 2a_2(k) + 2a_4(k) + \dots + 2a_{2n-2}(k) + a_{2n}(k) \\ \quad\quad\quad \text{for} \quad n = m. \end{cases}$$

The grand partition function of this system at magnetic field h and chemical potential A is written as follows,

$$\Xi = \sum_{N=0}^{\infty} \sum_{M=0}^{[N/2]} \left(\frac{\sinh\frac{(N-2M+1)h}{T}}{\sinh\frac{h}{T}}\right) \sum_{\{I,J^n\}} \exp\left(-\frac{1}{T}(E_{N,M,\{I,J^n\}} - AN)\right), \tag{12.22}$$

where $E_{N,M,\{I,J^n\}}$ are eigenvalues of the Hamiltonian

$$\mathcal{H} = -\sum_{i=1}^{n}\frac{\partial^2}{\partial x_i^2} + 2c\sum_{i<j}\delta(x_i - x_j),$$

with total spin $N/2 - M$. In other words the state belongs to the representation $[2^M 1^{N-2M}]$. By the steepest decent method the Gibbs free energy at a given temperature T, chemical potential A and magnetic field h is given by

$$G(T, A, h) = -T\ln\Xi = E(\rho) - AN(\rho)$$
$$-h(N(\rho) - 2M(\sigma_n)) - TS(\rho, \rho_h, \sigma_n, \sigma_n^h). \tag{12.23}$$

We have

$$\frac{E}{L} = \int k^2 \rho(k)dk, \quad \frac{N}{L} = \int \rho(k)dk,$$

$$\frac{M}{L} = \sum_{n=1}^{\infty} n \int \sigma_n(k)dk, \tag{12.24}$$

$$\frac{S}{L} = \int \{(\rho + \rho^h)\ln(\rho + \rho^h) - \rho \ln \rho - \rho^h \ln \rho^h\}dk$$

$$+ \sum_{n=1}^{\infty} \int \{(\sigma_n + \sigma_n^h)\ln(\sigma_n + \sigma_n^h) - \sigma_n \ln \sigma_n - \sigma_n^h \ln \sigma_n^h\}dk. \tag{12.25}$$

Thus the free energy is a functional of functions $\{\rho, \rho^h, \sigma_n, \sigma_n^h\}$. It must be minimized under the conditions (12.19) and (12.20). At the thermodynamic equilibrium the variation of G must be zero,

$$0 = \delta G/L = \int (k^2 - A - h)\delta \rho(k)dk + \sum_{n=1}^{\infty} 2nh \int \delta \sigma_n(k)dk$$

$$- T \int \left\{ \delta \rho \ln\left(\frac{\rho + \rho^h}{\rho}\right) + \delta \rho^h \ln\left(\frac{\rho + \rho^h}{\rho^h}\right) \right\}dk$$

$$- T \sum_{n=1}^{\infty} \int \left\{ \delta \sigma_n \ln\left(\frac{\sigma_n + \sigma_n^h}{\sigma_n}\right) + \delta \sigma_n^h \ln\left(\frac{\sigma_n + \sigma_n^h}{\sigma_n^h}\right) \right\}dk. \tag{12.26}$$

From equations (12.19) and (12.20) we have

$$\delta \rho^h = -\delta \rho + \sum_{n=1}^{\infty} a_n * \delta \sigma_n, \quad \delta \sigma^h = a_n * \delta \rho - \delta \sigma - \sum_{m=1}^{\infty} T_{nm} * \delta \sigma_m.$$

Substituting these into equation (12.26) we have

$$\frac{\delta G}{TL} = \int \left\{ \frac{k^2 - A - h}{T} - \ln\left(\frac{\rho^h}{\rho}\right) - \sum_{n=1}^{\infty} a_n * \ln\left(1 + \frac{\sigma_n}{\sigma_n^h}\right) \right\}\delta \rho dk$$

$$+ \sum_{n=1}^{\infty} \int \left\{ \frac{2nh}{T} - \ln\left(\frac{\sigma_n^h}{\sigma_n}\right) - a_n * \ln\left(1 + \frac{\rho}{\rho^h}\right) \right.$$

$$+ \sum_{m=1}^{\infty} T_{nm} * \ln\left(1 + \frac{\sigma_m^h}{\sigma_m}\right) \right\}\delta \sigma_n dk. \tag{12.27}$$

Thus we have a set of coupled non-linear integral equations for $\zeta(k) \equiv \rho^h(k)/\rho(k)$ and $\eta_n(k) \equiv \sigma_n^h(k)/\sigma_n(k)$ as follows:

$$\ln \zeta(k) = \frac{k^2 - A - h}{T} - \sum_{n=1}^{\infty} a_n * \ln(1 + \eta_n^{-1}(k)), \tag{12.28}$$

$$\ln \eta_n(k) = \frac{2nh}{T} - a_n * \ln(1 + \zeta^{-1}(k))$$

$$+ \sum_{m=1}^{\infty} T_{nm} * \ln(1 + \eta_m^{-1}(k)), \quad n = 1, 2, ... \tag{12.29}$$

Equations (12.19) and (12.20) are rewritten as

$$(1 + \zeta(k))\rho(k) = \frac{1}{2\pi} + \sum_{n=1}^{\infty} a_n * \sigma_n(k), \tag{12.30}$$

$$a_n * \rho(k) = (1 + \eta_n(k))\sigma_n(k) + \sum_{m=1}^{\infty} T_{nm} * \sigma_m(k). \tag{12.31}$$

From the thermodynamics the pressure is given by

$$p = -G/L. \tag{12.32}$$

Using equations (12.23), (12.25), (12.27) and (12.29) one obtains

$$p = T \int \ln(1 + \zeta^{-1}(k))\frac{dk}{2\pi}. \tag{12.33}$$

This is the same expression as the one for free energy of repulsive bosons in (7.20). The equations (12.27) can be written as follows:

$$\ln \zeta = \frac{k^2 - A}{T} - R * \ln(1 + \zeta^{-1}) - s * \ln(1 + \eta_1), \tag{12.34}$$

$$\ln \eta_1(k) = s * \ln((1 + \eta_2)/(1 + \zeta^{-1})), \tag{12.35}$$

$$\ln \eta_n(k) = s * \ln(1 + \eta_{n-1})(1 + \eta_{n+1}), \tag{12.36}$$

$$\lim_{n \to \infty} \frac{\ln \eta_n}{n} = \frac{2h}{T}, \tag{12.37}$$

where

$$s(x) = \frac{1}{4c'}\operatorname{sech}\frac{\pi x}{2c'}, \quad R(x) = a_1 * s(x) = \int_{-\infty}^{\infty} a_1(x - y)s(y)dy. \tag{12.38}$$

To derive these we should note the following relations:

$$s * (a_{n-1} + a_{n+1}) = a_n, \quad s * (T_{n-1,m} + T_{n+1,m}) = T_{nm} - (\delta_{n-1,m} + \delta_{n+1,m})s. \tag{12.39}$$

12.2 Some special limits

12.2.1 $c \to 0+$

At first we consider the limit $c \to 0+$. Here the width of the functions $s(x)$ and $R(x)$ becomes zero and we can replace them by $\frac{1}{2}\delta(x)$. Then equations

(12.34)–(12.37) are written as follows:

$$\ln \zeta(k) = \frac{k^2 - A}{T} - \frac{1}{2}\ln(1 + \zeta^{-1}(k))(1 + \eta_1(k)), \qquad (12.40)$$

$$\ln \eta_1(k) = \frac{1}{2}\ln((1 + \eta_2(k))/(1 + \zeta^{-1}(k))), \qquad (12.41)$$

$$\ln \eta_n(k) = \frac{1}{2}\ln(1 + \eta_{n-1}(k))(1 + \eta_{n+1}(k)), \qquad (12.42)$$

$$\lim_{n \to \infty} \frac{\ln \eta_n(k)}{n} = \frac{2h}{T}, \qquad (12.43)$$

Equation (12.42) is a difference equation. The general solution is

$$\eta_n = \left(\frac{a^n b - a^{-n} b^{-1}}{a - a^{-1}}\right)^2 - 1. \qquad (12.44)$$

a and b are arbitrary parameters which should be determined by other conditions. From (12.43) we have $a = \exp(|h|/T)$. Then the solution of (12.42) and (12.43) is written as

$$\eta_n(k) = \left(\frac{\sinh f(k) + nh/T}{\sinh h/T}\right)^2 - 1. \qquad (12.45)$$

Substituting this into (12.41) we have

$$\frac{1}{1 + \zeta^{-1}} = \left(\frac{\sinh f(k)}{h/T}\right)^2. \qquad (12.46)$$

Thus equation (12.40) becomes

$$\ln \frac{\sinh f(k)}{\sinh(h/T - f(k))} = \frac{k^2 - A}{T}. \qquad (12.47)$$

Solving this equation we have

$$f(k) = \frac{1}{2}\ln\left(\frac{1 + \exp[(k^2 - A + h)/T]}{1 + \exp[(k^2 - A - h)/T]}\right), \qquad (12.48)$$

$$1 + \zeta^{-1} = (1 + e^{-(k^2 - A + h)/T})(1 + e^{-(k^2 - A - h)/T}), \qquad (12.49)$$

$$p = T \int \frac{dk}{2\pi}\ln((1 + e^{-(k^2 - A + h)/T})(1 + e^{-(k^2 - A - h)/T})). \qquad (12.50)$$

The last equation coincides with (2.56) at $h = 0$. Thus we can reproduce the known results for the $c = 0$ case.

12.2.2 $c \to \infty$

In this case the $\eta_j(k)$ are independent of k, because $s(x)$ and $R(k)$ are infinitely wide functions. Then equations (12.35)–(12.37) become

$$\ln \eta_1 = \frac{1}{2} \ln(1 + \eta_2), \quad \ln \eta_j = \frac{1}{2} \ln(1 + \eta_{j-1})(1 + \eta_{j+1}),$$

$$\lim_{n \to \infty} \frac{\ln \eta_n}{n} = \frac{2h}{T}. \tag{12.51}$$

The solution of this set of equations is

$$\eta_n = \left(\frac{\sinh(n+1)h/T}{\sinh h/T} \right)^2 - 1. \tag{12.52}$$

From equation (12.34) we have

$$\zeta(k) = e^{\frac{k^2 - A}{T}} \frac{1}{e^{h/T} + e^{-h/T}},$$

$$p = T \int \frac{dk}{2\pi} \ln(1 + 2\cosh(h/T)e^{-\frac{k^2 - A}{T}}). \tag{12.53}$$

13

$S = 1/2$ fermions with an attractive potential

13.1 Derivation of the thermodynamic equations

In the ground state, each Λ is accompanied by two complex ks. This is regarded as a pair of up-spin and down-spin electrons,

$$k_\alpha^1 = \Lambda_\alpha' + i|c'|, \quad k_\alpha^2 = \Lambda_\alpha' - i|c'|, \quad \alpha = 1, 2, ..., M'.$$

At finite temperature there appear Λs which are not accompanied by complex ks. These Λs form strings similar to those in the repulsive case,

$$\Lambda_{\alpha,j}^n = \Lambda_\alpha^n + (n+1-2j)|c'|, \quad j = 1, 2, ..., n, \quad \alpha = 1, 2, ..., M_n.$$

Thus an eigenstate is characterized by $N - 2M'$ real ks, M' real Λ', and M_n Λ^n. We have $M = M' + \sum_{n=1}^\infty M_n$. From equation (12.1) we have

$$\exp i(k_\alpha^1 + k_\alpha^2)L = \prod_{\beta \neq \alpha} e\Big(\frac{\Lambda_\alpha' - \Lambda_\beta'}{-2|c'|}\Big)\Big\{\prod_{m,\beta,j} e\Big(\frac{\Lambda_\alpha' - \Lambda_\beta^{n,j}}{-2|c'|}\Big)\Big\}$$
$$\times e\Big(\frac{k_\alpha^1 - \Lambda_\alpha'}{-|c'|}\Big) e\Big(\frac{k_\alpha^2 - \Lambda_\alpha'}{-|c'|}\Big). \tag{13.1}$$

From equation (12.2) we have

$$e\Big(\frac{k_\alpha^1 - \Lambda_\alpha'}{-|c'|}\Big) e\Big(\frac{k_\alpha^2 - \Lambda_\alpha'}{-|c'|}\Big) = \prod_{j=1}^{N-2M'} e\Big(\frac{\Lambda_\alpha' - k_j}{-|c'|}\Big)\Big\{\prod_{m,\beta,j} e\Big(\frac{\Lambda_\alpha' - \Lambda_\beta^{n,j}}{2|c'|}\Big)\Big\}. \tag{13.2}$$

Substituting this into equation (13.1) we have

$$\exp(2i\Lambda_\alpha L) = -\prod_{j=1}^{N-2M'} e\Big(\frac{\Lambda_\alpha' - k_j}{-|c'|}\Big) \prod_{\beta=1}^{M'} e\Big(\frac{\Lambda_\alpha' - \Lambda_\beta'}{-2|c'|}\Big). \tag{13.3}$$

Here we have represented an unpaired k as k_j. From equation (12.1) we have

$$\exp(ik_j L) = \prod_{\alpha=1}^{M'} e\left(\frac{k_j - \Lambda'_\alpha}{-|c'|}\right) \prod_{n=1}^{\infty} \prod_{\alpha=1}^{M_n} e\left(\frac{k_j - \Lambda^n_\alpha}{-n|c'|}\right). \tag{13.4}$$

And from equation (12.2),

$$\prod_{j=1}^{N-2M'} e\left(\frac{\Lambda_\alpha - k_j}{n|c'|}\right) = -\prod_{m=1}^{\infty} \prod_{\beta=1}^{M_m} E_{nm}\left(\frac{\Lambda_\alpha - \Lambda^m_\beta}{|c'|}\right). \tag{13.5}$$

The logarithms of equations (13.3)–(13.5) are

$$2\Lambda_\alpha L = 2\pi J'_\alpha + \sum_{j=1}^{N-2M'} \theta\left(\frac{\Lambda'_\alpha - k_j}{-|c'|}\right) + \sum_{\beta=1}^{M'} \theta\left(\frac{\Lambda'_\alpha - \Lambda'_\beta}{-2|c'|}\right), \tag{13.6}$$

$$k_j L = 2\pi I_j + \sum_{\alpha=1}^{M'} \theta\left(\frac{k_j - \Lambda'_\alpha}{-|c'|}\right) + \sum_{n=1}^{\infty} \sum_{\alpha=1}^{M_n} \theta\left(\frac{k_j - \Lambda^n_\alpha}{-n|c'|}\right), \tag{13.7}$$

$$\sum_{j=1}^{N-2M'} \theta\left(\frac{\Lambda^n_\alpha - k_j}{n|c'|}\right) = 2\pi J^n_\alpha + \sum_{m=1}^{\infty} \sum_{\beta=1}^{M_m} \Theta_{nm}\left(\frac{\Lambda^n_\alpha - \Lambda^m_\beta}{|c'|}\right). \tag{13.8}$$

Here J'_α is an integer (half-odd integer) for $N - M'$ odd (even), I_j is an integer (half-odd integer) for $M' + M_1 + M_2 + \dots$ even (odd). J^n_α is an integer (half-odd integer) for $N - M_n$ odd (even) and satisfies:

$$|J^n_\alpha| \le \frac{1}{2}\left(N - 2M - \sum_{m=1}^{\infty} t_{nm} M_m\right).$$

We define functions $j'(\Lambda')$, $h(k)$, $j_n(\Lambda^n)$,

$$j'(\Lambda') \equiv 2\Lambda' - \frac{1}{L}\left\{\sum_{j=1}^{N-2M'} \theta\left(\frac{\Lambda' - k_j}{-|c'|}\right) + \sum_{\beta=1}^{M'} \theta\left(\frac{\Lambda' - \Lambda'_\beta}{-2|c'|}\right)\right\}, \tag{13.9}$$

$$h(k) \equiv k - \frac{1}{L}\left\{\sum_{\alpha=1}^{M'} \theta\left(\frac{k - \Lambda'_\alpha}{-|c'|}\right) + \sum_{n=1}^{\infty} \sum_{\alpha=1}^{M_n} \theta\left(\frac{k - \Lambda^n_\alpha}{-n|c'|}\right)\right\}, \tag{13.10}$$

$$j_n(\Lambda^n) \equiv \frac{1}{L}\left\{\sum_{j=1}^{N-2M'} \theta\left(\frac{\Lambda^n - k_j}{n|c'|}\right) - \sum_{m=1}^{\infty} \sum_{\beta=1}^{M_m} \Theta_{nm}\left(\frac{\Lambda^n - \Lambda^m_\beta}{|c'|}\right)\right\}.$$

$$\tag{13.11}$$

Holes of Λ', k and Λ^n are defined as solutions of

$$j'(\Lambda') = 2\pi \times (\text{omitted } J'),$$
$$h(k) = 2\pi \times (\text{omitted } I),$$
$$j_n(\Lambda^n) = 2\pi \times (\text{omitted } J^n). \tag{13.12}$$

In the limit of a very large system we define the distribution functions of Λ', k, Λ^n as $\sigma'(k)$, $\rho(k)$, $\sigma_n(k)$ and those of the holes as $\sigma'^h(k)$, $\rho^h(k)$ and $\sigma_n^h(k)$. Using the relations:

$$\frac{dj'(k)}{dk} = 2\pi(\sigma'(k) + \sigma'^h(k)),$$

$$\frac{dh(k)}{dk} = 2\pi(\rho'(k) + \rho'^h(k)),$$

$$\frac{dj_n(k)}{dk} = 2\pi(\sigma_n(k) + \sigma_n^h(k)),$$

we have equations for σ', σ'^h, ρ, ρ^h, σ_n, σ_n^h:

$$\frac{1}{\pi} = \sigma' + \sigma'^h + a_2 * \sigma' + a_1 * \rho, \tag{13.13}$$

$$\frac{1}{2\pi} = \rho + \rho^h + a_1 * \sigma' + \sum_n a_n * \sigma_n, \tag{13.14}$$

$$a_n * \rho = \sigma_n + \sigma_n^h + \sum_m T_{nm} * \sigma_m. \tag{13.15}$$

Here

$$a_n(k) = \frac{n|c'|}{\pi(k^2 + (nc')^2)}, \quad a_0(k) = \delta(k),$$

$$T_{nm}(k) \equiv \begin{cases} a_{|n-m|}(k) + 2a_{|n-m|+2}(k) + 2a_{|n-m|+4}(k) + \ldots \\[2mm] \quad +2a_{n+m-2}(k) + a_{n+m}(k)) \quad \text{for} \quad n \neq m, \\[2mm] 2a_2(k) + 2a_4(k) + \ldots + 2a_{2n-2}(k) + a_{2n}(k) \\[2mm] \quad\quad\quad \text{for} \quad n = m. \end{cases}$$
$$\tag{13.16}$$

The grand partition function is

$$\Xi = \sum_{N=0}^{\infty} \sum_{M=0}^{[N/2]} \left(\frac{\sinh \frac{(N-2M+1)h}{T}}{\sinh \frac{h}{T}} \right) \sum_{\{J',I,J^n\}} \exp\left(\frac{AN - E_{N,M,\{J',I,J^n\}}}{T} \right). \tag{13.17}$$

The Gibbs free energy is given by

$$G(T, A, h) = -T \ln \Xi = E(\rho, \sigma') - AN(\rho, \sigma')$$
$$-h(N(\rho, \sigma') - 2M(\sigma', \sigma_n)) - TS(\sigma', \sigma'^h, \rho, \rho_h, \sigma_n, \sigma_n^h), \tag{13.18}$$

where

$$E(\rho, \sigma')/L = \int k^2 \rho(k) + 2(k^2 - c'^2)\sigma'(k)dk, \tag{13.19}$$

$$N(\rho, \sigma')/L = \int \rho(k) + 2\sigma'(k)dk, \tag{13.20}$$

$$M(\sigma', \sigma_n)/L = \int \sigma'(k) + \sum_{n=1}^{\infty} n\sigma_n(k)dk, \tag{13.21}$$

$$S(\sigma', \sigma'^h, \rho, \rho_h, \sigma_n, \sigma_n^h)/L$$
$$= \int \{(\sigma' + \sigma'^h)\ln(\sigma' + \sigma'^h) - \sigma'\ln\sigma' - \sigma'^h\ln\sigma'^h\}dk$$
$$+ \int \{(\rho + \rho^h)\ln(\rho + \rho^h) - \rho\ln\rho - \rho^h\ln\rho^h\}dk$$
$$+ \sum_n \int \{(\sigma_n + \sigma_n^h)\ln(\sigma_n + \sigma_n^h) - \sigma_n\ln\sigma_n - \sigma_n^h\ln\sigma_n^h\}dk.$$

$$\tag{13.22}$$

By the condition of minimizing the free energy, we have

$$0 = \delta G/L = \int 2(k^2 - c'^2)\delta\sigma'(k) + (k^2 - A - h)\delta\rho(k)dk$$

$$+ \sum_{n=1}^{\infty} 2nh \int \delta\sigma_n(k)dk$$

$$- T \int \left\{ \delta\sigma' \ln\left(\frac{\sigma' + \sigma'^h}{\sigma'}\right) + \delta\sigma'^h \ln\left(\frac{\sigma' + \sigma'^h}{\sigma'^h}\right) \right.$$

$$\left. + \delta\rho \ln\left(\frac{\rho + \rho^h}{\rho}\right) + \delta\rho^h \ln\left(\frac{\rho + \rho^h}{\rho^h}\right) \right\}dk$$

$$- T \sum_{n=1}^{\infty} \int \left\{ \delta\sigma_n \ln\left(\frac{\sigma_n + \sigma_n^h}{\sigma_n}\right) + \delta\sigma_n^h \ln\left(\frac{\sigma_n + \sigma_n^h}{\sigma_n^h}\right) \right\}dk. \tag{13.23}$$

From equations (13.13), (13.14) and (13.15) we have

$$\delta\sigma'^h = -\delta\sigma' - a_2 * \delta\sigma' - a_1 * \delta\rho,$$

$$\delta\rho^h = -\delta\rho - a_1 * \delta\sigma' - \sum_{n=1}^{\infty} a_n * \delta\sigma_n,$$

$$\delta\sigma_n^h = a_n * \delta\rho - \delta\sigma_n - \sum_{m=1}^{\infty} T_{nm} * \delta\sigma_m.$$

Substituting these into equation (13.23) we have a set of coupled non-linear integral equations for $\eta' = \sigma'^h/\sigma'$, $\zeta = \rho^h/\rho$, and $\eta_n = \sigma_n^h/\sigma_n$ as follows:

$$\ln \eta' = \frac{2(k^2 - c'^2 - A)}{T} + a_2 * \ln(1 + \eta'^{-1}) + a_1 * \ln(1 + \zeta^{-1}),$$

$$\ln \zeta = \frac{k^2 - A - h}{T} + a_1 * \ln(1 + \eta'^{-1}) - \sum_{n=1}^{\infty} a_n * \ln(1 + \eta_n^{-1}),$$

$$\ln \eta_n = \frac{2nh}{T} + a_n * \ln(1 + \zeta^{-1}) + \sum_{m=1}^{\infty} T_{nm} * \ln(1 + \eta_m^{-1}). \qquad (13.24)$$

The pressure p and Gibbs free energy G are given by

$$p = -\frac{G}{L} = T \int \ln(1 + \eta'^{-1}) \frac{dk}{\pi} + T \int \ln(1 + \zeta^{-1}) \frac{dk}{2\pi}. \qquad (13.25)$$

The equations (13.24) can be written as follows:

$$\ln \eta' = \frac{2(k^2 - c'^2 - A)}{T} + a_2 * \ln(1 + \eta'^{-1}) + a_1 * \ln(1 + \zeta^{-1}),$$
$$\ln \zeta = s * \ln(1 + \eta') - s * \ln(1 + \eta_1),$$
$$\ln \eta_1 = s * (\ln(1 + \zeta^{-1}) + \ln(1 + \eta_2)),$$
$$\ln \eta_n = s * (\ln(1 + \eta_{n-1}) + \ln(1 + \eta_{n+1})), \quad n = 2, 3, ...,$$
$$\lim_{n \to \infty} \frac{\ln \eta_n}{n} = \frac{2h}{T}. \qquad (13.26)$$

From equations (13.13), (13.14) and (13.15) we have a set of linear equations for σ', ρ, σ_n,

$$\frac{1}{\pi} = \sigma'(1 + \eta') + a_2 * \sigma' + a_1 * \rho, \qquad (13.27)$$

$$\frac{1}{2\pi} = \rho(1 + \zeta) + a_1 * \sigma' + \sum_n a_n * \sigma_n, \qquad (13.28)$$

$$a_n * \rho = (1 + \eta_n)\sigma_n + \sum_m T_{nm} * \sigma_m. \qquad (13.29)$$

13.2 Some special limits
13.2.1 $c \to 0-$

In this limit we can put $a_1 = a_2 = \delta(x)$, $s = \frac{1}{2}\delta(x)$. Then equation (13.26) becomes

$$\eta' = \exp\left(\frac{2(k^2 - A)}{T}\right)(1 + \eta'^{-1})(1 + \zeta^{-1}),$$

$$\zeta = \sqrt{\frac{1 + \eta'}{1 + \eta_1}},$$

$$\eta_1 = \sqrt{(1 + \zeta^{-1})(1 + \eta_2)},$$
$$\eta_n = \sqrt{(1 + \eta_{n-1})(1 + \eta_{n+1})},$$
$$\lim_{n \to \infty} \frac{\ln \eta_n}{n} = \frac{2h}{T}. \tag{13.30}$$

The general solution of the second, third, fourth and fifth equations is

$$\eta_n = f^2(n) - 1, \quad \zeta^{-1} = f^2(0) - 1, \quad \frac{1}{1 + \eta'} = f^2(-1),$$

$$f(n) \equiv \frac{bz^n - b^{-1}z^{-n}}{z - z^{-1}}. \tag{13.31}$$

Here b is a function of k and $z \equiv e^{-h/T}$. Substituting these into the first equation we can determine $b(k)$,

$$b(k) = z^2 \sqrt{\frac{1 + z^{-1}e^{(k^2-A)/T}}{1 + ze^{(k^2-A)/T}}}$$

Substituting these into (13.25) we have

$$p = T \int \frac{dk}{2\pi} \ln(1 + e^{-(k^2-A+h)/T})(1 + e^{-(k^2-A-h)/T}). \tag{13.32}$$

At $h = 0$ this result coincides with the pressure of non-interacting fermions (2.56).

13.2.2 $T \to 0+$

In the low-temperature limit we introduce the following quantities:

$$\epsilon'(k) = T \ln \eta'(k), \quad \kappa(k) = T \ln \zeta(k), \quad \epsilon_n(k) = T \ln \eta_n(k).$$

We can show that $\epsilon_n(k) > 0$. Then in the limit of $T \to 0$, $\epsilon'(k)$ and $\kappa(k)$ are determined by:

$$\epsilon'(k) = 2(k^2 - A - c'^2) - a_2 * \epsilon'^-(k) - a_1 * \kappa^-(k),$$
$$\kappa(k) = k^2 - A - h - a_1 * \epsilon'^-(k). \tag{13.33}$$

Here we used (13.24). From these equations we can show that ϵ' and κ are monotonically increasing functions of k^2. We define parameters B and Q by $\epsilon'(B) = 0$ and $\kappa(Q) = 0$. η' and ζ are zero in the region $[B, -B]$ and $[Q, -Q]$, respectively, and infinity outside these regions. From equations (13.13)–(13.15) we have a set of linear integral equations in the limit $T \to 0$,

$$\frac{1}{\pi} = \sigma'(k) + \int_{-B}^{B} a_2(k - k')\sigma'(k')\mathrm{d}k'$$

$$+ \int_{-Q}^{Q} a_1(k - k')\rho(k')\mathrm{d}k',$$

$$\frac{1}{2\pi} = \rho(k) + \int_{-B}^{B} a_1(k - k')\sigma'(k')\mathrm{d}k'.$$

These are equivalent to equations (2.120) and (2.121).

14

Thermodynamics of the Hubbard model

14.1 Strings of the Hubbard model

The author obtained the thermodynamic Bethe-ansatz equation for the Hubbard model[95]. It is expected that there are strings of Λs of arbitrary length and real independent ks by analogy with the repulsive delta fermion problem. But for the thermodynamics we must consider pairs of fermions and bound states of pairs of fermions. A pair of fermions is given by

$$k_\alpha^1 = \pi - \sin^{-1}(\Lambda_\alpha' + U'i),$$
$$k_\alpha^2 = \pi - \sin^{-1}(\Lambda_\alpha' - U'i). \tag{14.1}$$

Here we take the branch of $\sin^{-1}$ as $-\pi/2 < \Re \sin^{-1} x \leq \pi/2$. The energy of this state is given by

$$E = -2t(\cos k_\alpha^1 + \cos k_\alpha^2) = 4t\Re\sqrt{1 - (\Lambda_\alpha' - U'i)^2}. \tag{14.2}$$

This value of this excitation is always positive, and is of the order of U when U is large. Thus for the low-temperature thermodynamics this excitation is not relevant. But in the high-temperature limit one cannot get the correct entropy per site ($\ln 4$) without these excitations. This excitation resembles the pairs of fermions which appear in attractive delta-fermions. In the Hubbard model there appear bound states of n pairs. This excitation has $2n$ ks and n Λs,

$$\Lambda_\alpha'^{nj} = \Lambda_\alpha'^n + (n+1-2j)U'i, \quad j = 1, 2, ..., n,$$
$$k_\alpha^1 = \pi - \sin^{-1}(\Lambda_\alpha'^n + niU'),$$
$$k_\alpha^2 = \sin^{-1}(\Lambda_\alpha'^n + (n-2)iU'),$$
$$k_\alpha^3 = \pi - k_\alpha^2,$$
$$k_\alpha^4 = \sin^{-1}(\Lambda_\alpha'^n + (n-4)iU'),$$
$$k_\alpha^5 = \pi - k_\alpha^4,$$

$$\cdots,$$
$$k_\alpha^{2n-2} = \sin^{-1}(\Lambda_\alpha'^n - (n-2)iU'),$$
$$k_\alpha^{2n-1} = \pi - k_\alpha^{2n-2},$$
$$k_\alpha^{2n} = \pi - \sin^{-1}(\Lambda_\alpha'^n - niU'). \tag{14.3}$$

The energy of this excitation is

$$E = -2t \sum_{j=1}^{2n} \cos k_\alpha^j = 4t\Re\sqrt{1 - (\Lambda_\alpha' - nU'i)^2}. \tag{14.4}$$

The other excitation is strings of n Λs,

$$\Lambda_\alpha^{nj} = \Lambda_\alpha^n + (n+1-2j)U'i. \tag{14.5}$$

Essler, Korepin and Schoutens showed that this classification of the Bethe-ansatz eigenstates gives 4^N states[24]. We should construct equations for the M_n' $\Lambda_\alpha'^n$ s, $N - 2\sum nM_n'$ k_js and M_n Λ_ns. The equations for these are derived from (6.29) and (6.30),

$$\exp(ik_j N_a) = \prod_{(n,\alpha)} e\left(\frac{\sin k_j - \Lambda_\alpha^n}{nU'}\right) \prod_{(n,\alpha)} e\left(\frac{\sin k_j - \Lambda_\alpha'^n}{nU'}\right), \tag{14.6}$$

$$\exp(iN_a \sum_{l=1}^{2n} k_\alpha^{n,l})$$
$$= \exp\left(-N_a(\sin^{-1}(\Lambda_\alpha'^n + inU') + \sin^{-1}(\Lambda_\alpha'^n - inU'))\right)$$
$$= -\prod_{j=1}^{N-2M'} e\left(\frac{\Lambda_\alpha'^n - \sin k_j}{nU'}\right) \prod_{(m,\beta)} E_{nm}\left(\frac{\Lambda_\alpha'^n - \Lambda_\beta'^m}{U'}\right), \tag{14.7}$$

$$\prod_{j=1}^{N-2M'} e\left(\frac{\Lambda_\alpha^n - \sin k_j}{nU'}\right) = -\prod_{(m,\beta)} E_{nm}\left(\frac{\Lambda_\alpha^n - \Lambda_\beta^m}{U'}\right). \tag{14.8}$$

Taking the logarithm of these equations we have

$$k_j N_a = 2\pi I_j - \sum_{n=1}^{\infty}\sum_{\alpha=1}^{M_n} \theta\left(\frac{\sin k_j - \Lambda_\alpha^n}{nU'}\right) - \sum_{n=1}^{\infty}\sum_{\alpha=1}^{M_n'} \theta\left(\frac{\sin k_j - \Lambda_\alpha'^n}{nU'}\right), \tag{14.9}$$

$$N_a(\sin^{-1}(\Lambda_\alpha'^n + inU') + \sin^{-1}(\Lambda_\alpha'^n - inU')) = 2\pi J_\alpha'^n$$
$$+ \sum_{j=1}^{N-2M'} \theta\left(\frac{\Lambda_\alpha'^n - \sin k_j}{nU'}\right) + \sum_{(m,\beta)} \Theta_{nm}\left(\frac{\Lambda_\alpha'^n - \Lambda_\beta'^m}{U'}\right), \tag{14.10}$$

$$\sum_{j=1}^{N-2M'} \theta\left(\frac{\Lambda_\alpha^n - \sin k_j}{nU'}\right) = 2\pi I_\alpha^n + \sum_{(m,\beta)} \Theta_{nm}\left(\frac{\Lambda_\alpha^n - \Lambda_\beta^m}{U'}\right). \tag{14.11}$$

Here I_j, J_α^n, $J_\alpha'^n$ should satisfy the following conditions:

$$I_j = \begin{cases} \text{integer}, & (\sum_{i=1}^\infty (M_i + M_i') \text{ even}) \\ \text{half} - \text{odd integer}, & (\sum_{i=1}^\infty (M_i + M_i') \text{ odd}) \end{cases}$$

$$J'^n = \begin{cases} \text{integer}, & (N - M_n' \text{ odd}) \\ \text{half} - \text{odd integer}, & (N - M_n' \text{ even}) \end{cases}$$

$$J^n = \begin{cases} \text{integer}, & (N - M_n \text{ odd}) \\ \text{half} - \text{odd integer}, & (N - M_n \text{ even}) \end{cases}$$

$$|J_\alpha'^n| < \frac{1}{2}\left(N_a - N + 2M' - \sum_{m=1}^\infty t_{nm} M_m'\right),$$

$$|J_\alpha^n| < \frac{1}{2}\left(N - 2M' - \sum_{m=1}^\infty t_{nm} M_m'\right), \tag{14.12}$$

where $t_{nm} \equiv 2\min(n,m) - \delta_{nm}$. A set of integers $\{I_j, J_\alpha^n, J_\alpha'^n\}$ satisfying this condition gives an eigenstate $|\Psi\rangle$ with particle number N and down-spin number M. We can generate $(N_a - N + 1)(N - 2M + 1)$ states using the η^- and S^- operators defined in (6.10) and (6.14). The grand partition function is given by

$$\Xi(T, A, h) = \sum_{I_j, J_\alpha^n, J_\alpha'^n} \exp\left(-\frac{E - AN}{T}\right)$$

$$\times (1 + e^{(2A-U)/T} + e^{2(2A-U)/T} + \dots + e^{(N_a-N)(2A-U)/T})$$

$$\times (1 + e^{-2h/T} + e^{-4h/T} + \dots + e^{-2(N-2M)h/T})$$

$$= \sum_{N=0}^{N_a} \frac{\sinh(N_a - N + 1)(A - U/2)/T}{\sinh(A - U/2)/T} \sum_{M=0}^{[N/2]} \frac{\sinh(N - 2M + 1)h/T}{\sinh h/T}$$

$$\times \sum_{I_j, J_\alpha^n, J_\alpha'^n} \exp\left(-\frac{E - AN_a + (N_a - N)U/2 + h(N - 2M)}{T}\right). \tag{14.13}$$

The energy E and momentum K of this state is given by

$$E = \sum_{j=1}^{N-2M'} (-2t\cos k_j - h) + \sum_{(n,\alpha)} 4t\Re\sqrt{1 - (\Lambda_\alpha'^n - inU')^2}$$

$$+2h \sum_{n=1}^{\infty} nM_n, \tag{14.14}$$

$$K = \frac{2\pi}{N_a} \Big(\sum_j I_j - \sum_{(n,\alpha)} J_\alpha'^n - \sum_{(n,\alpha)} J_\alpha^n \Big). \tag{14.15}$$

14.2 Thermodynamic Bethe-ansatz equation for the Hubbard model

In the case $A \le U/2$, $h \ge 0$ we have

$$G = -T \ln \Xi = E - AN - TS, \tag{14.16}$$

$$E/N_a = \int_{-\pi}^{\pi} (-2t \cos k - h)\rho(k)dk$$

$$+ \sum_{n=1}^{\infty} \int 4t\Re\sqrt{1 - (\Lambda - inU')^2}\sigma_n'(\Lambda)d\Lambda + 2h \sum_{n=1}^{\infty} n \int \sigma_n(\Lambda)d\Lambda,$$

$$N/N_a = \int_{-\pi}^{\pi} \rho(k)dk + \sum_{n=1}^{\infty} 2n \int \sigma_n'(\Lambda)d\Lambda,$$

$$S/N_a = \int_{-\pi}^{\pi} \{(\rho + \rho^h)\ln(\rho + \rho^h) - \rho \ln \rho - \rho^h \ln \rho^h\}dk$$

$$+ \sum_n \int \{(\sigma_n + \sigma_n^h)\ln(\sigma_n + \sigma_n^h) - \sigma_n \ln \sigma_n - \sigma_n^h \ln \sigma_n^h\}d\Lambda$$

$$+ \sum_n \int \{(\sigma_n' + \sigma_n'^h)\ln(\sigma_n' + \sigma_n'^h) - \sigma_n' \ln \sigma_n' - \sigma_n'^h \ln \sigma_n'^h\}d\Lambda. \tag{14.17}$$

From equations (14.9)–(14.11) we get

$$\frac{1}{2\pi} = \rho(k) + \rho^h(k)$$

$$- \cos k \Big(\sum_{n=1}^{\infty} \int_{-\infty}^{\infty} a_n(\sin k - \Lambda)(\sigma_n'(\Lambda) + \sigma_n(\Lambda))d\Lambda \Big), \tag{14.18}$$

$$\int_{-\pi}^{\pi} a_n(\sin k - \Lambda)\rho(k)dk = \sigma_n(\Lambda) + \sigma_n^h(\Lambda)$$

$$+ \sum_{m=1}^{\infty} T_{nm} * \sigma_m(\Lambda), \tag{14.19}$$

$$\frac{1}{\pi}\Re\frac{1}{\sqrt{1 - (\Lambda - nU'i)^2}} - \int_{-\pi}^{\pi} a_n(\sin k - \Lambda)\rho(k)dk$$

$$= \sigma_n'(\Lambda) + \sigma_n'^h(\Lambda) + \sum_{m=1}^{\infty} T_{nm} * \sigma_m'(\Lambda). \tag{14.20}$$

The condition $\delta G = 0$ gives the equations for $\zeta \equiv \rho^h/\rho$, $\eta_n \equiv \sigma_n^h/\sigma_n$ and $\eta_n' \equiv \sigma_n'^h/\sigma_n'$,

$$
\ln \zeta(k) = \frac{-2t \cos k - h - A}{T}
$$
$$
+ \sum_{n=1}^{\infty} \int_{-\infty}^{\infty} a_n(\sin k - \Lambda) \ln \left(\frac{1 + \eta_n'^{-1}(\Lambda)}{1 + \eta_n^{-1}(\Lambda)} \right) d\Lambda, \tag{14.21}
$$

$$
\ln \eta_n(\Lambda) = \frac{2nh}{T} - \int_{-\pi}^{\pi} dk \cos k \, a_n(\sin k - \Lambda) \ln(1 + \zeta^{-1}(k))
$$
$$
+ \sum_{m=1}^{\infty} T_{nm} * \ln(1 + \eta_m^{-1}(\Lambda)), \tag{14.22}
$$

$$
\ln \eta_n'(\Lambda) = \frac{4t\Re\sqrt{1 - (\Lambda - nU'i)^2} - 2nA}{T}
$$
$$
- \int_{-\pi}^{\pi} dk \cos k \, a_n(\sin k - \Lambda) \ln(1 + \zeta^{-1}(k))
$$
$$
+ \sum_{m=1}^{\infty} T_{nm} * \ln(1 + \eta_m'^{-1}(\Lambda)). \tag{14.23}
$$

Substituting (14.18), (14.19), (14.20), (14.21), (14.22) and (14.23) we have an expression for the Gibbs free energy per site,

$$
G/N_a = -T \int_{-\pi}^{\pi} \ln(1 + \zeta^{-1}(k)) \frac{dk}{2\pi}
$$
$$
- T \sum_{n=1}^{\infty} \int_{-\infty}^{\infty} \Re \frac{1}{\sqrt{1 - (\Lambda - nU'i)^2}} \ln(1 + \eta_n'^{-1}(\Lambda)) \frac{d\Lambda}{\pi}, \tag{14.24}
$$

Using the $n = 1$ equation of (14.23) this is transformed as

$$
g = \frac{E_0}{N_a} - A - T \left\{ \int_{-\pi}^{\pi} \rho_0(k) \ln(1 + \zeta(k)) dk \right.
$$
$$
\left. + \int_{-\infty}^{\infty} \sigma_0(\Lambda) \ln(1 + \eta_1(\Lambda)) d\Lambda \right\}. \tag{14.25}
$$

Here $E_0, \rho_0(k)$ and $\sigma_0(\Lambda)$ are the energy, $\rho(k)$ and $\sigma_1(\Lambda)$ at $T = h = U/2 - A = 0$, (half-filled, zero-field ground state),

$$
E_0/N_a = -4t \int_0^{\infty} \frac{J_0(\omega) J_1(\omega) d\omega}{1 + \exp(2U'\omega)},
$$
$$
\sigma_0(\Lambda) = \int_{-\pi}^{\pi} s(\Lambda - \sin k) \frac{dk}{2\pi},
$$
$$
\rho_0(k) = \frac{1}{2\pi} + \cos k \int_{-\infty}^{\infty} a_1(\Lambda - \sin k) \sigma_0(\Lambda) d\Lambda, \tag{14.26}
$$

and

$$s(x) \equiv \frac{1}{4U'}\operatorname{sech}\frac{\pi x}{2U'}. \tag{14.27}$$

Equations (14.21), (14.22) and (14.23) are transformed as follows:

$$\ln \zeta(k) = \frac{\kappa_0(k)}{T} + \int_{-\infty}^{\infty} d\Lambda s(\Lambda - \sin k) \ln\left(\frac{1 + \eta_1'(\Lambda)}{1 + \eta_1(\Lambda)}\right), \tag{14.28}$$

$$\ln \eta_1(\Lambda) = s * \ln(1 + \eta_2(\Lambda))$$
$$- \int_{-\pi}^{\pi} s(\Lambda - \sin k) \ln(1 + \zeta^{-1}(k)) \cos k dk, \tag{14.29}$$

$$\ln \eta_1'(\Lambda) = s * \ln(1 + \eta_2'(\Lambda))$$
$$- \int_{-\pi}^{\pi} s(\Lambda - \sin k) \ln(1 + \zeta(k)) \cos k dk, \tag{14.30}$$

$$\ln \eta_n(\Lambda) = s * \ln\{(1 + \eta_{n-1}(\Lambda))(1 + \eta_{n+1}(\Lambda))\}, \quad n \geq 2, \tag{14.31}$$

$$\ln \eta_n'(\Lambda) = s * \ln\{(1 + \eta_{n-1}'(\Lambda))(1 + \eta_{n+1}'(\Lambda))\}, \quad n \geq 2, \tag{14.32}$$

$$\lim_{n\to\infty} \frac{\ln \eta_n(\Lambda)}{n} = \frac{2h}{T}, \tag{14.33}$$

$$\lim_{n\to\infty} \frac{\ln \eta_n'(\Lambda)}{n} = \frac{U - 2A}{T}. \tag{14.34}$$

Here $\kappa_0(k)$ is defined by

$$\kappa_0(k) \equiv -2t \cos k - 4t \int_{-\infty}^{\infty} d\Lambda s(\Lambda - \sin k)\Re\sqrt{1 - (\Lambda - U'i)^2}. \tag{14.35}$$

The last terms of equations (14.29) and (14.30) are rewritten as follows:

$$- \int_{-\pi/2}^{\pi/2} s(\Lambda - \sin k) \ln\left(\frac{1 + \zeta^{-1}(k)}{1 + \zeta^{-1}(\pi - k)}\right) \cos k dk,$$

$$- \int_{-\pi/2}^{\pi/2} s(\Lambda - \sin k) \ln\left(\frac{1 + \zeta(k)}{1 + \zeta(\pi - k)}\right) \cos k dk. \tag{14.36}$$

From equation (14.35) we have $\kappa_0(k) - \kappa_0(\pi - k) = -4t \cos k$. Then at $|k| < \pi/2$ we have $\ln \zeta(k) < \ln \zeta(\pi - k)$. Thus the last term of (14.29) is negative and that of (14.30) is positive. The equation for η_j resembles that for the antiferromagnetic XXX model, and that for η_j' resembles the ferromagnetic XXX model.

14.3 Some special limits

14.3.1 The limit $U \to \infty$

In this limit we get from (14.23) that

$$\eta_n' = \infty.$$

From (14.22) we have

$$\ln \eta_n(\Lambda) = \frac{2nH}{T} + O(U^{-1}) + \sum_{m=1}^{\infty} T_{nm} * \ln(1 + \eta_m^{-1}(\Lambda)). \qquad (14.37)$$

This equation is equivalent to the $J = 0$ case of the XXX model, and we have

$$\eta_n(\Lambda) = f^2(n) - 1, \quad f(n) = \sinh((n+1)h/T)/\sinh(h/T). \qquad (14.38)$$

Substituting this into (14.21),

$$\ln \zeta(k) = \frac{-2t \cos k - h - A}{T} - \sum_{n=1}^{\infty} \ln(1 + \eta_n^{-1})$$

$$= \frac{-2t \cos k - A}{T} - \ln(2 \cosh h/T). \qquad (14.39)$$

Using (14.24),

$$G/N_a = -T \int_{-\pi}^{\pi} \ln(1 + 2 \cosh(h/T) \exp[(2t \cos k + A)/T]) \frac{dk}{2\pi}. \qquad (14.40)$$

The particle number and magnetization are

$$N = -\frac{\partial G(T, A, h)}{\partial A} = \cosh(h/T)X,$$

$$2S_z = -\frac{\partial G(T, A, h)}{\partial h} = \sinh(h/T)X$$

$$X = N_a \int \frac{2e^{(2t \cos k + A)/T}}{1 + 2 \cosh(h/T)e^{(2t \cos k + A)/T}} \frac{dk}{2\pi}. \qquad (14.41)$$

Thus $S_z/N = \frac{1}{2} \tanh(h/T)$. This means that the magnetization is the same as for free spins with $S = 1/2$.

14.3.2 The limit $U \to 0$

ρ and ζ are functions of k. But we define $\rho_-(\Lambda)$, $\rho_+(\Lambda)$, $\zeta_-(\Lambda)$ and $\zeta_+(\Lambda)$ as follows:

$$\rho_-(\Lambda) \equiv \frac{1}{\sqrt{1 - \Lambda^2}} \rho(\sin^{-1} \Lambda), \quad \zeta_-(\Lambda) \equiv \zeta(\sin^{-1} \Lambda),$$

$$\rho_+(\Lambda) \equiv \frac{1}{\sqrt{1 - \Lambda^2}} \rho(\pi - \sin^{-1} \Lambda), \quad \zeta_+(\Lambda) \equiv \zeta(\pi - \sin^{-1} \Lambda),$$

$$(14.42)$$

where $|\sin^{-1}\Lambda| \leq \pi/2$, $|\Lambda| \leq 1$. Then the integral equations (14.28)–(14.35) become

$$\ln \zeta_\pm = t(2 \pm 2)T^{-1}\sqrt{1-\Lambda^2} + \frac{1}{2}\ln\left(\frac{1+\eta_1'}{1+\eta_1}\right), \qquad (14.43)$$

$$\ln \eta_1 = \frac{1}{2}\ln\left(\frac{1+\zeta_+^{-1}}{1+\zeta_-^{-1}}\right) + \frac{1}{2}\ln(1+\eta_2), \qquad (14.44)$$

$$\ln \eta_1' = \frac{1}{2}\ln\left(\frac{1+\zeta_+}{1+\zeta_-}\right) + \frac{1}{2}\ln(1+\eta_2'), \qquad (14.45)$$

$$\ln \eta_n = \frac{1}{2}\ln(1+\eta_{n-1})(1+\eta_{n+1}), \qquad (14.46)$$

$$\ln \eta_n' = \frac{1}{2}\ln(1+\eta_{n-1}')(1+\eta_{n+1}'), \qquad (14.47)$$

$$\lim_{n\to\infty} \frac{\ln \eta_n}{n} = \frac{2h}{T}, \qquad (14.48)$$

$$\lim_{n\to\infty} \frac{\ln \eta_n'}{n} = \frac{-2A}{T}. \qquad (14.49)$$

The solution of equations (14.46)–(14.49) is

$$\eta_n = f^2(n) - 1, \quad f(n) = \frac{az^n - a^{-1}z^{-n}}{z - z^{-1}}, \quad z = e^{-h/T},$$

$$\eta_n' = f^2(n) - 1, \quad f'(n) = \frac{bw^n - b^{-1}w^{-n}}{w - w^{-1}}, \quad w = e^{A/T}. \qquad (14.50)$$

a and b are functions of Λ at $|\Lambda| \leq 1$. Substituting this into (14.43)–(14.45) we have

$$\zeta_+ = f'(1)/f(1), \quad \zeta_- = x^{-2}f'(1)/f(1),$$

$$f^2(0) = \frac{1+\zeta_+^{-1}}{1+\zeta_-^{-1}}, \quad f'^2(0) = \frac{1+\zeta_+}{1+\zeta_-}, \quad x \equiv \exp\left(\frac{2t\sqrt{1-\Lambda^2}}{T}\right). \qquad (14.51)$$

From these equations, one can show

$$f(-1) + f'(-1) = 0, \quad xf(0) = f'(0).$$

We obtain the relation

$$b^2 = \frac{a^2(w + x^{-1}z^{-1}) - (w + x^{-1}z)}{a^2(w^{-1} + x^{-1}z^{-1}) - (w^{-1} + x^{-1}z)}.$$

We can eliminate b from this equation and obtain an equation for a^2. The final result is

$$a = \sqrt{\frac{(1 + x^{-1}zw)(1 + x^{-1}zw^{-1})}{(1 + x^{-1}z^{-1}w)(1 + x^{-1}z^{-1}w^{-1})}},$$

$$b = \sqrt{\frac{(1 + xzw)(1 + xz^{-1}w)}{(1 + xzw^{-1})(1 + xz^{-1}w^{-1})}}. \tag{14.52}$$

Using equation (14.26) we have

$$G/N_a = -\frac{4}{\pi} - A$$

$$-T \int_{-1}^{1} \frac{1}{\sqrt{1 - \Lambda^2}} \ln(1 + \zeta_-(\Lambda)) + \frac{1}{2} \ln(1 + \eta_1(\Lambda)) \frac{d\Lambda}{\pi}.$$

From (14.50), (14.51) and (14.52),

$$(1 + \zeta_-(\Lambda))\sqrt{1 + \eta_1(\Lambda)} = f(1) + x^{-2}f'(1) = \{(1 + x^{-1}wz)$$
$$(1 + x^{-1}wz^{-1})(1 + x^{-1}w^{-1}z)(1 + x^{-1}w^{-1}z^{-1})\}^{1/2}.$$

Thus the thermodynamic potential per site is

$$G/N_a = -T \int_{-\pi}^{\pi} \frac{dk}{2\pi} \ln(1 + e^{(2t\cos k + A - h)/T})(1 + e^{(2t\cos k + A + h)/T}) \tag{14.53}$$

This result coincides with the result for free band electrons.

14.3.3 The limit $T \to 0$

We write

$$\kappa(k) = T \ln \zeta(k), \quad \epsilon_j(k) = T \ln \eta_j(\Lambda), \quad \epsilon'_j(\Lambda) = T \ln \eta'_j(\Lambda).$$

From the equations we find that $\epsilon'_1, \epsilon'_2, \dots$ and $\epsilon_2, \epsilon_3, \dots$ are always positive. Then in the limit $T \to 0$ these equations become

$$\kappa(k) = -2t \cos k - h - A + \int a_1(\sin k - \Lambda)\epsilon_1^-(\Lambda)d\Lambda$$

$$\epsilon_1(\Lambda) = 2h + \int dk \cos k a_1(\sin k - \Lambda)\kappa^-(k)$$

$$- \int d\Lambda' a_2(\Lambda - \Lambda')\epsilon_1^-(\Lambda'). \tag{14.54}$$

Functions $\sigma_2(\Lambda), \sigma_3(\Lambda)\dots$ and $\sigma'_1(\Lambda), \sigma'_2(\Lambda), \dots$ are zero. We denote points where $\kappa(k)$ changes sign by $\pm Q$, and points for $\epsilon_1(\Lambda)$ by $\Lambda = \pm B$. The equation for $\rho(k)$ and $\sigma_1(\Lambda)$ becomes

$$\rho(k) = \frac{1}{2\pi} + \cos k \int_{-B}^{B} a_1(\sin k - \Lambda)\sigma_1(\Lambda)d\Lambda,$$

$$\int_{-Q}^{Q} a_1(\sin k - \Lambda)dk = \sigma_1(\Lambda) + \int_{-B}^{B} a_2(\Lambda - \Lambda')d\Lambda'. \quad (14.55)$$

This is equivalent to the Fredholm-type integral equation in (6.37)–(6.39)

14.3.4 The limit $t \to 0$

In this limit the first term on the r.h.s. of (14.28) is a constant, $U/(2T)$. Then $\zeta(k)$ has no k dependence and is a constant. The last terms of equations (14.29) and (14.30) are zero. Thus η_j and η'_j are also constants. The integral equations become difference equations which appeared in the thermodynamics of the XXX model. The solutions are

$$\eta_n = \Big(\frac{\sinh(n+1)h/T}{\sinh h/T}\Big)^2 - 1,$$

$$\eta'_n = \Big(\frac{\sinh[(n+1)(U/2-A)/T]}{\sinh[(U/2-A)/T]}\Big)^2 - 1,$$

$$\zeta = \frac{U}{2T} + \frac{\cosh[(U/2-A)/T]}{\cosh h/T}. \quad (14.56)$$

Substituting these into (14.25) we have

$$G/(N_a T) = \frac{(U/2-A)}{T} - \ln[2\cosh h/T + 2\cosh(U/2-A)/T]$$

$$= -\ln[1 + e^{(2A-U)/T} + 2e^{(A-U/2)/T}\cosh(h/T)]. \quad (14.57)$$

The low-temperature thermodynamics of the Hubbard model was investigated by the author[98]. Actual numerical calculation of this set of integral equations was done by Kawakami, Usuki and Okiji[49].

Part four

The quantum transfer matrix and recent developments

15

The transfer matrix and correlation length

15.1 The transfer matrix for the Ising chain

We consider the simplest case of a one-dimensional system,

$$\mathcal{H} = -J_z \sum_l S_l^z S_{l+1}^z - 2h \sum_l S_l^z. \tag{15.1}$$

The partition function of this model is

$$Z = \mathrm{Tr} \exp\left(-\frac{\mathcal{H}}{T}\right) = \sum_{\sigma_l = \pm 1} \exp\left(\frac{K}{4}\sigma_l \sigma_{l+1} + H\sigma_l\right) = \mathrm{Tr}\mathbf{T}^N,$$

$$K = J/T, \quad H = h/T. \tag{15.2}$$

Here $\mathbf{T}$ is the transfer matrix,

$$\mathbf{T} = \begin{bmatrix} \exp(K/4 + H) & \exp(-K/4 + H) \\ \exp(-K/4 - H) & \exp(K/4 - H) \end{bmatrix}. \tag{15.3}$$

The partition function is given by

$$Z = \mathrm{Tr} \exp(-\mathcal{H}/T) = \lambda_1^N + \lambda_2^N,$$

$$\lambda_{1,2} = e^{K/4}(\cosh H \pm \sqrt{\sinh^2 H + e^K}). \tag{15.4}$$

λ_1 and λ_2 are eigenvalues of this transfer matrix. The column corresponds to the state of σ_l and the row corresponds to the state of σ_{l+1}. This is an example of the well-known transfer matrix method for classical systems. We can calculate the free energy per site of an infinite system by finding the largest eigenvalue,

$$f = -T \ln(e^{K/4}(\cosh H + \sqrt{\sinh^2 H + e^K})).$$

185

The two-point correlation function $\langle S_j^z S_{j+n}^z \rangle$ is given by

$$\frac{\mathrm{Tr}\exp(-\mathcal{H}/T)S_j^z S_{j+n}^z}{\mathrm{Tr}\exp(-\mathcal{H}/T)} = \frac{\mathrm{Tr}\mathbf{T}^j S^z \mathbf{T}^n S^z \mathbf{T}^{N-n-j}}{\mathrm{Tr}\mathbf{T}^N}$$

$$= \frac{\lambda_1^N (S^z)_{11}^2 + \lambda_2^N (S^z)_{22}^2 + (\lambda_1^{N-n}\lambda_2^n + \lambda_2^{N-n}\lambda_1^n)|(S^z)_{12}|^2}{\lambda_1^N + \lambda_2^N}, \qquad (15.5)$$

where $(S^z)_{ab} = \langle a|S^z|b\rangle$. In the limit of $N \to \infty$ this becomes

$$\langle 1|S^z|1\rangle^2 + |\langle 1|S^z|2\rangle|^2 (\lambda_2/\lambda_1)^n. \qquad (15.6)$$

Then the two-point correlation function decays exponentially as $A + B \exp(-n/\xi)$. The correlation length ξ is given by $(\ln(\lambda_1/\lambda_2))^{-1}$. Generally speaking the largest eigenvalue of the transfer matrix gives the free energy per site and the correlation length is given by the ratio of the second largest eigenvalue and the largest eigenvalue. This method is applicable to the two-dimensional Ising model.

15.2 The transfer matrix for the classical Heisenberg model

The classical Heisenberg model is given by the following Hamiltonian:

$$\mathcal{H} = -J\sum_{l=1}^{N}(n_l^x n_{l+1}^x + n_l^y n_{l+1}^y + n_l^z n_{l+1}^z),$$

$$(n_l^x)^2 + (n_l^y)^2 + (n_l^z)^2 = 1. \qquad (15.7)$$

Nakamura and Fisher investigated this problem[73, 25]. The partition function is given by an angular integral over θ_l and φ_l,

$$Z = \int \prod_l \frac{\mathrm{d}\varphi_l \mathrm{d}\cos\theta_l}{4\pi}$$

$$\exp\left[\frac{J}{T}\sum_l \sin\theta_l \sin\theta_{l+1}\cos(\varphi_l - \varphi_{l+1}) + \cos\theta_l \cos\theta_{l+1}\right]. \qquad (15.8)$$

Then the partition function is given by:

$$Z = \lambda_1^N[1 + (\lambda_2/\lambda_1)^N + (\lambda_3/\lambda_1)^N + ...], \qquad (15.9)$$

where $|\lambda_1| \geq |\lambda_2| \geq ...$ are eigenvalues of an integral operator $\mathbf{T}$,

$$\mathbf{T}f = \frac{1}{4\pi}\int_0^{2\pi}\mathrm{d}\varphi'\int_0^{\pi}\mathrm{d}\cos\theta' K(\theta,\varphi,\theta',\varphi')f(\theta',\varphi') = \lambda_l f(\theta,\varphi),$$

$$K(\theta,\varphi,\theta',\varphi') \equiv \exp\frac{J}{T}\left[\sin\theta\sin\theta'\cos(\varphi-\varphi') + \cos\theta\cos\theta'\right].$$

$$(15.10)$$

This operator commutes with the angular momentum operators,

$$l_x = -i\left[\sin\varphi\frac{\partial}{\partial\theta} - \cot\theta\cos\varphi\frac{\partial}{\partial\varphi}\right],$$

$$l_y = i\left[\cos\varphi\frac{\partial}{\partial\theta} + \cot\theta\sin\varphi\frac{\partial}{\partial\varphi}\right],$$

$$l_z = -i\frac{\partial}{\partial\varphi}, \tag{15.11}$$

and the total angular momentum operator $l^2 = l_x^2 + l_y^2 + l_z^2$. Thus this operator should be diagonalized by spherical harmonics. The kernel K can be expanded in terms of spherical harmonics,

$$K(\theta, \varphi, \theta', \varphi') = \sum_{l=0}^{\infty}(-i)^l\sum_{m=-l}^{l} j_l\left(\frac{iJ}{T}\right) Y_{lm}(\theta, \varphi)\overline{Y_{lm}(\theta', \varphi')}. \tag{15.12}$$

Here the spherical Bessel functions $j_l(z)$ and spherical harmonics $Y_{lm}(\theta, \varphi)$ are defined by

$$j_l(z) = (-1)^l z^l\left(\frac{1}{z}\frac{d}{dz}\right)^l\frac{\sin z}{z},$$

$$Y_{lm}(\theta, \varphi) = e^{im\varphi}(-1)^{\frac{m+|m|}{2}}\sqrt{\left(l + \frac{1}{2}\right)\frac{(l - |m|)!}{(l + |m|)!}}P_l^m(\cos\theta),$$

$$P_l^m(z) = (1 - z^2)^{\frac{|m|}{2}}\frac{d^{|m|}}{dz^{|m|}}P_l(z). \tag{15.13}$$

The Legendre polynomials $P_l(z)$ are defined by

$$P_l(z) = \frac{1}{2^l l!}\frac{d^l}{dz^l}(z^2 - 1)^l.$$

The largest eigenvalue λ_1 of $\mathbf{T}$ is $\sinh(J/T)/(J/T)$ and its eigenvector is Y_{00}. The free energy per site is

$$f = -\frac{T}{N}\ln\lambda_1^N = -T\ln(\sinh(J/T)/(J/T)). \tag{15.14}$$

The energy per site is given by

$$e = -T^2\frac{\partial}{\partial T}\frac{f}{T} = -J\left[\coth\frac{J}{T} - \frac{T}{J}\right]. \tag{15.15}$$

The specific heat per site is

$$C = \frac{\partial e}{\partial T} = 1 - \left[\frac{J/T}{\sinh(J/T)}\right]^2. \tag{15.16}$$

The two-point function is given as follows:

$$s(k) = \langle n_l^z n_{l+k}^z \rangle = \frac{\mathrm{Tr}\mathbf{T}^{N-k}n^z\mathbf{T}^k n^z}{\mathrm{Tr}\mathbf{T}^N}. \tag{15.17}$$

For $k \ll N$ this is given by

$$s(k) = \sum_{l,m} |\langle l, m | n^z | 00 \rangle|^2 \left(\frac{\lambda_{lm}}{\lambda_{00}}\right)^k. \tag{15.18}$$

From the recursion relation

$$z P_l^m(z) = \frac{l-m+1}{2l+1} P_{l+1}^m(z) + \frac{l+m}{2l+1} P_{l-1}^m(z), \tag{15.19}$$

we get

$$\cos\theta\, Y_{lm}(\theta, \varphi) = \sqrt{\frac{(l+1)^2 - m^2}{4(l+1)^2 - 1}}\, Y_{l+1,m}(\theta, \varphi) + \sqrt{\frac{l^2 - m^2}{4l^2 - 1}}\, Y_{l-1,m}(\theta, \varphi). \tag{15.20}$$

Thus

$$\langle l, m | n^z | 00 \rangle = \sqrt{\frac{1}{3}}\delta_{m,0}\delta_{l,1},$$

$$\lambda_{10} = \cosh(J/T)/(J/T) - \sinh(J/T)/(J/T)^2. \tag{15.21}$$

and

$$s(k) = \frac{1}{3}(\coth(J/T) - (T/J))^k. \tag{15.22}$$

Thus the magnetic susceptibility per site is given by

$$\chi = \frac{1}{T}\left[s(0) + 2\sum_{k=1}^{\infty} s(k)\right] = \frac{1}{3T}\frac{1 + \coth(J/T) - (T/J)}{1 - \coth(J/T) + (T/J)}. \tag{15.23}$$

16

The spin 1/2 XXZ model in a magnetic field

16.1 The diagonal-to-diagonal transfer matrix

Here we give a set of equations which determines the free energy and correlation length of the $S = 1/2$ XXZ chain in a magnetic field in the z-direction. The Hamiltonian is given as follows:

$$\mathcal{H} = -J \sum_{j=1}^{N} \{S_j^x S_{j+1}^x + S_j^y S_{j+1}^y + \Delta S_j^z S_{j+1}^z\} - 2h \sum_{j=1}^{N} S_j^z,$$
$$J > 0, \quad -\infty < \Delta < \infty. \tag{16.1}$$

The quantum transfer matrix for this model is equivalent to that of the diagonal-to-diagonal transfer matrix of the six-vertex model. This eigenvalue problem is a special case of the inhomogeneous six-vertex model[7]. The XYZ chain in a zero magnetic field is also solvable,

$$\mathcal{H} = -\sum_{j=1}^{N} \{J_x S_j^x S_{j+1}^x + J_y S_j^y S_{j+1}^y + J_z S_j^z S_{j+1}^z\}, \quad |J_x| \leq J_y \leq J_z. \tag{16.2}$$

The thermodynamics of quantum 1D systems is transformed to some kind of classical 2D strip by adding an imaginary time axis. The width of the strip is proportional to the inverse of the temperature.

The most popular method is the checker board decomposition. The Monte Carlo method is applicable and one can calculate low-temperature properties of such systems[6,38]. Betsuyaku defined the quantum transfer matrix in the real space direction[14,15]. The numerical treatment of the matrix is useful for the calculation of thermodynamic quantities[14]. The transfer matrix can be used for unsolvable models like spin 1 chains[88]. The dimension of the matrix increases exponentially as the Trotter number M increases. The Trotter number is the number of slices along the imaginary time axis. So in this numerical method, it is impossible to treat the very low-temperature

189

case. Koma[51, 52] showed that the eigenvalue problem of the quantum transfer matrix for the XXZ chain in a magnetic field is solvable by the Bethe-ansatz method. It is equivalent to that of the diagonal-to-diagonal transfer matrix of the six-vertex model. Usually the six-vertex model is treated by the row-to-row transfer matrix. Bariev[5], Truong and Schotte[113] had showed that the six-vertex model can be treated also by the diagonal-to-diagonal transfer matrix. Koma calculated the Bethe-ansatz equation numerically and got the free energy and the susceptibility of the ferromagnetic XXX chain. The results coincide completely with those of the thermodynamic integral equation method[109]. Suzuki, Akutsu and Wadati showed that the XYZ problem is reduced to the diagonal-to-diagonal transfer matrix of the eight-vertex model and this can be treated by the inhomogeneous eight-vertex model[86].

The partition function Z of the system described by (16.1) is approximated as follows:

$$Z = \text{Tr}\{\exp(-\mathscr{H}_1/(MT))\exp(-\mathscr{H}_2/(MT))\}^M,$$

$$\mathscr{H}_1 = -\sum_{j=\text{odd}} J\{S_j^x S_{j+1}^x + S_j^y S_{j+1}^y + \Delta S_j^z S_{j+1}^z\} + h(S_j^z + S_{j+1}^z),$$

$$\mathscr{H}_2 = -\sum_{j=\text{even}} J\{S_j^x S_{j+1}^x + S_j^y S_{j+1}^y + \Delta S_j^z S_{j+1}^z\} + h(S_j^z + S_{j+1}^z).$$

$$(16.3)$$

Then the partition function is given by the six-vertex model with $2M \times N$ bonds,

$$Z = \sum_{\{\sigma\}} \prod_{j=1}^{N} \prod_{i=1}^{M} B(\sigma_{2i+j,j}\sigma_{2i+j+1,j}; \sigma_{2i+j,j+1}\sigma_{2i+j+1,j+1}).$$

Here $\sigma_{l,m}; l = 1, ..., 2M$, $m = 1, ..., N$ gives the direction of bonds and takes values $+1$ and -1. B is given by the 4×4 matrix,

$$B(\sigma_1\sigma_2; \sigma_1'\sigma_2') = \begin{bmatrix} b & 0 & 0 & c \\ 0 & 0 & a & 0 \\ 0 & a & 0 & 0 \\ c & 0 & 0 & b' \end{bmatrix}, \qquad (16.4)$$

$$a = \exp\left(-\frac{J\Delta}{4MT}\right)\sinh\left(\frac{J}{2MT}\right), \quad b = \exp\left(\frac{J\Delta - 4h}{4MT}\right),$$

$$b' = \exp\left(\frac{J\Delta + 4h}{4MT}\right), \quad c = \exp\left(-\frac{J\Delta}{4MT}\right)\cosh\left(\frac{J}{2MT}\right). \quad (16.5)$$

If we put $\sigma_{l,m} \rightarrow (-1)^{l+m+1}\sigma_{l,m}$, Z is given by

$$Z = \sum_{\{\sigma\}} \prod_{j=1}^{N} \prod_{i=1}^{M} A(\sigma_{2i+j,j}\sigma_{2i+j+1,j}; \sigma_{2i+j,j+1}\sigma_{2i+j+1,j+1}),$$

$$A(\sigma_1\sigma_2;\sigma_1'\sigma_2') = \begin{bmatrix} a & 0 & 0 & 0 \\ 0 & c & b' & 0 \\ 0 & b & c & 0 \\ 0 & 0 & 0 & a \end{bmatrix}.$$

Then in the case $N = 2M \times$ integer, we have

$$Z = \mathrm{Tr}\mathbf{T}^N, \quad \mathbf{T}(\sigma_1, \sigma_2, ..., \sigma_{2M}; \sigma_1', \sigma_2', ..., \sigma_{2M}')$$

$$\equiv A(\sigma_1\sigma_2; \sigma_{2M}'\sigma_1')A(\sigma_3\sigma_4; \sigma_2'\sigma_3')...A(\sigma_{2M-1}\sigma_{2M}; \sigma_{2M-2}'\sigma_{2M-1}').$$

This $2^{2M} \times 2^{2M}$ matrix is the diagonal-to-diagonal transfer matrix of the six-vertex model. The eigenvalue problem of this transfer matrix is a special case of the inhomogeneous six-vertex model on the square lattice which was treated by Baxter[7]. The relation with the diagonal-to-diagonal six-vertex model and inhomogeneous six-vertex model is shown in Fig. (16.1).

Consider an inhomogeneous six-vertex model with the following column-dependent Boltzman weights:

$$a_l = \rho_l h(v + v_l + \eta)$$
$$b_l = \rho_l \omega^{-1} h(v + v_l - \eta)$$
$$b_l' = \rho_l \omega h(v + v_l - \eta)$$
$$c_l = \rho_l h(2\eta), \quad l = 1, ..., L. \tag{16.6}$$

Here L is the number of columns. $h(u)$ is u, $\sin(u)$ or $\sinh(u)$. In this case the transfer matrix $\mathbf{T}(v)$ is defined in 2^L-dimensional space,

$$\mathbf{T} = \mathrm{Tr}\mathbf{R}_1(\sigma_1, \sigma_1')\mathbf{R}_2(\sigma_2, \sigma_2')...\mathbf{R}_L(\sigma_L, \sigma_L'),$$
$$\mathbf{R}_l(++) = \begin{pmatrix} a_l & 0 \\ 0 & b_l \end{pmatrix}, \quad \mathbf{R}_l(+-) = \begin{pmatrix} 0 & 0 \\ c_l & 0 \end{pmatrix},$$
$$\mathbf{R}_l(-+) = \begin{pmatrix} 0 & c_l \\ 0 & 0 \end{pmatrix}, \quad \mathbf{R}_l(--) = \begin{pmatrix} b_l' & 0 \\ 0 & a_l \end{pmatrix}. \tag{16.7}$$

The space is divided into subspaces by the number of down-spins k. Without loss of generality we can put $k \le L/2$. In this subspace we can construct a

Bethe-ansatz wave function with k parameters $u_1, ..., u_k$,

$$|\Psi\rangle = \sum f(y_1, y_2, ..., y_k)\sigma_{y_1}^- \sigma_{y_2}^- ... \sigma_{y_k}^- |0\rangle,$$

$$f(y_1, y_2, ..., y_k) = \sum_P A(P) \prod_{j=1}^{k} F(y_j; u_{Pj}),$$

$$F(y; u) \equiv \omega^y \prod_{l=1}^{y-1} \mathsf{h}(u + v_l + \eta) \prod_{l=y+1}^{L} \mathsf{h}(u + v_l - \eta),$$

$$A(P) = \epsilon(P) \sum_{j<l} \mathsf{h}(u_{Pj} - u_{Pl} - 2\eta). \tag{16.8}$$

The periodic boundary condition is satisfied by

$$\frac{\phi(u_j + \eta)}{\phi(u_j - \eta)} = -\omega^{-L} \prod_{m=1}^{k} \frac{\mathsf{h}(u_j - u_m + 2\eta)}{\mathsf{h}(u_j - u_m - 2\eta)},$$

$$\phi(v) = \prod_{l=1}^{L} \rho_l \mathsf{h}(v + v_l). \tag{16.9}$$

The corresponding eigenvalue is given by

$$T(v) = \omega^{-L+k} \phi(v - \eta)\frac{Q(v + 2\eta)}{Q(v)} + \omega^k \phi(v + \eta)\frac{Q(v - 2\eta)}{Q(v)},$$

$$Q(v) = \prod_{j=1}^{k} \mathsf{h}(v - u_j). \tag{16.10}$$

To solve the diagonal-to-diagonal transfer matrix we should consider an inhomogeneous six-vertex model, the Boltzman weights of which are given by

$$a_l = c_l = 1, \quad b_l = b_l' = 0 \quad \text{for even } l,$$

$$a_l = \exp\left(-\frac{J\Delta}{4MT}\right) \sinh\left(\frac{J}{2MT}\right), \quad b_l = \exp\left(\frac{J\Delta - 4h}{4MT}\right),$$

$$b_l' = \exp\left(\frac{J\Delta + 4h}{4MT}\right), c_l = \exp\left(-\frac{J\Delta}{4MT}\right) \cosh\left(\frac{J}{2MT}\right)$$

$$\text{for odd } l. \tag{16.11}$$

The conditions (16.6) are satisfied if we put

$$L = 2M, \quad \omega = \pm\exp\left(\frac{h}{MT}\right), \quad v = 0, \quad \frac{\mathsf{h}'(2\eta)}{\mathsf{h}'(0)} = \pm\frac{\sinh(\frac{J\Delta}{MT})}{\sinh(\frac{J}{MT})}, \tag{16.12}$$

and

$$\rho_l = 1/\mathsf{h}(2\eta), \quad v_l = \eta \quad \text{for even } l,$$

$$\rho_l = \pm \frac{\sqrt{bb'}}{\mathsf{h}(v_l - \eta)}, \quad \pm \frac{\mathsf{h}(v_l + \eta)}{\mathsf{h}(v_l - \eta)}$$

$$= \frac{a}{\sqrt{bb'}} = \exp\left(-\frac{J\Delta}{2MT}\right) \sinh\left(\frac{J}{2MT}\right) \quad \text{for odd } l. \quad (16.13)$$

Then we have the Bethe-ansatz equation for u_j, $j = 1, ..., k$,

$$\left(\frac{\mathsf{h}(u_j + 2\eta)\mathsf{h}(u_j + \eta + v_1)}{\mathsf{h}(u_j)\mathsf{h}(u_j - \eta + v_1)}\right)^M = -\omega^{-2M} \prod_{m=1}^{k} \frac{\mathsf{h}(u_j - u_m + 2\eta)}{\mathsf{h}(u_j - u_m - 2\eta)}. \quad (16.14)$$

The eigenvalue is given by

$$\Lambda = T(0) = \omega^k \phi(\eta) \frac{Q(-2\eta)}{Q(0)} = a^M \omega^k \prod_{m=1}^{k} \frac{\mathsf{h}(u_m - 2\eta)}{\mathsf{h}(u_m)}. \quad (16.15)$$

We put $\eta + v_1 = 2\alpha$ and $x_j = u_j + \alpha$. By these transformations equations (16.14) and (16.15) become

$$\left[\frac{\mathsf{h}(x_l - \alpha)\mathsf{h}(x_l + \alpha + 2\eta)}{\mathsf{h}(x_l + \alpha)\mathsf{h}(x_l - \alpha - 2\eta)}\right]^M = -\exp\left(-\frac{2h}{T}\right) \prod_{j=1}^{k} \frac{\mathsf{h}(x_l - x_j + 2\eta)}{\mathsf{h}(x_l - x_j - 2\eta)},$$

$$l = 1, 2, ..., k, \quad (16.16)$$

$$\Lambda = \exp\left(\frac{hk}{MT} + \frac{J\Delta}{4T}\right) \left(\frac{\mathsf{h}(2\alpha)}{\mathsf{h}(2\alpha + 2\eta)}\right)^M \prod_{l=1}^{k} \frac{\mathsf{h}(x_l - \alpha - 2\eta)}{\mathsf{h}(x_l - \alpha)}. \quad (16.17)$$

For $|\Delta| > 1$ we put

$$\mathsf{h}(x) = \sin x, \quad \eta = i\theta/2, \quad \theta = \text{sign}(\Delta)\cosh^{-1}|\frac{\sinh(J\Delta/2MT)}{\sinh(J/2MT)}|,$$

$$\alpha = \frac{i}{2}\tanh^{-1}\left(\tanh\theta \tanh\frac{J\Delta}{2MT}\right), \quad (16.18)$$

for $\Delta = \pm 1$ we put

$$\mathsf{h}(x) = x, \quad \eta = i\,\text{sign}(\Delta)/2, \quad \alpha = \frac{i}{2}\tanh\frac{J}{2MT}, \quad (16.19)$$

and for $|\Delta| < 1$ we put

$$\mathsf{h}(x) = \sinh x, \quad \eta = i\theta/2, \quad \theta = \cos^{-1}\left(\frac{\sinh(J\Delta/2MT)}{\sinh(J/2MT)}\right),$$

$$\alpha = \frac{i}{2}\tan^{-1}\left(\tan\theta \tanh\frac{J\Delta}{2MT}\right). \quad (16.20)$$

Without loss of generality we can put $k \leq M$, because the case of $k > M$ is reduced to the $k < M$ case by reversing spin directions and putting

$h \rightarrow -h$. As all of the elements of the matrix $\mathbf{T}$ are non-negative, the largest eigenvalue (with respect to the absolute value) is real and positive. As the matrix $\mathbf{T}$ is asymmetric, the eigenvalues are not necessarily real. For the largest eigenvalue, k is equal to M. We write this eigenvalue as Λ_M^1. The exact free energy per site is given by the limiting value of Λ_M^1 as $M \rightarrow \infty$,

$$f = -T \ln(\lim_{M \rightarrow \infty} \Lambda_M^1). \tag{16.21}$$

Next we consider two-point functions $S^{xx}(n) \equiv \langle S_j^x S_{j+n}^x \rangle$ and $S^{zz}(n) \equiv \langle S_j^z S_{j+n}^z \rangle$. $S^{yy}(n) \equiv \langle S_j^y S_{j+n}^y \rangle$ is the same as $S^{xx}(n)$. In the case of an infinite chain, these are written as

$$S^{xx}(n) = \frac{\langle 0|\mathbf{U}^x \mathbf{T}^n \mathbf{R}^{-n} \mathbf{U}^x \mathbf{R}^n|0\rangle}{\langle 0|\mathbf{T}^n|0\rangle},$$

$$S^{zz}(n) = \frac{\langle 0|\mathbf{U}^z \mathbf{T}^n \mathbf{R}^{-n} \mathbf{U}^z \mathbf{R}^n|0\rangle}{\langle 0|\mathbf{T}^n|0\rangle}.$$

Here $\mathbf{R}$ is the shift operator,

$$\mathbf{U}^x \equiv \frac{1}{2}(\sigma^x)_{\sigma_1,\sigma_1'} \prod_{j=2}^{2M} \delta_{\sigma_j,\sigma_j'},$$

$$\mathbf{U}^z \equiv \frac{1}{2}(\sigma^z)_{\sigma_1,\sigma_1'} \prod_{j=2}^{2M} \delta_{\sigma_j,\sigma_j'},$$

$$\mathbf{R} \equiv \delta_{\sigma_1,\sigma_2'} \delta_{\sigma_2,\sigma_3'} \ldots \delta_{\sigma_{2M},\sigma_1'}.$$

σ^x and σ^y are Pauli matrices. $\langle 0|$ and $|0\rangle$ are the left and right eigenstates of the largest eigenvalue. $\mathbf{R}^2$ and the transfer matrix $\mathbf{T}$ commute. So we put the corresponding eigenvalue of $\mathbf{R}^2$ as $\exp(iK_j)$ of the eigenstate $|j\rangle$. K_j is given by $2\pi l/M$ and is the total momentum of the state j. $S^{xx}(n)$ at $n = $ even is given by

$$S^{xx}(n) = \sum_j \langle 0|\mathbf{U}^x|j\rangle \langle j|\mathbf{U}^x|0\rangle (\frac{\Lambda_j}{\Lambda_0})^n \exp(i(K_0 - K_j)n/2).$$

Then if $|j\rangle$ is the largest eigenvalue which satisfies $\langle 0|\mathbf{U}^x|j\rangle \langle j|\mathbf{U}^x|0\rangle \neq 0$, the correlation length ξ_{xx} is given by $1/\ln(\Lambda_0/|\Lambda_j|)$. ξ_{xx} is determined by the largest eigenvalue in the sector $k = M \pm 1$, Λ_{M-1}^1. While ξ_{zz} is determined by the second eigenvalue in the subspace $k = M$, Λ_M^2. Λ_{M-1}^1 is always real. But Λ_M^2 is not necessarily real,

$$\xi_{xx} = \lim_{M \rightarrow \infty} 1/\ln(\Lambda_M^1/|\Lambda_{M-1}^1|), \quad \xi_{zz} = \lim_{M \rightarrow \infty} 1/\ln(\Lambda_M^1/|\Lambda_M^2|). \tag{16.22}$$

It seems that Λ_M^2 becomes complex in some regions for $\Delta < 1$ and $h \neq 0$. The second eigenstate belongs always to the $k = M - 1$ subspace for the $|\Delta| \leq 1$ case. But for $|\Delta| > 1$, it belongs to the $k = M$ subspace if $|h|$ is sufficiently small. At $k = M$, equation (16.16) can also be written as

$$M \ln\left\{\frac{h(x_l - \alpha)h(x_l + \alpha + 2\eta)h(x_l - 2\eta)}{h(x_l + \alpha)h(x_l - \alpha - 2\eta)h(x_l + 2\eta)}\right\} = -2\pi\left(1 - \frac{1}{2}\right)i$$

$$-\frac{2h}{T} + \sum_{j=1}^{M} \ln\left\{\frac{h(x_l - x_j + 2\eta)h(x_l - 2\eta)}{h(x_l - x_j - 2\eta)h(x_l + 2\eta)}\right\}. \tag{16.23}$$

We write the largest eigenvalue as Λ_M^1 and

$$\mathfrak{R}\, x_1 > \mathfrak{R}\, x_2 > ... > \mathfrak{R}\, x_M,$$

$$x_l = -\bar{x}_{M+1-l}, \quad l = 1, 2, ..., M. \tag{16.24}$$

We put the solution of (16.23) as y_l to discriminate from solutions of the largest eigenvalue. For Λ_{M-1}^1 we remove y_M. The corresponding equation is

$$\mathfrak{R}y_1 > ... > \mathfrak{R}y_{M-1}, \quad y_l = -\bar{y}_{M-l}, \quad l = 1, 2, ..., M - 1, \tag{16.25}$$

$$M \ln\left\{\frac{h(y_l - \alpha)h(y_l + \alpha + 2\eta)h(y_l - 2\eta)}{h(y_l + \alpha)h(y_l - \alpha - 2\eta)h(y_l + 2\eta)}\right\} = -2\pi l i - \frac{2h}{T}$$

$$+ \ln\left\{\frac{h(2\eta - y_l)}{h(2\eta + y_l)}\right\} + \sum_{j=1}^{M-1} \ln\left\{\frac{h(y_l - y_j + 2\eta)h(y_l - 2\eta)}{h(y_l - y_j - 2\eta)h(y_l + 2\eta)}\right\}. \tag{16.26}$$

For the second largest eigenvalue in the $k = M$ subspace Λ_M^2, we have

$$M \ln\left\{\frac{h(z_l - \alpha)h(z_l + \alpha + 2\eta)h(z_l - 2\eta)}{h(z_l + \alpha)h(z_l - \alpha - 2\eta)h(z_l + 2\eta)}\right\}$$

$$= -2\pi(l - 1)i - \frac{2h}{T} + \ln\left\{\frac{h(z_l - z_1 + 2\eta)h(2\eta - z_l)}{h(z_l - z_1 - 2\eta)h(2\eta + z_l)}\right\}$$

$$+ \sum_{j=2}^{M} \ln\left\{\frac{h(z_l - z_j + 2\eta)h(z_l - 2\eta)}{h(z_l - z_j - 2\eta)h(z_l + 2\eta)}\right\}. \tag{16.27}$$

In the case $J > 0$, $\Delta > 1$, and h sufficiently small, z_1 lies on the $\mathfrak{R}z_1 = \pi/2$ axis,

$$\pi/2 = \mathfrak{R}z_1 > \mathfrak{R}z_2 > ... > \mathfrak{R}z_M > -\pi/2,$$

$$z_l = -\bar{z}_{M+2-l}, \quad l = 2, 3, ..., M, \tag{16.28}$$

If the magnetic field h is sufficiently large and $\Delta < 1$, Λ_M^2 is not necessarily real. In this case the correlation function $S^{zz}(n)$ is oscillating depending on the phase of Λ_M^2 and decays exponentially with the correlation length ξ_{zz}.

Thus we can calculate Λ^1_M, Λ^1_{M-1} and Λ^2_M using these equations and equation (16.17). The correct eigenvalue is obtained in the limit of infinite M. In Koma's[51] and Yamada's[115] calculation this limit is taken via numerical extrapolation. Here we take the limit of $M \to \infty$ in equations (16.17) and (16.16) analytically.

α is of the order of J/MT. Then in the limit of $M \to \infty$, equations (16.23)–(16.28) become

$$\frac{2J_s}{T}g(x_l) = 2\pi\left(l - \frac{1}{2}\right) + \frac{2hi}{T} + \frac{1}{i}\sum_{j=1}^{\infty} \ln[f(x_l, x_j)f(x_l, -\bar{x}_j)],$$
$$l = 1, 2, ... \tag{16.29}$$

for the largest eigenvalue Λ^1_M and

$$\frac{2J_s}{T}g(y_l) = 2\pi l + \frac{2hi}{T} + \frac{1}{i}\left[\ln\left(\frac{h(2\eta + y_l)}{h(2\eta - y_l)}\right)\right.$$
$$\left. + \sum_{j=1}^{\infty} \ln(f(y_l, y_j)f(y_l, -\bar{y}_j))\right], \quad l = 1, 2, 3, ... \tag{16.30}$$

for Λ^1_{M-1} and

$$\frac{2J_s}{T}g(z_l) = 2\pi l + \frac{2hi}{T} + \frac{1}{i}\left[\ln(-f(z_l, z_0))\right.$$
$$\left. + \sum_{j=1}^{\infty} \ln(f(z_l, z_j)f(z_l, z_{-j}))\right], \quad l = ..., -1, 0, 1, 2, ... \tag{16.31}$$

for Λ^2_M. Here

$$J_s \equiv \lim_{M \to \infty} -i\alpha MT = \begin{cases} \dfrac{J}{4}\sqrt{\Delta^2 - 1}, & \text{for } |\Delta| > 1, \\[2mm] \dfrac{J}{4}, & \text{for } |\Delta| = 1, \\[2mm] \dfrac{J}{4}\sqrt{1 - \Delta^2}, & \text{for } |\Delta| < 1, \end{cases} \tag{16.32}$$

and

$$g(x) \equiv \frac{h'(x)}{h(x)} - \frac{1}{2}\frac{h'(x + 2\eta)}{h(x + 2\eta)} - \frac{1}{2}\frac{h'(x - 2\eta)}{h(x - 2\eta)}$$

$$
= \begin{cases}
\cot x - \dfrac{\sin 2x}{\cosh 2\theta - \cos 2x}, & \text{for} \quad |\Delta| > 1, \\[3mm]
\dfrac{1}{x} - \dfrac{x}{x^2 + 1}, & \text{for} \quad |\Delta| = 1, \\[3mm]
\coth x - \dfrac{\sinh 2x}{\cosh 2x - \cos 2\theta}, & \text{for} \quad |\Delta| < 1,
\end{cases}
\tag{16.33}
$$

$$
f(x, y) \equiv \frac{h(x - y - 2\eta)h(x + 2\eta)}{h(x - y + 2\eta)h(x - 2\eta)}.
\tag{16.34}
$$

When M becomes large, α becomes small as $O(J/(4MT))$. Then the form of equation (16.17) is not appropriate to take the limit $M \to \infty$. For this purpose the following equations are useful:

$$
\prod_{l=1}^{M}\left(h'(-2\eta) + \frac{h(-2\eta)}{\delta_l^+ - \alpha}\right) = \left(\frac{h(2\eta) + 2\alpha h'(2\eta)}{2\alpha}\right)^M + \left(\frac{h(2\eta)}{2\alpha}\right)^M,
$$

$$
h'(-2\eta)\prod_{l=2}^{M}\left(h'(-2\eta) + \frac{h(-2\eta)}{\delta_l^- - \alpha}\right)
$$
$$
= \left(\frac{h(2\eta) + 2\alpha h'(2\eta)}{2\alpha}\right)^M - \left(\frac{h(2\eta)}{2\alpha}\right)^M,
\tag{16.35}
$$

$$
\delta_l^+ \equiv -i\alpha \cot\left(\frac{\pi(l - \tfrac{1}{2})}{M}\right), \quad \delta_l^- \equiv -i\alpha \cot\left(\frac{\pi l}{M}\right).
\tag{16.36}
$$

Using these we rewrite equation (16.17) as follows:

$$
\Lambda_M^1 = CD^+ \prod_{l=1}^{M} \frac{h(x_l - \alpha - 2\eta)}{h(x_l - \alpha)\{h'(-2\eta) + h(-2\eta)/(\delta_l^+ - \alpha)\}},
\tag{16.37}
$$

$$
\Lambda_{M-1}^1 = \exp\left(\frac{-h}{MT}\right)\frac{CD^-}{h'(-2\eta)}
$$
$$
\times \prod_{l=1}^{M-1} \frac{h(y_l - \alpha - 2\eta)}{h(y_l - \alpha)\{h'(-2\eta) + h(-2\eta)/(\delta_l^- - \alpha)\}},
\tag{16.38}
$$

$$
\Lambda_M^2 = CD^- \frac{h(z_M - \alpha - 2\eta)}{h(z_M - \alpha)h'(-2\eta)}
$$
$$
\times \prod_{l=1}^{M-1} \frac{h(z_l - \alpha - 2\eta)}{h(z_l - \alpha)\{h'(-2\eta) + h(-2\eta)/(\delta_l^- - \alpha)\}},
\tag{16.39}
$$

Table 16.1. *Boltzmann weights of vertices for inhomogeneous six- or (eight-) vertex models which correspond to Hamiltonians (16.1) and (16.2)*

		even l	XYZ odd l	XXZ odd l
a_l		1	$e^{-\frac{J_z}{4MT}}\sinh\frac{J_x+J_y}{4MT}$	$e^{-\frac{J\Delta}{4MT}}\sinh\frac{J}{2MT}$
b_l, b_l'		0	$e^{\frac{J_z}{4MT}}\cosh\frac{J_x-J_y}{4MT}$	$e^{\frac{J\Delta-4h}{4MT}}, e^{\frac{J\Delta+4h}{4MT}}$
c_l		1	$e^{-\frac{J_z}{4MT}}\cosh\frac{J_x+J_y}{4MT}$	$e^{-\frac{J\Delta}{4MT}}\cosh\frac{J}{2MT}$
d_l		0	$e^{\frac{J_z}{4MT}}\sinh\frac{J_x-J_y}{4MT}$	0

where

$$C = e^{J\Delta/4T+h/T}\left(\frac{h(2\alpha)(h(2\eta)+2\alpha h'(2\eta))}{2\alpha h(2\eta+2\alpha)}\right)^M,$$

$$D^{\pm} = \left[1 \pm \left(\frac{h(2\eta)}{h(2\eta)+2\alpha h'(2\eta)}\right)^M\right]. \tag{16.40}$$

16.2 The limit of an infinite Trotter number

From equations (16.29)–(16.31) we have

$$x_l \simeq \delta_l^+ \quad \text{and} \quad y_l \simeq \delta_l^-. \tag{16.41}$$

Then terms in the product of (16.37) and (16.39) approach unity if $l \gg 1$, $M - l \gg 1$ and $M \gg 1$. In the limit of infinite M equation (16.37) becomes

$$\Lambda_M^1 = 2e^{h/T}\cosh\left(\frac{J\Delta}{4T}\right)\prod_{l=1}^{\infty}\frac{h(2\eta-x_l)h(2\eta+\overline{x}_l)J_s^2}{h(x_l)h(\overline{x}_l)h^2(2\eta)((T\pi(l-\frac{1}{2}))^2+(\frac{J\Delta}{4})^2)}. \tag{16.42}$$

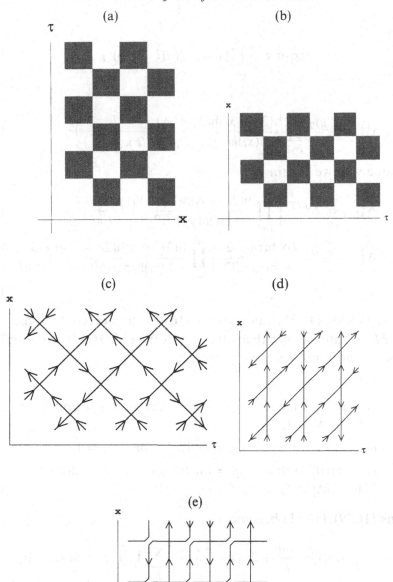

Fig. 16.1. Equivalence of two-dimensional lattices. (a) The thermodynamics of a chain is reduced to that of a checkerboard. (b) The x axis and τ axis are exchanged. (c) This problem is equivalent to the six-vertex (or eight-vertex) problem with a diagonal-to-diagonal transfer matrix. (d) This is equivalent to the slanted lattice. (e) The slanted lattice is equivalent to an inhomogeneous six-vertex (or eight-vertex) model where even columns have special Boltzmann weights ($a = c = 1$, $b = d = 0$).

Using

$$\cosh x = \prod_{l=1}^{\infty}(1 + (x/(\pi(l - \tfrac{1}{2})))^2),$$

we have

$$\Lambda_M^1 = 2e^{h/T}\prod_{l=1}^{\infty}\frac{h(2\eta - x_l)h(2\eta + \bar{x}_l)}{h(x_l)h(\bar{x}_l)h^2(2\eta)}\left(\frac{J_s}{T\pi(l - \tfrac{1}{2})}\right)^2. \tag{16.43}$$

In the same way we obtain

$$\Lambda_{M-1}^1 = 2e^{h/T}\frac{J}{4}\prod_{l=1}^{\infty}\frac{h(2\eta - y_l)h(2\eta + \bar{y}_l)}{h(y_l)h(\bar{y}_l)h^2(2\eta)}\left(\frac{J_s}{T\pi l}\right)^2. \tag{16.44}$$

$$\Lambda_M^2 = 2e^{h/T}\frac{J\Delta}{4}\frac{h(z_0 - 2\eta)}{h(z_0)h'(2\eta)}\prod_{l=1}^{\infty}\frac{h(2\eta - z_l)h(2\eta - z_{-l})}{h(z_l)h(z_{-l})h^2(2\eta)}\left(\frac{J_s}{T\pi l}\right)^2. \tag{16.45}$$

Equations (16.30)–(16.31) and (16.43)–(16.45) do not contain the Trotter number M and are simpler than (16.23)–(16.28) and (16.17). We should note that $z_l = -\bar{z}_l$ does not necessarily hold.

If we set

$$p_l = \tanh\theta/\tan x_l, \quad q_l = \tanh\theta/\tan y_l, \quad r_l = \tanh\theta/\tan z_l$$
$$\text{for} \quad |\Delta| > 1,$$
$$p_l = 1/x_l, \quad q_l = 1/y_l, \quad r_l = 1/z_l \quad \text{for} \quad |\Delta| = 1,$$
$$p_l = \tan\theta/\tanh x_l, \quad q_l = \tan\theta/\tanh y_l, \quad r_l = \tan\theta/\tanh z_l$$
$$\text{for} \quad |\Delta| < 1, \tag{16.46}$$

equations (16.29)–(16.31) become

$$\frac{J}{2T}G(p_l) = \frac{2hi}{T} + 2\pi\left(l - \frac{1}{2}\right) + \frac{1}{i}\sum_{j=1}^{\infty}\ln\left[L(p_l, p_j)L(p_l, -\bar{p}_j)\right], \tag{16.47}$$

$$\frac{J}{2T}G(q_l) = \frac{2hi}{T} + 2\pi l + \frac{1}{i}\left[\ln\left(\frac{1 + iq_l}{1 - iq_l}\right)\right.$$
$$\left. + \sum_{j=1}^{\infty}\ln(L(q_l, q_j)L(q_l, -\bar{q}_j))\right], \tag{16.48}$$

$$\frac{J}{2T}G(r_l) = \frac{2hi}{T} + 2\pi l + \frac{1}{i}\left[\ln(-L(r_l, r_0))\right.$$
$$\left. + \sum_{j=1}^{\infty}\ln(L(r_l, r_j)L(r_l, r_{-j}))\right], \tag{16.49}$$

where

$$G(x) = \Delta\frac{x(x^2 + 1 - \Delta^{-2})}{x^2 + 1}, \quad L(x, y) \equiv \frac{iy + (1 - \Delta^{-2}/(1 - ix))}{-iy + (1 - \Delta^{-2}/(1 + ix))}. \quad (16.50)$$

Equations (16.43)–(16.45) become

$$\Lambda_M^1 = 2e^{h/T} \prod_{l=1}^{\infty}\Big(\frac{J\Delta}{4\pi T(l - 1/2)}\Big)^2 (1 + ip_l)(1 - i\bar{p}_l), \quad (16.51)$$

$$\Lambda_{M-1}^1 = 2e^{h/T}\Big(\frac{J}{4T}\Big) \prod_{l=1}^{\infty}\Big(\frac{J\Delta}{4\pi Tl}\Big)^2 (1 + iq_l)(1 - i\bar{q}_l), \quad (16.52)$$

$$\Lambda_M^2 = 2e^{h/T}\Big(\frac{J\Delta}{4T}\Big)(1 + iq_0) \prod_{l=1}^{\infty}\Big(\frac{J\Delta}{4\pi Tl}\Big)^2 (1 + iq_l)(1 + iq_{-l}). \quad (16.53)$$

These equations are convenient for numerical calculation because they do not contain complex functions.

16.3 Analytical solutions for special cases
16.3.1 The Ising limit

In this limit we have

$$\Delta \to \infty, \quad J = J_z/\Delta, \quad \tanh\theta = 1. \quad (16.54)$$

Then equation (16.47) becomes

$$p_l = \frac{4\pi T(l - 1/2)}{|J_z|} - \frac{(4h + 2TD)i}{J_z}, \quad (16.55)$$

$$D \equiv \ln\Big[\prod_{j=1}^{\infty} \frac{(p_j - i)(\bar{p}_j + i)}{(p_j + i)(\bar{p}_j - i)}\Big]. \quad (16.56)$$

Substituting (16.55) into (16.56), we find that

$$e^D = \frac{\cosh((J_z + 4h + 2TD)/(4T))}{\cosh((J_z - 4h - 2TD)/(4T))}. \quad (16.57)$$

Solving this equation with respect to e^D, we have

$$e^D = \exp\Big(\frac{J_z - 2h}{2T}\Big)\Big(\sqrt{\sinh^2\frac{h}{T} + e^{-J_z/T}} + \sinh\frac{h}{T}\Big). \quad (16.58)$$

Using this and equations (16.51) and (16.54), we have

$$\Lambda_M^1 = \exp\Big(\frac{J_z}{4T}\Big)\Big(\cosh\frac{h}{T} + \sqrt{\sinh^2\frac{h}{T} + e^{-J_z/T}}\Big). \quad (16.59)$$

So the free energy per site is

$$f = -T \ln\left[\exp\left(\frac{J_z}{4T}\right)\left(\cosh\frac{h}{T} + \sqrt{\sinh^2\frac{h}{T} + e^{-J_z/T}}\right)\right]. \qquad (16.60)$$

Equation (16.48) gives

$$r_l = \frac{4\pi T l}{|J_z|} - \frac{(4h + 2TD')i}{J_z},$$

$$D' \equiv \ln\left[\left(\frac{1+ir_0}{1-ir_0}\right)\prod_{j=1}^{\infty}\frac{(r_j - i)(r_{-j} - i)}{(r_j + i)(r_{-j} + i)}\right]. \qquad (16.61)$$

Then we have

$$e^{D'} = \frac{\sinh((J_z + 4h + 2TD')/4T)}{\sinh((J_z - 4h - 2TD')/4T)}, \qquad (16.62)$$

and therefore

$$e^{D'} = \exp\left(\frac{J_z - 2h}{2T}\right)\left(\sqrt{\sinh^2\frac{h}{T} + e^{-J_z/T}} - \sinh\frac{h}{T}\right). \qquad (16.63)$$

Equation (16.52) gives

$$\Lambda_M^2 = \exp\left(\frac{J_z}{4T}\right)\left(\cosh\frac{h}{T} - \sqrt{\sinh^2\frac{h}{T} + e^{-J_z/T}}\right). \qquad (16.64)$$

Using (16.22), (16.59) and (16.64) we find that the correlation length is

$$\xi_{zz}^{-1} = \ln\left[\frac{\cosh h/T + \sqrt{\sinh^2 h/T + e^{-J_z/T}}}{\cosh h/T - \sqrt{\sinh^2 h/T + e^{-J_z/T}}}\right]. \qquad (16.65)$$

Equations (16.60) and (16.65) match the results for the Ising chain, (15.4) and (15.6).

16.3.2 The XY limit

From (16.16) we have $\theta = \pi/2$, $h(x) = \sinh x$. Then all logarithms in equations (16.29)–(16.31) vanish. Equation (16.29) gives

$$\frac{1}{\tanh x_l} - \tanh x_l = \frac{4\pi T(l - \frac{1}{2})}{J} + \frac{4hi}{J}$$

and

$$\frac{1}{\tanh x_l} = \left(\frac{2\pi T(l - \frac{1}{2}) + 2hi}{J}\right) + \sqrt{\left(\frac{2\pi T(l - \frac{1}{2}) + 2hi}{J}\right)^2 + 1}.$$

Substituting this into equation (16.43) we have

$$\Lambda_M^1 = 2\exp\Big(\sum_{l=1}^{\infty} \frac{1}{2\pi}\int_0^{2\pi} dx\ln\Big[1 + \frac{(J\cos x + 2h)^2}{(2\pi T(l-\frac{1}{2}))^2}\Big]\Big)$$
$$= \exp\Big[\frac{1}{2\pi}\int_0^{2\pi} dx\ln\Big(2\cosh\Big(\frac{J\cos x + 2h}{2T}\Big)\Big)\Big]. \qquad (16.66)$$

Here we used the identity

$$\ln(a + \sqrt{a^2 - b^2}) = \frac{1}{2\pi}\int_0^{2\pi} dx\ln(2a + 2b\cos x). \qquad (16.67)$$

Thus

$$f = -T\frac{1}{2\pi}\int_0^{2\pi} dx\ln\Big(2\cosh\Big(\frac{J\cos x + 2h}{2T}\Big)\Big). \qquad (16.68)$$

This is equivalent to (1.19). In the same way,

$$\frac{1}{\tanh y_l} = \Big(\frac{2\pi Tl + 2hi}{J}\Big) + \sqrt{\Big(\frac{2\pi Tl + 2hi}{J}\Big)^2 + 1},$$
$$\Lambda_{M-1}^1 = \frac{J}{2T}\exp\Big(\frac{1}{2\pi}\int_0^{2\pi} dx\ln\Big(\frac{\sinh[(J\cos x + 2h)/2T]}{(J\cos x + 2h)/2T}\Big)\Big).$$

The inverse of the correlation length is

$$\xi_{xx}^{-1} = \frac{1}{2\pi}\int_0^{2\pi} dx\ln(\coth|(J\cos x + 2h)/2T|), \qquad (16.69)$$

for $2|h| \le J$ and

$$\xi_{xx}^{-1} = \ln\Big(\Big|\frac{2h}{J}\Big| + \sqrt{\Big(\frac{2h}{J}\Big)^2 - 1}\Big) + \frac{1}{2\pi}\int_0^{2\pi} dx\ln\Big(\coth\Big|\frac{J\cos x + 2h}{2T}\Big|\Big), \qquad (16.70)$$

for $2|h| > J$. Results (16.68)–(16.70) are consistent with known results.

16.3.3 The $T = h = 0$ case

For very small T, the distribution of x_ls becomes dense. The mean distance is of the order of T/J_s. At $h = 0$ all the x_ls are on the real axis. Denote the distribution function of the x_ls as $\rho(x)$. $\rho(x)$ obeys the equation

$$\frac{\pi T}{J_s}\int_x^K \rho(t)dt + \frac{T}{J_s}\int_0^K \frac{1}{2i}\ln[f(x, y)f(x, -y)]\rho(y)dy = g(x). \qquad (16.71)$$

We define the function $F(x)$,

$$
F(x) \equiv \begin{cases}
\dfrac{\pi T}{J_s} \displaystyle\int_x^K \rho(t)\mathrm{d}t, & \text{at } x > 0, \\[4mm]
-F(-x), & \text{at } x < 0.
\end{cases}
\tag{16.72}
$$

By partial differentiation of equation (16.71), one has

$$
F(x) - \int_{-K}^K q(x - y)F(y)\mathrm{d}y = g(x).
\tag{16.73}
$$

$$
q(x) \equiv \frac{1}{2\pi i}\left(\frac{h'(x - 2\eta)}{h(x - 2\eta)} - \frac{h'(x + 2\eta)}{h(x + 2\eta)}\right)
$$

$$
= \begin{cases}
\dfrac{1}{\pi}\dfrac{\sinh 2\theta}{\cosh 2\theta - \cos 2x}, & \text{for } |\Delta| > 1, \\[4mm]
\dfrac{1}{\pi}\dfrac{\mathrm{sign}(\Delta)}{x^2 + 1}, & \text{for } |\Delta| = 1, \\[4mm]
\dfrac{1}{\pi}\dfrac{\sin 2\theta}{\cosh 2x - \cos 2\theta}, & \text{for } |\Delta| < 1.
\end{cases}
\tag{16.74}
$$

The Fourier transform of $g(x)$ and $q(x)$ is

$$
\begin{cases}
\tilde{g}(n) = -\pi i \; \mathrm{sign}(n)(1 - e^{-|2n\theta|}), & \text{for } |\Delta| > 1, \\[4mm]
\tilde{g}(\omega) = -\pi i \; \mathrm{sign}(\omega)(1 - e^{-|\omega|}), & \text{for } |\Delta| = 1, \\[4mm]
\tilde{g}(\omega) = -\pi i\left(\coth\dfrac{\pi\omega}{2} - \dfrac{\cosh(\pi - 2\theta)\omega/2}{\sinh \pi\omega/2}\right), & \text{for } |\Delta| < 1,
\end{cases}
\tag{16.75}
$$

$$
\begin{cases}
\tilde{q}(n) = \; \mathrm{sign}(\Delta)e^{-|2n\theta|}, & \text{for } |\Delta| > 1, \\[4mm]
\tilde{q}(\omega) = \; \mathrm{sign}(\Delta)e^{-|\omega|}, & \text{for } |\Delta| = 1, \\[4mm]
\tilde{q}(\omega) = \; \dfrac{\sinh[(\pi - 2\theta)\omega/2]}{\sinh(\pi\omega/2)}, & \text{for } |\Delta| < 1.
\end{cases}
\tag{16.76}
$$

The function $g(x)$ has a pole at $x = 0$, but we can define the Fourier integral by the principal integral. One should note that the Fourier transform becomes discrete at $|\Delta| > 1$ because the interval of integration is finite. Thus we have

$$\tilde{F}(n) = -\pi i \text{sign}(n), \qquad F(x) = \cot x,$$
$$\text{for} \quad \Delta > 1,$$

$$\tilde{F}(\omega) = -\pi i \text{sign}(\omega), \qquad F(x) = \frac{1}{x},$$
$$\text{for} \quad \Delta = 1,$$

$$\tilde{F}(\omega) = -\pi i \tanh\left(\frac{(\pi - \theta)\omega}{2}\right), F(x) = \frac{\pi/(\pi - \theta)}{\sinh(\pi x/(\pi - \theta))},$$
$$\text{for} \quad |\Delta| < 1,$$

$$\tilde{F}(\omega) = -\pi i \tanh(\omega/2), \qquad F(x) = \frac{\pi}{\sinh \pi x},$$
$$\text{for} \quad \Delta = -1,$$

$$\tilde{F}(n) = -\pi i \tanh n\theta, \qquad F(x) = \sum_{n=-\infty}^{\infty} \frac{\pi\theta}{\sinh[\pi\theta(x - n\pi)]},$$
$$\text{for} \quad \Delta < -1. \tag{16.77}$$

From equation (16.43) the ground state energy per site e is

$$e = -\lim_{T \to 0} T \ln \Lambda_M^1 = \frac{J_s}{\pi} \int_0^K \ln\left\{\frac{h(2\eta - x)h(x + 2\eta)}{F^2(x)h^2(x)h^2(2\eta)}\right\} F'(x) dx$$

$$= \frac{J_s}{\pi} \int_{-K}^K F'(x) + g(x)F(x) dx. \tag{16.78}$$

Thus

$$e = \begin{cases} -\dfrac{J\Delta}{4} & \text{for} \quad \Delta \geq 1, \\[2ex] -J\left(\dfrac{\Delta}{4} + \sqrt{1 - \Delta^2} \displaystyle\int_0^\infty \dfrac{\sinh(\theta\omega)}{\sinh(\pi\omega)\cosh[(\pi - \theta)\omega]} d\omega\right) \\ \qquad \text{for} \quad |\Delta| < 1, \\[2ex] J\left(\dfrac{1}{4} - \ln 2\right) & \text{for} \quad \Delta = -1, \\[2ex] J\left(\dfrac{\Delta}{4} - \sqrt{\Delta^2 - 1} \displaystyle\sum_{n=1}^{\infty} e^{-2n|\theta|} \tanh|n\theta|\right) & \text{for} \quad \Delta < -1. \end{cases} \tag{16.79}$$

These results coincide with the known exact ground state energy per site of the XXZ model in a zero field, as expected.

16.4 Numerical calculations of the XXZ model

In equations (16.47)–(16.49) it is possible to calculate each term of the sum, but in each term one must calculate a logarithmic function. So we would like to take the product of each term and after that take the logarithm of it. The logarithm is, however, a multi-valued function. In actual numerical calculation the imaginary part of $\ln x$ is fixed at $[\pi, -\pi]$, but there is no guarantee that this choice of branch is appropriate. So we transform equations (16.47)–(16.49) as follows:

$$
\begin{aligned}
p_l = V\Big(\frac{4hi}{J\Delta} + \frac{4T}{J\Delta}\Big\{\pi(l - \tfrac{1}{2}) \\
+ \frac{1}{2i} \ln\Big[\exp\Big(\frac{Ji}{2T}(\Delta V^{-1}(p_l) - G(p_l))\Big) \prod_{j=1}^{\infty} L(p_l, p_j)L(p_l, -\bar{p}_j)\Big]\Big\}\Big),
\end{aligned}
$$

(16.80)

$$
\begin{aligned}
q_l = V\Big(\frac{4hi}{J\Delta} + \frac{4T}{J\Delta}\Big\{\pi l + \frac{1}{2i} \ln\Big[\exp\Big(\frac{Ji}{2T}(\Delta V^{-1}(q_l) - G(q_l))\Big) \\
\times \Big(\frac{1 + iq_l}{1 - iq_l}\Big) \prod_{j=1}^{\infty} L(q_l, q_j)L(q_l, -\bar{q}_j)\Big]\Big\}\Big), \quad l \geq 1,
\end{aligned}
$$

(16.81)

$$
\begin{aligned}
r_l = V\Big(\frac{4hi}{J\Delta} + \frac{4T}{J\Delta}\Big\{\pi l + \frac{1}{2i} \ln\Big[\exp\Big(\frac{Ji}{2T}(\Delta V^{-1}(r_l) - G(r_l))\Big) \\
\times (-L(r_l, r_0)) \prod_{j=1}^{\infty} L(r_l, r_j)L(r_l, r_{-j})\Big]\Big\}\Big).
\end{aligned}
$$

(16.82)

The function $V(x)$ is chosen as follows:

$$
\begin{cases}
V(x) = V^{-1}(x) = x & \text{at} \quad \Delta \geq 1 \\[2ex]
V(x) = \tan\theta / \tanh\Big(\frac{\pi - \theta}{\pi} \sinh^{-1}\frac{\pi}{(\pi - \theta)x}\Big), \\
V^{-1}(x) = \frac{\pi/(\pi - \theta)}{\sinh\big(\frac{\pi}{\pi - \theta} \tanh^{-1}\frac{\tan\theta}{x}\big)}, & \text{at} \quad |\Delta| < 1 \\[2ex]
V(x) = \pi / \sinh^{-1}(\pi/x), \quad V^{-1}(x) = \pi / \sinh(\pi/x) & \text{at} \quad \Delta = -1 \\[2ex]
V(x) = \tanh\theta / \tanh\Big(\frac{2K(u)}{\pi} \mathrm{tn}^{-1}\big(\frac{\pi}{2K(u)x}, u\big)\Big), \\
V^{-1}(x) = \frac{\pi/2K}{\mathrm{tn}((\pi/2K)\tanh^{-1}(\tanh\theta/x), u)}, \\
K(u')/K(u) = \theta/\pi, & \text{at} \quad \Delta < -1.
\end{cases}
$$

(16.83)

Here $\mathrm{tn}(x, u) \equiv \mathrm{sn}(x, u)/\mathrm{cn}(x, u)$ is an elliptic function. $K(u)$ is a complete elliptic integral of the first kind with modulus u. u' is $\sqrt{1 - u^2}$. In these equations the imaginary part of the logarithms is fixed at $[\pi, -\pi]$. These functions are obtained from the distribution of roots (16.77) at zero temperature. In actual numerical calculation we determine the first several hundred p_ls. The higher p_ls are approximated by

$$p_l = 4T\pi(l - 1/2)/J\Delta + \text{const.} \times i$$

and the sum is estimated accurately.

Iterative calculations of equations (16.80) and (16.81) converges very rapidly for arbitrary h, Δ and T. This yields Λ_M^1 and Λ_{M-1}^1. In contrast, equation (16.82) for Λ_M^2 is sometimes unstable for $h \neq 0$. It stands $|\Lambda_M^2| > |\Lambda_{M-1}^1|$ at $|\Delta| > 1$ and $h_{\text{critical}} > |h|$. h_{critical} is dependent on Δ and temperature. It seems that equation (16.82) gives a stable solution at least in this region.

We give the susceptibility and specific heat as functions of temperature for $\Delta = 1$ and $\Delta = -1$ in a zero field in Tables (16.2) and (16.3).

Table 16.2. *Susceptibility and specific heat of the S = 1/2 ferromagnetic XXX chain*

T/J	$J\chi$	specific heat
0.025	441.5	0.0847726
0.050	132.56	0.103884
0.075	67.218	0.1144228
0.100	42.028	0.1210856
0.1250	29.41	0.1255503
0.1500	22.08	0.1285785
0.1750	17.39	0.1305907
0.2000	14.180	0.1319037
0.2250	11.867	0.1327730
0.2500	10.137	0.1333757
0.2750	8.802	0.1338067
0.3000	7.746	0.1340965
0.3250	6.894	0.1342357
0.3500	6.1935	0.1341973
0.3750	5.6094	0.1339523
0.4000	5.1160	0.1334778
0.4250	4.69470	0.1327611
0.4500	4.33125	0.1317999
0.4750	4.01505	0.1306011
0.5000	3.7378	0.1291786
0.5250	3.4930	0.1275514
0.5500	3.2756	0.1257419
0.5750	3.0813	0.1237738
0.6000	2.90686	0.1216711
0.6250	2.74945	0.1194573
0.6500	2.6068	0.1171547
0.6750	2.4770	0.1147838
0.7000	2.35856	0.1123635
0.7250	2.25000	0.1099107
0.7500	2.15023	0.1074401
0.7750	2.05825	0.1049650
0.8000	1.97323	0.1024966
0.8250	1.89445	0.1000445
0.8500	1.82126	0.09761703
0.8750	1.75312	0.09522095
0.9000	1.68953	0.09286195
0.9250	1.63009	0.09054464
0.9500	1.57440	0.08827272

Table 16.3. *Susceptibility and specific heat of the S = 1/2 antiferromagnetic XXX chain*

T/J	$J\chi$	specific heat
0.0250	0.4419	0.0167
0.0500	0.4484	0.03386
0.0750	0.45393	0.05148
0.1000	0.45925	0.06998
0.1250	0.4647	0.0898
0.1500	0.4705	0.1117
0.1750	0.4770	0.1359
0.2000	0.48412	0.16234
0.2250	0.4920	0.18999
0.2500	0.5004	0.21776
0.2750	0.5093	0.244473
0.3000	0.51839	0.269086
0.3250	0.52746	0.290837
0.3500	0.53630	0.309245
0.3750	0.54472	0.324084
0.4000	0.55256	0.335345
0.4250	0.55970	0.343176
0.4500	0.566066	0.347838
0.4750	0.571606	0.349658
0.5000	0.57629	0.3489939
0.5250	0.580142	0.3462115
0.5500	0.583156	0.3416629
0.5750	0.585372	0.3356767
0.6000	0.586830	0.3285502
0.6250	0.587577	0.3205466
0.6500	0.587663	0.3118945
0.6750	0.587142	0.3027889
0.7000	0.586066	0.2933943
0.7250	0.584487	0.2838467
0.7500	0.582454	0.2742577
0.7750	0.580016	0.2647172
0.8000	0.5772157	0.2552966
0.8250	0.5740957	0.2460515
0.8500	0.5706945	0.2370242
0.8750	0.5670477	0.2282464
0.9000	0.5631878	0.2197403
0.9250	0.5591444	0.2115209
0.9500	0.5549446	0.2035972

17

The XYZ model with no magnetic field

17.1 The transfer matrix for the XYZ model

The partition function of (16.2) at temperature T is approximated by that of the eight-vertex model on a $2M \times N$ checker board lattice, as shown in Fig. (16.1). The transfer matrix in the real direction is a $2^{2M} \times 2^{2M}$ matrix. It is defined by:

$$
\begin{aligned}
&\mathbf{T}(\sigma_1, \sigma_2, ..., \sigma_{2M}; \sigma_1', \sigma_2', ..., \sigma_{2M}') \\
&= A(\sigma_1\sigma_2; \sigma_{2M}'\sigma_1')A(\sigma_3\sigma_4; \sigma_2'\sigma_3')...A(\sigma_{2M-1}\sigma_{2M}; \sigma_{2M-2}'\sigma_{2M-1}'), \\
&A(++++) = A(----) = a, \quad A(++--) = A(--++) = d, \\
&A(+-+-) = A(-+-+) = c, \quad A(+--+) = A(-++-) = b, \\
&a = e^{-J_z/4MT}\sinh\left(\frac{J_x + J_y}{4MT}\right), \quad b = e^{J_z/4MT}\cosh\left(\frac{J_y - J_x}{4MT}\right), \\
&c = e^{-J_z/4MT}\cosh\left(\frac{J_x + J_y}{4MT}\right), \quad d = -e^{J_z/4MT}\sinh\left(\frac{J_y - J_x}{4MT}\right).
\end{aligned}
$$

$$(17.1)$$

The other elements of A are zero. The free energy per site and the correlation length are given by:

$$ f = -T\ln\Lambda_0, \quad \xi = 1/\ln(|\Lambda_0/\Lambda_1|), \tag{17.2} $$

where Λ_0 and Λ_1 are the largest and the second largest eigenvalues of $\mathbf{T}$. This transfer matrix is that of an inhomogeneous eight-vertex model on a square lattice with $2M$ columns,

$$
\begin{aligned}
\mathbf{T} &= \mathrm{Tr}\mathbf{R}_1(\sigma_1, \sigma_1')\mathbf{R}_2(\sigma_2, \sigma_2')...\mathbf{R}_{2M}(\sigma_{2M}, \sigma_{2M}'), \\
\mathbf{R}_l(++) &= \begin{pmatrix} a_l & 0 \\ 0 & b_l \end{pmatrix}, \quad \mathbf{R}_l(+-) = \begin{pmatrix} 0 & d_l \\ c_l & 0 \end{pmatrix}, \\
\mathbf{R}_l(-+) &= \begin{pmatrix} 0 & c_l \\ d_l & 0 \end{pmatrix}, \quad \mathbf{R}_l(--) = \begin{pmatrix} b_l & 0 \\ 0 & a_l \end{pmatrix},
\end{aligned}
$$

$$(17.3)$$

$$a_{2l} = c_{2l} = 1, \quad b_{2l} = d_{2l} = 0 \quad \text{and}$$

$$a_{2l-1} = a, \quad b_{2l-1} = b, \quad c_{2l-1} = c, \quad d_{2l-1} = d. \tag{17.4}$$

The weights of this square lattice are given in Table (16.1).

17.1.1 Baxter's theory for the inhomogeneous eight-vertex model

According to Baxter[8] the inhomogeneous eight-vertex model is soluble if the Boltzmann weights on the l-th column are given by

$$a_l = \rho_l \Theta(2\eta)\Theta(v_l - \eta)H(v_l + \eta),$$
$$b_l = \rho_l \Theta(2\eta)H(v_l - \eta)\Theta(v_l + \eta),$$
$$c_l = \rho_l H(2\eta)\Theta(v_l - \eta)\Theta(v_l + \eta),$$
$$d_l = \rho_l H(2\eta)H(v_l - \eta)H(v_l + \eta). \tag{17.5}$$

Here H and Θ are Jacobian elliptic theta functions. ρ_l and v_l are dependent on l but η and the modulus k should be common for all columns. In his theory, the matrix $\mathbf{T}(v)$ with the parameter v is defined by changing $v_l \rightarrow v_l + v$ in equation (17.5). Matrices $\mathbf{T}(v)$ and $\mathbf{T}(v')$ commute with each other for any v and v'. Define the function

$$\phi(v) = \prod_{l=1}^{2M} \rho_l \Theta(0)H(v - v_l)\Theta(v - v_l). \tag{17.6}$$

One can construct a matrix $\mathbf{Q}(v)$ with parameter v which satisfies

$$[\mathbf{T}(v), \mathbf{Q}(v)] = 0, \quad [\mathbf{Q}(v), \mathbf{Q}(v')] = 0,$$
$$\mathbf{T}(v)\mathbf{Q}(v) = \phi(v + \eta)\mathbf{Q}(v - 2\eta) + \phi(v - \eta)\mathbf{Q}(v + 2\eta). \tag{17.7}$$

$\mathbf{T}(v)$ and $\mathbf{Q}(v)$ are simultaneously diagonalized. An eigenvalue of $\mathbf{Q}(v)$ is factorized as follows:

$$Q(v) = \exp\left(-\frac{1}{2}iv\pi v/K\right) \prod_{j=1}^{M} \mathsf{h}(v - w_j), \quad \mathsf{h}(x) \equiv H(x)\Theta(x). \tag{17.8}$$

The sum of the w_js should satisfy

$$\left\{ 2\sum_{j=1}^{M} w_j - \sum_{l=1}^{2M} v_l \right\} = 2K(v'' + M + \text{even integer}) + ivK'. \tag{17.9}$$

v and v'' are integers. v'' is 0 or 1 in the symmetric or antisymmetric subspace, respectively, with respect to reversing all arrows. K and K' are the complete elliptic integrals of the first kind with modulus k and $\sqrt{1-k^2}$, respectively. The corresponding eigenvalue of $\mathbf{T}(v)$ is given by

$$T(v) = \phi(v+\eta)e^{iv\pi\eta/K} \prod_{l=1}^{M} \frac{h(v-w_l-2\eta)}{h(v-w_l)}$$

$$+ \phi(v-\eta)e^{-iv\pi\eta/K} \prod_{l=1}^{M} \frac{h(v-w_l+2\eta)}{h(v-w_l)}. \tag{17.10}$$

17.1.2 Transcendental equations

We apply the above theory to the eigenvalue problem of the quantum transfer matrix given in equation (17.4). For even columns we have

$$v_{2l} = \eta, \quad \rho_{2l} = (\Theta(0)H(2\eta)\Theta(2\eta))^{-1}. \tag{17.11}$$

For the odd columns we have

$$v_{2l-1} = v_1, \quad \rho_{2l-1} = b(\Theta(2\eta)H(v_1-\eta)\Theta(v_1+\eta))^{-1},$$
$$a/c = \operatorname{sn}(v_1+\eta)/\operatorname{sn}(2\eta), \quad b/c = \operatorname{sn}(v_1-\eta)/\operatorname{sn}(2\eta),$$
$$d/b = k\operatorname{sn}(2\eta)\operatorname{sn}(v_1+\eta). \tag{17.12}$$

The modulus k, η and v_1 are determined by

$$k = \frac{\sqrt{A_z^2 - A_x^2} - \sqrt{A_z^2 - A_y^2}}{\sqrt{A_z^2 - A_x^2} + \sqrt{A_z^2 - A_y^2}},$$

$$\operatorname{sn}^2(2\eta) = cd/(kab),$$
$$\operatorname{sn}(\eta+v_1) = \operatorname{sn}(2\eta)a/c. \tag{17.13}$$

Here we put

$$A_z \equiv a^2 + b^2 - c^2 - d^2 = 2\sinh\left(\frac{J_z}{2MT}\right),$$

$$A_y \equiv 2(ab - cd) = 2\sinh\left(\frac{J_y}{2MT}\right)$$

$$A_x \equiv 2(ab + cd) = 2\sinh\left(\frac{J_x}{2MT}\right). \tag{17.14}$$

From equation (17.2) we find that d is negative and therefore $\operatorname{sn}(2\eta)$ and $\operatorname{sn}(\eta+v_1)$ are pure imaginary,

$$\operatorname{sn}(2\eta) = i\frac{\sqrt{A_z^2 - A_x^2} + \sqrt{A_z^2 - A_y^2}}{A_x + A_y}, \quad 0 \le 2\eta/i \le K'.$$

By this parametrization equation (17.5) is satisfied. The function ϕ in equation (17.6) becomes

$$\phi(v) = \left\{ \frac{b\Theta(0)h(v-\eta)h(v-v_1)}{\Theta(2\eta)H(v_1-\eta)\Theta(v_1+\eta)h(2\eta)} \right\}^M. \tag{17.15}$$

Putting $v = 0$ in equation (17.10), we have the eigenvalue of **T**,

$$\Lambda = T(0) = e^{-iv\pi\eta/K}\phi(-\eta)\prod_{l=1}^{M} \frac{h(-w_l+2\eta)}{h(-w_l)} = e^{-iv\pi\eta/K}$$

$$\times \left[\frac{b\Theta(0)\Theta(2\eta-(\eta+v_1))h(v_1+\eta)}{\Theta(\eta+v_1)\Theta(2\eta)h(v_1-\eta)} \right]^M \prod_{l=1}^{M} \frac{h(-w_l+2\eta)}{h(-w_l)}. \tag{17.16}$$

The w_ls are solutions of the following equation:

$$\left(\frac{h(w_l)h(w_l+\eta-v_1)}{h(w_l-2\eta)h(w_l-\eta-v_1)} \right)^M = -e^{-2iv\pi\eta/K}\prod_{j=1}^{M} \frac{h(w_l-w_j+2\eta)}{h(w_l-w_j-2\eta)}. \tag{17.17}$$

If we put $\alpha \equiv -\frac{1}{2}(\eta+v_1)$ and $x_l \equiv w_l + \alpha$, equations (17.16) and (17.17) become

$$\left(\frac{h(x_l-\alpha)h(x_l+\alpha+2\eta)}{h(x_l+\alpha)h(x_l-\alpha-2\eta)} \right)^M = -e^{-2iv\pi\eta/K}\prod_{j=1}^{M} \frac{h(x_l-x_j+2\eta)}{h(x_l-x_j-2\eta)}, \tag{17.18}$$

$$\Lambda = \left[\frac{b\Theta(0)\Theta(2\eta+2\alpha)h(2\alpha)}{\Theta(2\alpha)\Theta(2\eta)h(2\alpha+2\eta)} \right]^M e^{-iv\pi\eta/K}\prod_{l=1}^{M} \frac{h(x_l-\alpha-2\eta)}{h(x_l-\alpha)}. \tag{17.19}$$

In the case $J_x = 0$, A_x is zero and we have

$$k = \frac{1-\sqrt{1-(A_y/A_z)^2}}{1+\sqrt{1-(A_y/A_z)^2}}, \quad sn(2\eta) = \frac{i}{\sqrt{k}}, \quad 2\eta = iK'/2. \tag{17.20}$$

Using the quasi-periodicity of $h(x)$,

$$h(x+iK') = -\exp(-\pi(K'+2ix)/2K)h(x),$$

and equation (17.9), we write equation (16.36) as follows:

$$\left(\frac{h(x_l-\alpha)h(x_l+\alpha+2\eta)}{h(x_l+\alpha)h(x_l-\alpha-2\eta)} \right)^M = -(-1)^{v''+M}. \tag{17.21}$$

The largest eigenvalue belongs to $v'' + M =$ even, and the second largest eigenvalue belongs to $v'' + M =$ odd. So we have

$$\frac{h(x_l - \alpha)h(x_l + \alpha + 2\eta)}{h(x_l + \alpha)h(x_l - \alpha - 2\eta)} = \exp[2\pi(l - 1/2)i/M], \qquad (17.22)$$

$$\frac{h(y_l - \alpha)h(y_l + \alpha + 2\eta)}{h(y_l + \alpha)h(y_l - \alpha - 2\eta)} = \exp[2\pi(l - 1)i/M]. \qquad (17.23)$$

Hereafter we write the y_ls as solutions of (17.18) for the second eigenvalue. The x_ls and y_ls are all real and satisfy the following conditions:

$$x_l = -x_{M+1-l}, \quad l = 1, 2, ..., M,$$

$$y_1 = K, \quad y_l = -y_{M+2-l}, \quad l = 2, 3, ..., M. \qquad (17.24)$$

We can set $v = 0$ for these eigenvalues. These properties hold in the region $|J_x| \le J_y$. So one can determine the x_ls and y_ls for fixed M and T. In the case of even M, equation (17.18) is written as follows,

$$\frac{2J_s}{T}g_M(x_l) = 2\pi(l - 1/2) - i\sum_{j=1}^{M/2}\ln[f(x_l, x_j)f(x_l, -x_j)],$$

$$\frac{2J_s}{T}g_M(y_l) = 2\pi(l - 1) + \frac{1}{i}\Big[\ln(-f(y_l, K))$$

$$+ \sum_{j=2}^{M/2-1}\ln(f(y_l, y_j)f(y_l, -y_j))\Big], \qquad (17.25)$$

where

$$g_M(x) \equiv \frac{1}{2\alpha}\ln\Big\{\frac{h(x_l + \alpha)h(x_l - \alpha - 2\eta)h(x_l + 2\eta)}{h(x_l - \alpha)h(x_l + \alpha + 2\eta)h(x_l - 2\eta)}\Big\},$$

$$f(x, y) \equiv \frac{h(x - y - 2\eta)h(x + 2\eta)}{h(x - y + 2\eta)h(x - 2\eta)},$$

$$J_s \equiv -2i\alpha M. \qquad (17.26)$$

Using the Liouville theorem one can show

$$f(x, y)f(x, -y) = \frac{1 - \text{tdn}^2(y)/\text{tdn}^2(x - 2\eta)}{1 - \text{tdn}^2(y)/\text{tdn}^2(x + 2\eta)}, \qquad (17.27)$$

$$f(x, K) = -\frac{\text{tdn}(2\eta + x)}{\text{tdn}(2\eta - x)}, \qquad (17.28)$$

$$\frac{h'^2(0)h(x - y)h(x + y)}{h^2(x)h^2(y)} = \frac{1}{\text{tdn}^2(y)} - \frac{1}{\text{tdn}^2(x)}, \qquad (17.29)$$

$$\text{tdn}(x) \equiv \text{sn}(x)/(\text{cn}(x)\text{dn}(x)). \qquad (17.30)$$

17.1.3 The limit of $M \to \infty$

The exact free energy per site is given by the limiting value of Λ as $M \to \infty$. The correlation length is given by the ratio of the largest and the second eigenvalues. We take the limit $M \to \infty$ in equations (17.25)–(17.26) and (17.19) analytically. In this limit we have

$$k \to \frac{\sqrt{J_z^2 - J_x^2} - \sqrt{J_z^2 - J_y^2}}{\sqrt{J_z^2 - J_x^2} + \sqrt{J_z^2 - J_y^2}},$$

$$\mathrm{sn}(2\eta) \to i \frac{\sqrt{J_z^2 - J_x^2} + \sqrt{J_z^2 - J_y^2}}{J_x + J_y},$$

$$J_s \to \frac{\sqrt{J_z^2 - J_x^2} + \sqrt{J_z^2 - J_y^2}}{8} = \frac{J_z \mathrm{tdn}(2\eta)}{4i}.$$

$$(17.31)$$

As α is $O(M^{-1})$, the function $g_M(x)$ in equations (17.25)–(17.26) becomes

$$\lim_{M \to \infty} g_M(x) = g(x) \equiv \frac{h'(x)}{h(x)} - \frac{1}{2} \frac{h'(x + 2\eta)}{h(x + 2\eta)} - \frac{1}{2} \frac{h'(x - 2\eta)}{h(x - 2\eta)}$$

$$= \frac{2}{\mathrm{sn}(2x_l)} \frac{\mathrm{tdn}^2(2\eta)}{\mathrm{tdn}^2(2\eta) - \mathrm{tdn}^2(x_l)}. \tag{17.32}$$

Then equations (17.25)–(17.26) are written as follows:

$$\frac{2J_s}{T} g(x_l) = 2\pi \left(l - \frac{1}{2}\right) + \frac{1}{i} \sum_{j=1}^{\infty} \ln\left[\frac{1 - \mathrm{tdn}^2(x_j)/\mathrm{tdn}^2(x_l - 2\eta)}{1 - \mathrm{tdn}^2(x_j)/\mathrm{tdn}^2(x_l + 2\eta)}\right],$$

$$\frac{2J_s}{T} g(y_l) = 2\pi(l - 1) + \frac{1}{i}\left[\ln\left(\frac{\mathrm{tdn}(2\eta + y_l)}{\mathrm{tdn}(2\eta - y_l)}\right)\right.$$

$$\left. + \sum_{j=2}^{\infty} \ln\left(\frac{1 - \mathrm{tdn}^2(y_j)/\mathrm{tdn}^2(y_l - 2\eta)}{1 - \mathrm{tdn}^2(y_j)/\mathrm{tdn}^2(y_l + 2\eta)}\right)\right]. \tag{17.33}$$

From these equations we can calculate the limiting value of x_l and y_l of equations (17.25)–(17.26) as $M \to \infty$.

The form of equation (17.19) is not appropriate to take the limit $M \to \infty$. For this purpose the following identities are useful:

$$\prod_{l=1}^{M}\left(h'(-2\eta) + \frac{h(-2\eta)}{\delta_l^+ - \alpha}\right) = \left(\frac{h(2\eta) + 2\alpha h'(2\eta)}{2\alpha}\right)^M + \left(\frac{h(2\eta)}{2\alpha}\right)^M,$$

$$h'(-2\eta) \prod_{l=2}^{M}\left(h'(-2\eta) + \frac{h(-2\eta)}{\delta_l^- - \alpha}\right)$$

$$= \left(\frac{h(2\eta) + 2\alpha h'(2\eta)}{2\alpha}\right)^M - \left(\frac{h(2\eta)}{2\alpha}\right)^M, \tag{17.34}$$

$$\delta_l^+ \equiv -i\alpha \cot\left(\frac{\pi(l - \frac{1}{2})}{M}\right), \quad \delta_l^- \equiv -i\alpha \cot\left(\frac{\pi(l - 1)}{M}\right). \tag{17.35}$$

Using these we rewrite equation (17.19) as follows:

$$\Lambda_0 = C\left[1 + \left(\frac{h(2\eta)}{h(2\eta) + 2\alpha h'(2\eta)}\right)^M\right]$$

$$\times \prod_{l=1}^{M} \frac{h'(0)h(x_l - \alpha - 2\eta)}{h(x_l - \alpha)\{h'(-2\eta) + h(-2\eta)/(\delta_l^+ - \alpha)\}}, \tag{17.36}$$

$$\Lambda_1 = C\left[1 - \left(\frac{h(2\eta)}{h(2\eta) + 2\alpha h'(2\eta)}\right)^M\right] \frac{h'(0)h(y_1 - \alpha - 2\eta)}{h(y_1 - \alpha)h'(-2\eta)}$$

$$\times \prod_{l=2}^{M} \frac{h'(0)h(y_l - \alpha - 2\eta)}{h(y_l - \alpha)\{h'(-2\eta) + h(-2\eta)/(\delta_l^- - \alpha)\}}, \tag{17.37}$$

where

$$C \equiv \left[\frac{b\Theta(0)\Theta(2\eta + 2\alpha)h(2\alpha)(h(2\eta) + 2\alpha h'(2\eta))}{\Theta(2\alpha)\Theta(2\eta)h'(0)2\alpha h(2\eta + 2\alpha)}\right]^M.$$

From equations (17.25)–(17.26) we have

$$x_l = \delta_l^+ + O(l^{-3}) + O((M - l)^{-3}), \quad \text{and} \quad y_l \simeq \delta_l^- + O(l^{-3}) + O((M - l)^{-3}).$$

Then terms in the product of (17.36) and (17.37) approach unity if $l \gg 1$, $M - l \gg 1$ and $M \gg 1$. In the limit of infinite M, the equations (17.36) become

$$\Lambda_0 = 2\exp\left(\frac{J_z}{4T} + \frac{J_s i}{T}\left(2\frac{\Theta'(2\eta)}{\Theta(2\eta)} - \frac{h'(2\eta)}{h(2\eta)}\right)\right) \cosh\left(\frac{J_s i}{T}\frac{h'(2\eta)}{h(2\eta)}\right)$$

$$\times \prod_{l=1}^{\infty}\left(\frac{h'^2(0)h(2\eta - x_l)h(2\eta + x_l)J_s^2}{h^2(x_l)h^2(2\eta)(T^2\pi^2(l - \frac{1}{2})^2 + (J_s i h'(2\eta)/h(2\eta))^2)}\right).$$

Using

$$H'(2\eta)/H(2\eta) - \Theta'(2\eta)/\Theta(2\eta) = 1/\text{tdn}(2\eta),$$

$$\cosh x = \prod_{l=1}^{\infty}(1 + x^2/(\pi(l - \frac{1}{2}))^2),$$

and equation (17.29) we have

$$\Lambda_0 = 2\prod_{l=1}^{\infty}\Big(\frac{h'^2(0)h(2\eta - x_l)h(2\eta + x_l)J_s^2}{h^2(x_l)h^2(2\eta)T^2\pi^2(l - \frac{1}{2})^2}\Big)$$

$$= 2\prod_{l=1}^{\infty}\Big(\frac{J_z}{4\pi T(l - 1/2)}\Big)^2\Big(1 - \frac{\text{tdn}^2 2\eta}{\text{tdn}^2 x_l}\Big). \tag{17.38}$$

Using

$$\sinh x = x\prod_{l=1}^{\infty}(1 + x^2/(\pi l)^2),$$

we find in a similar way that

$$\Lambda_1 = 2\Big(\frac{J_z}{4T}\Big)\prod_{l=2}^{\infty}\Big(\frac{h'^2(0)h(2\eta - y_l)h(2\eta + y_l)J_s^2}{h^2(y_l)h^2(2\eta)T^2\pi^2(l - 1)^2}\Big)$$

$$= 2\Big(\frac{J_z}{4T}\Big)\prod_{l=2}^{\infty}\Big(\frac{J_z}{4\pi T(l - 1)}\Big)^2\Big(1 - \frac{\text{tdn}^2 2\eta}{\text{tdn}^2 y_l}\Big). \tag{17.39}$$

The free energy and the correlation length are given by

$$\begin{aligned}
f &= -T\ln\Lambda_0 \\
&= -T\ln 2 - T\sum_{l=1}^{\infty}\ln\Big[\Big(\frac{J_z}{4\pi T(l - 1/2)}\Big)^2\Big(1 - \frac{\text{tdn}^2 2\eta}{\text{tdn}^2 x_l}\Big)\Big],
\end{aligned}$$

$$\tag{17.40}$$

$$\begin{aligned}
\xi^{-1} &= \ln\Big|\frac{\Lambda_0}{\Lambda_1}\Big| = \sum_{l=1}^{\infty}\ln\Big[\Big(\frac{J_z}{4\pi T(l - 1/2)}\Big)^2\Big(1 - \frac{\text{tdn}^2 2\eta}{\text{tdn}^2 x_l}\Big)\Big] \\
&\quad - \ln\Big(\frac{J_z}{4T}\Big) - \sum_{l=2}^{\infty}\ln\Big[\Big(\frac{J_z}{4\pi T(l - 1)}\Big)^2\Big(1 - \frac{\text{tdn}^2 2\eta}{\text{tdn}^2 y_l}\Big)\Big]. \tag{17.41}
\end{aligned}$$

If we put $p_l = -i\text{tdn}2\eta/\text{tdn}x_l$, $q_l = -i\text{tdn}2\eta/\text{tdn}y_l$, equations (17.33) become as follows:

$$\frac{J_z}{2T}G(p_l) = 2\pi(l - 1/2) + \frac{1}{i}\sum_{j=1}^{\infty}\ln\Big[L(p_l, p_j)L(p_l, -p_j)\Big],$$

$$\frac{J_z}{2T}G(q_l) = 2\pi(l - 1) + \frac{1}{i}\Big(\ln(L(q_l, 0))$$

$$+ \sum_{j=2}^{\infty}\ln[L(q_l, q_j)L(q_l, -q_j)]\Big),$$

$$G(x) \equiv x\frac{\sqrt{(x^2 + D_x^2)(x^2 + D_y^2)}}{x^2 + 1},$$

$$L(x, y) \equiv \frac{\sqrt{(x^2 + D_x^2)(x^2 + D_y^2)} - ix\Delta_x\Delta_y - iy(1 + x^2)}{\sqrt{(x^2 + D_x^2)(x^2 + D_y^2)} + ix\Delta_x\Delta_y + iy(1 + x^2)},$$

$$\Delta_x = J_x/J_z, \quad \Delta_y = J_y/J_z,$$
$$D_x = \sqrt{1 - \Delta_x^2}, \quad D_y = \sqrt{1 - \Delta_y^2}. \tag{17.42}$$

We find that the solution of (17.42) by iteration converges very rapidly. For $j \gg J_z/(4\pi T)$, p_j is about $4\pi T(j - 1/2)/J_z$. So one can replace the sum for large j by an integral. Λ_0 and Λ_1 are

$$\Lambda_0 = 2\prod_{j=1}^{\infty}\left(\frac{J_z}{4\pi T(j - \frac{1}{2})}\right)^2(1 + p_j^2),$$

$$\Lambda_1 = 2\frac{J_z}{4T}\prod_{j=2}^{\infty}\left(\frac{J_z}{4\pi T(j - 1)}\right)^2(1 + q_j^2). \tag{17.43}$$

Equations (17.42)–(17.43) are very convenient for numerical calculations because they do not contain complicated functions such as elliptic functions.

17.2 Special cases and numerical methods

17.2.1 The $T \to 0$ limit

At very small T, the distribution of x_ls becomes dense. The mean distance is of the order of T/J_s. Put the distribution function of x_ls as $\rho(x)$. Equation (17.33) becomes:

$$\frac{\pi T}{J_s}\int_x^K \rho(t)dt + \frac{T}{J_s}\int_0^K \frac{1}{2i}\ln[f(x, y)f(x, -y)]\rho(y)dy = g(x). \tag{17.44}$$

The function $g(x)$ has a pole at $x = 0$, so the inhomogeneous term of the integral equation for $\rho(x)$ has a second order pole. It is difficult to apply the Fourier transform technique for $\rho(x)$, so we define the function $F(x)$,

$$F(x) \equiv \begin{cases} \dfrac{\pi T}{J_s}\displaystyle\int_x^K \rho(t)dt, & \text{at } x > 0, \\[4mm] -F(-x), & \text{at } x < 0. \end{cases} \tag{17.45}$$

By partial differentiation, equation (17.44) becomes

$$F(x) - \int_{-K}^K q(x - y)F(y)dy = g(x),$$

$$q(x) \equiv \frac{1}{2\pi i}\left(\frac{h'(x - 2\eta)}{h(x - 2\eta)} - \frac{h'(x + 2\eta)}{h(x + 2\eta)}\right) + \frac{2i\eta}{KK'}.$$

Using the imaginary transformation and the infinite product representation of the elliptic theta functions, we have

$$
h(x) = \sqrt{k}\Theta^2(0)\exp\left(-\frac{\pi x^2}{2KK'}\right)
$$

$$
\times \sinh\frac{\pi x}{K'}\prod_{j=1}^{\infty}\frac{\sinh(\pi(2jK-x)/K')\sinh(\pi(2jK+x)/K')}{\sinh^2((2\pi jK)/K')}.
$$

Thus the functions $q(x)$ and $g(x)$ in equation (17.32) become

$$
q(x) = \sum_{m=-\infty}^{\infty}\overline{q}(x-2mK), \quad \overline{q}(x) = \frac{1}{K'}\frac{\sin(2\theta)}{\cosh(2\pi x/K')+\cos 2\theta},
$$

$$
g(x) = \sum_{m=-\infty}^{\infty}\overline{g}(x-2mK),
$$

$$
\overline{g}(x) = \frac{\pi}{K'}\left\{\coth(\pi x/K') - \frac{\sinh(2\pi x/K')}{\cosh(2\pi x/K')+\cos 2\theta}\right\}.
$$

Here we put $\theta \equiv \pi(1+2i\eta/K')$. The Fourier transform of these equations is

$$
\tilde{q}_n = \int_{-K}^{K}q(x)e^{in\pi x/K}\,\mathrm{d}x = \int_{-\infty}^{\infty}\overline{q}(x)e^{in\pi x/K}\,\mathrm{d}x
$$

$$
= -\frac{\sinh(n(\tau-2\lambda))}{\sinh(n\tau)},
$$

$$
\tilde{g}_n = \int_{-K}^{K}g(x)e^{in\pi x/K}\,\mathrm{d}x = -\pi i\left(\coth(n\tau) - \frac{\cosh(n(\tau-2\lambda))}{\sinh(n\tau)}\right),
$$

$$
\tau \equiv \frac{\pi K'}{2K}, \quad \lambda \equiv \frac{K'\theta}{2K}. \tag{17.46}
$$

The function $g(x)$ has a pole at $x = 0$. We define its Fourier transform by a principal integral. Then we have the Fourier transform of $F(x)$,

$$
\tilde{F}_n = \tilde{g}_n/(1-\tilde{q}_n) = -\pi i\tanh(n\lambda). \tag{17.47}
$$

Thus,

$$
F(x) = \sum_{m=-\infty}^{\infty}\overline{F}(x-2mK), \quad \overline{F}(x) = \frac{\pi^2}{K'\theta}\mathrm{cosech}\frac{\pi^2 x}{K'\theta}.
$$

Introducing a new modulus u,

$$
\frac{\theta K_k'}{2\pi K_k} = \frac{K_u'}{K_u}, \tag{17.48}
$$

we write $F(x)$ and its derivative in terms of elliptic functions,

$$F(x) = \frac{K_u}{K_k} \text{ctn}\left(\frac{K_u x}{K_k}, u\right), \quad \text{ctn}(x) \equiv \frac{\text{cn}(x)}{\text{sn}(x)},$$

$$\rho(x) = \frac{J_s}{\pi T}\left(\frac{K_u}{K_k}\right)^2 \text{dn}\left(\frac{K_u x}{K_k}, u\right)\left(\text{sn}\left(\frac{K_u x}{K_k}, u\right)\right)^{-2}. \tag{17.49}$$

From equation (17.38) we get the energy per site:

$$e = -\lim_{T \to 0} T \ln \Lambda_0 = \frac{J_s}{\pi}\int_0^K \ln\left\{\frac{h'^2(0)h(2\eta - x)h(2\eta + x)}{F^2(x)h^2(x)h^2(2\eta)}\right\}F'(x)dx.$$

By partial integration we have

$$e = \frac{2J_s}{\pi}\int_0^K F'(x) + g(x)F(x)dx.$$

Here we used $F(x)h(x) = h'(0) + O(x^2)$ and $h(2\eta - x)h(2\eta + x) = h^2(2\eta) + O(x^2)$ near $x = 0$. As F and g have a pole at $x = 0$ we put

$$u(x) \equiv F(x) - w(x), \quad v(x) \equiv g(x) - w(x), \quad w(x) \equiv \frac{\pi}{2K}\cot\frac{\pi x}{2K}.$$

We find $\tilde{w}_n = -\pi i \text{sign}(n)$ and that $u(x)$ and $v(x)$ have no singularity. Thus

$$e = \frac{J_s}{\pi}\int_{-K}^K u(x)v(x) + u'(x) + w(x)(u(x) + w(x)) + w^2(x) + w'(x)dx.$$

As $w^2(x) + w'(x) = -(\pi/2K)^2$ we have

$$e = -\frac{J_s \pi}{2K} + \frac{J_s}{K\pi}\sum_{n=1}^\infty \left[\tilde{u}_n \tilde{v}_{-n} + \tilde{w}_n(\tilde{u}_{-n} + \tilde{v}_{-n})\right]$$

$$= -\frac{J_s \pi}{2K}\left[1 + 2\sum_{n=1}^\infty -\frac{\sinh(n(\tau - 2\lambda))}{\sinh(n\tau)} + 2\frac{\sinh(n(\tau - \lambda))}{\sinh(n\tau)\cosh(n\lambda)}\right]. \tag{17.50}$$

This coincides with the exact ground state energy per site of the XYZ chain obtained in (5.80).

17.2.2 The $J_x = 0$ case (anisotropic XY chain)

In equation (17.42) all interaction terms vanish and it becomes:

$$p_l \sqrt{\frac{p_l^2 + 1 - \Delta_y^2}{p_l^2 + 1}} = \alpha_l, \quad \alpha_l \equiv 4\pi T(l - 1/2)/J_z.$$

So we have

$$p_l^2 = (\Delta_y^2 + \alpha_l^2 - 1 + \sqrt{(\Delta_y^2 + \alpha_l^2 - 1)^2 + 4\alpha_l^2})/2$$

and equations (17.43) yield:

$$\Lambda_0 = 2 \prod_{l=1}^{\infty} \left\{ \frac{1 + (1 + \Delta_y^2)\alpha_l^{-2} + \sqrt{(1 + (1 + \Delta_y^2)\alpha_l^{-2})^2 - 4\alpha_l^{-4}\Delta_y^2}}{2} \right\}. \qquad (17.51)$$

Using the formula

$$\ln(a + \sqrt{a^2 - b^2}) = \frac{1}{2\pi} \int_0^{2\pi} dx \ln(2a + 2b \cos x),$$

we have

$$\ln \Lambda_0 = \ln 2 + \frac{1}{2\pi} \int_0^{2\pi} dx \sum_{l=1}^{\infty} \ln[1 + (1 + \Delta_y^2 + 2\Delta_y \cos x)\alpha_l^{-2}]$$

$$= \frac{1}{2\pi} \int_0^{2\pi} \ln\left\{ 2 \cosh\left(\frac{1}{4T} \sqrt{J_z^2 + J_y^2 + 2J_z J_y \cos x} \right) \right\} dx.$$

In the same way we obtain Λ_1,

$$\ln \Lambda_1 = \frac{1}{2\pi} \int_0^{2\pi} \ln\left\{ 2 \sinh\left(\frac{1}{4T} \sqrt{J_z^2 + J_y^2 + 2J_z J_y \cos x} \right) \right\} dx.$$

The free energy per site and correlation length are

$$f = -\frac{T}{2\pi} \int_0^{2\pi} \ln\left\{ 2 \cosh\left(\frac{1}{4T} \sqrt{J_z^2 + J_y^2 + 2J_z J_y \cos x} \right) \right\} dx, \qquad (17.52)$$

$$\xi^{-1} = \frac{1}{2\pi} \int_0^{2\pi} \ln\left\{ \coth\left(\frac{1}{4T} \sqrt{J_z^2 + J_y^2 + 2J_z J_y \cos x} \right) \right\} dx. \qquad (17.53)$$

Equation (17.52) coincides with equation (1.47) at $h = 0$.

17.3 Numerical calculations

Using equations (17.42) and (17.43) we calculate the free energy and correlation length for arbitrary $(J_x/J_z, J_y/J_z)$. In Table (17.1) we give these quantities for $(J_x/J_z, J_y/J_z)$ values of $(0.8, 0.8)$, $(0.5, 0.5)$, $(0.5, 1.0)$ in tables A, B, C, respectively. We take p_l and q_l up to $l = 1024$. The sums for larger l in equations are approximated by integrals.

It seems that the correlation length ξ becomes longer as the temperature goes down,

$$\xi \sim T^c \exp(\Delta_\xi / T). \qquad (17.54)$$

Table 17.1. *Free energy and correlation length according to equations* *(17.42)–(17.43) for different anisotropy parameters* $(J_x/J_z, J_y/J_z)$

(A) (0.8,0.8)

T/J_z	$(e-f)/J_z$	ξ
0.03	4.35 E-6	11019.5
0.04	4.688 E-5	906.813
0.05	2.1972 E-4	203.643
0.07	1.4456 E-3	37.7358
0.10	6.37579E-3	11.1653
0.20	4.07554E-2	2.87522
0.40	0.139038	1.32996

(B) (0.5,0.5)

T/J_z	$(e-f)/J_z$	ξ
0.03	2 E-8	9.28 E5
0.04	8.2 E-7	2.514 E4
0.05	8.96 E-6	2884.67
0.07	1.522 E-4	243.033
0.10	1.4141 E-3	38.1380
0.20	2.28744E-2	4.52924
0.40	0.115028	1.537295

(C) (0.5,1.0)

T/J_z	$-f/J_z$	ξ
0.03	0.2752456	20.61248
0.04	0.2758122	15.42187
0.05	0.2765430	12.29748
0.07	0.2785067	8.700840
0.10	0.2827654	5.954982
0.20	0.3085682	2.738899
0.40	0.3964288	1.352457

(D) (-0.5,0.6)

T/J_z	$-f/J_z$	ξ
0.03	0.3201312	139.6721
0.04	0.3203167	63.6121
0.05	0.3206097	37.96712
0.07	0.3215180	19.46736
0.10	0.3236850	10.59361
0.20	0.3383583	3.831694
0.40	0.4041916	1.609889

(E) (1.0,1.0)

T/J_z	$-f/J_z$	ξ
0.03	0.2546415	10.04573
0.04	0.2569907	7.741370
0.05	0.2595858	6.341317
0.07	0.2653749	4.717322
0.10	0.2752379	3.473658
0.20	0.3149567	1.971316
0.40	0.4126261	1.175348

(F) (-1.0,1.0)

T/J_z	$-f/J_z$	ξ
0.03	0.4434480	18.3789
0.04	0.4436825	13.8750
0.05	0.4439843	11.16418
0.07	0.4447916	8.054780
0.10	0.4465170	5.71082
0.20	0.456958	2.94813
0.40	0.5028667	1.535774

It is expected that spin-wave excitation will not affect the long range correlation. But spinon excitation drastically destroys the long range order. Then it is expected that $\Delta_\xi = \Delta_{\text{spinon}}$. On the contrary, the energy is determined by the short range order. So we can expect that

$$\Delta_f = \min(\Delta_{\text{spinon}}, \Delta_{\text{spin-wave}}).\tag{17.55}$$

On the line AF–F of Fig. (11.2) the system is gapless and the correlation length behaves as $\xi \sim 1/T$.

18
Recent developments and related topics

18.1 Numerical analysis of the $S = 1$ chain

From analogy with the $S = 1/2$ Heisenberg antiferromagnet, people expected the general S antiferromagnet to have gapless excitations. But Haldane predicted that for integer S the system has an energy gap, and is gapless for half-odd integer S, based on field theoretical considerations[33, 34]. Several numerical calculations were performed on the Hamiltonian,

$$\mathscr{H} = J \sum_{i=1}^{N} \mathbf{S}_i \cdot \mathbf{S}_{i+1}, \quad \mathbf{S}_{N+1} \equiv \mathbf{S}_1. \tag{18.1}$$

The diagonalization method up to $N = 14$[17] for $S = 1$ could not give a decisive answer regarding the existence of the gap. Nightingale and Blöte[75] calculated the energy gap of the $N = 32$ chain using the diffusion Monte Carlo method. They estimated the gap as $0.41J$ in the limit $N = \infty$. The author calculated the correlation functions using the world line Monte Carlo method[104]. He showed that it decays algebraically for the $S = 1/2$ chain and exponentially for the $S = 1$ chain. The estimation of the correlation length is $\xi = 4.5 \pm 2$. Nomura[76] did a more elaborate calculation and estimated as $\xi = 6.1$. Monte Carlo calculation of the elementary excitation was done for the $N = 32$ chain[105]. The asymmetry of this spectrum with respect to $Q = \pi/2$ axis was found. This excitation spectrum is justified by the neutron scattering experiments[64]. Recent diagonalization calculations[53] support the existence of a gap. The actual $S = 1$ Heisenberg chain is modified by an anisotropy effect.

Thus the general Hamiltonian is

$$\mathscr{H} = \sum_{i=1}^{N} J(S_i^x S_{i+1}^x + S_i^y S_{i+1}^y + \lambda S_i^z S_{i+1}^z) + D(S_i^z)^2, \quad J > 0. \tag{18.2}$$

For $\lambda = 1$ and $D = 0$, the existence of a gap was established by the quantum Monte Carlo method. The next problem to solve is how the gap is destroyed by the anisotropy. On the line $D = 0$, the gap vanishes at $\lambda = 1.17$. The inter-chain coupling also destroys the gap and induces long range order. The critical value is estimated by a simple mean field theory. It must be emphasized that the $S = 1/2$ quasi one-dimensional system always has three-dimensional order at low temperatures.

In the analysis of the $S = 1$ case described by the Hamiltonian (18.2), the Monte Carlo method and the diagonalization method are both very useful. The estimation of physical quantities as $N = \infty$ can be done through Shank's transformation. The determination of λ_2, where the system becomes gapless, is very difficult because the gap becomes small very slowly near the critical point. This transition is of the Kosterlitz and Thouless type.

In the $S = 1/2$ case the two-point function $\langle S_i^z S_{i+r}^z \rangle$ decays algebraically. But for the $S = 1$ case the two-point function decays exponentially. This state is far from the ordered state. Thus a weak perturbation of interchain coupling cannot make the ordered state. There should be a critical inter-chain coupling which separates the ordered and disordered ground state. $Ni(C_2H_8N_2)_2NO_2ClO_4$, abbreviated NENP, is expected in the disordered phase because of small interchain coupling. On the other hand $CsNiCl_3$ has a bigger interchain coupling and thus is ordered at low temperatures. This critical interchain coupling is estimated by Sakai and Takahashi[80].

At zero temperature the magnetization is zero in a weak magnetic field. Magnetization appears suddenly at the critical field. The magnetization curve and differential magnetic susceptibility were calculated by the exact diagonalization method[107]. These results coincide very well with those of experiments[111].

In the magnetized state the system should be described by the conformal field theory or Luttinger liquid theory because the system is gapless. In this region, spin correlation functions decay algebraically. It is shown that the central charge is unity and the correlation exponents are functions of the magnetization. Numerical calculation of exponents was done at $\lambda = 1$ and $D = 0$[107].

The $S = 1$ Heisenberg model is not soluble. But if a biquadratic term is added the system becomes mathematically tractable in some cases,

$$\mathcal{H} = \sum_{i=1}^{N} \mathbf{S}_i \cdot \mathbf{S}_{i+1} - \beta(\mathbf{S}_i \cdot \mathbf{S}_{i+1})^2. \tag{18.3}$$

The point $\beta = -1$ was solved by Sutherland[85] using the Bethe-ansatz method. The point $\beta = 1$ was solved by Takhtajan and Babujian[112]. In both cases the system is gapless. The point $\beta = -1/3$ was investigated by Affleck, Kennedy, Lieb and Tasaki[1]. The ground state is known at this point and the system has a gap. The excited states are not soluble. This point is different from the Bethe-ansatz soluble point. It is expected that the state has a gap in the region $-1 < \beta < 1$.

The elementary excitations of the $S = 1$ Heisenberg chain were calculated by the quantum Monte Carlo method and the diagonalization method. The lowest excitation energy is at momentum π. This spectrum is not doubly periodic nor symmetric with respect to the axis $Q = \pi/2$. These points are different from the $S = 1/2$ chain. The scattering intensity of these lowest energy states is calculated. In the region $0.3\pi < Q \leq \pi$, there is a strong delta function peak in the dynamical structure factor. On the other hand for $0 < Q < 0.3\pi$ the lowest energy state is the lower edge of the continuum. It is expected that in this region the lowest energy state is a scattering state of two elementary excitations near $Q = \pi$. Then it is expected that the gap at $Q = 0$ is twice the gap at $Q = \pi$. In the actual NENP system there is anisotropy and the situation becomes more complicated. Details are discussed in Ref. 106.

Appendix A

The Young tableau and the theorem of Lieb and Mattis

In this section we consider a system of one dimensional particles which interact via a many-body potential. Consider

$$\mathscr{H} = -\sum \frac{\partial^2}{\partial x_i^2} + V(x_1, x_2, ..., x_N). \tag{A.1}$$

The potential is a symmetric function with respect to the exchange of particles. There are $N!$ permutation operators X_P of variables. It is evident that these operators commute with the Hamiltonian and form a group. Generally speaking, if a Hamiltonian commutes with each element of a group one can choose a complete set of eigenfunctions, each of which transforms like an irreducible representation of that group. A group formed by all N-order permutations is called a symmetry group of order N and abbreviated as S_N. The irreducible representation of S_N is represented by a Young tableau. Consider a function of N variables,

$$\phi(x_1, x_2, ..., x_{n_1} | x_{n_1+1}, ..., x_{n_1+n_2} | x_{n_1+n_2+1}, ..., x_N);$$

$$n_1 \geq n_2 \geq ..., \tag{A.2}$$

where we mean that ϕ is separately antisymmetric in the variables $x_1, x_2, ..., x_n$, in the variables $x_{n_1+1}, ..., x_{n_1+n_2}$ and so on. If the bars cannot be moved to the right, i.e.

$$\left\{ 1 - \sum_{k=1}^{n_j} P(n_1 + n_2 + ... + n_{j-1} + k, n_1 + n_2 + ... + n_{l-1} + 1) \right\} \phi$$

$$= 0, \tag{A.3}$$

for arbitrary $l > j$, we say that the function ϕ transforms like an irreducible representation which corresponds to the Young tableau

$$[1^{n_1-n_2} 2^{n_2-n_3} ...].$$

226

For example if $\phi(x_1, x_2 | x_3)$ satisfies (A.3), two functions $\phi_1(x_1, x_2, x_3) = \phi(x_1, x_2 | x_3)$ and $\phi_2(x_1, x_2, x_3) = \phi(x_1, x_3 | x_2)$ form a base of an irreducible representation $[1, 2]$ of S_3. In fact we show

$$I \begin{pmatrix} \phi_1 \\ \phi_2 \end{pmatrix} = \begin{pmatrix} 1 & 0 \\ 0 & 1 \end{pmatrix} \begin{pmatrix} \phi_1 \\ \phi_2 \end{pmatrix}, P_{12} \begin{pmatrix} \phi_1 \\ \phi_2 \end{pmatrix} = \begin{pmatrix} -1 & 0 \\ -1 & 1 \end{pmatrix} \begin{pmatrix} \phi_1 \\ \phi_2 \end{pmatrix},$$

$$P_{13} \begin{pmatrix} \phi_1 \\ \phi_2 \end{pmatrix} = \begin{pmatrix} 1 & -1 \\ 0 & -1 \end{pmatrix} \begin{pmatrix} \phi_1 \\ \phi_2 \end{pmatrix}, P_{23} \begin{pmatrix} \phi_1 \\ \phi_2 \end{pmatrix} = \begin{pmatrix} 0 & 1 \\ 1 & 0 \end{pmatrix} \begin{pmatrix} \phi_1 \\ \phi_2 \end{pmatrix},$$

$$P \begin{pmatrix} 1 & 2 & 3 \\ 3 & 1 & 2 \end{pmatrix} \begin{pmatrix} \phi_1 \\ \phi_2 \end{pmatrix} = \begin{pmatrix} 0 & -1 \\ 1 & -1 \end{pmatrix} \begin{pmatrix} \phi_1 \\ \phi_2 \end{pmatrix},$$

$$P \begin{pmatrix} 1 & 2 & 3 \\ 2 & 3 & 1 \end{pmatrix} \begin{pmatrix} \phi_1 \\ \phi_2 \end{pmatrix} = \begin{pmatrix} 1 & -1 \\ -1 & 0 \end{pmatrix} \begin{pmatrix} \phi_1 \\ \phi_2 \end{pmatrix}.$$

This is an irreducible representation by 2×2 matrices. The many-body wave function of N particles with spin value s is a function of spatial coordinates $x_1, ..., x_N$ and spin coordinates $s_1, s_2, ..., s_N$. Spin coordinates take the discrete values $s, s - 1, ..., -s$. We put the number of particles with the same spin value as n_i with $n_1 \geq n_2 \geq ... \geq n_{2s+1}$. The eigenfunctions of (A.1) can be written as

$$\psi(x_1 s_1, x_2 s_2, ..., x_N s_N) = \sum_{j=1}^{\frac{N!}{n_1! n_2! ... n_{2s+1}!}} \phi_j(x_1, x_2, ..., x_N) G_j. \tag{A.4}$$

A typical spin function for particles with spin value s is

$$G_1 = (s, s, ..., s | s - 1, s - 1, ..., s - 1 | ... | - s, -s, ..., -s). \tag{A.5}$$

The spatial wave function ϕ_1 is separately symmetric (antisymmetric) for bosons (fermions) for variables $x_1, x_2, ..., x_{n_1}$ and $x_{n_1+1}, ..., x_{n_1+n_2}$ and so on. The other ϕ_js are obtained if we require that (A.4) is totally symmetric or antisymmetric.

Theorem: Let us define the lowest eigenvalue of (A.1) as $E(Y)$ when the wave function transform like an irreducible representation defined by a Young tableau Y. If two Young tableaux $Y = (1^{n_1-n_2}, 2^{n_2-n_3}, ...)$ and $Y = (1^{n_1'-n_2'}, 2^{n_2'-n_3'}, ...)$ have relations:

$$n_1 \geq n_1', n_1 + n_2 \geq n_1' + n_2', n_1 + n_2 + n_3 \geq n_1' + n_2' + n_3', ...,$$

then $E(Y) \geq E(Y')$.

This is a famous theorem due to Lieb and Mattis. From this theorem one can show that the one-dimensional fermions with $1/2$ spin described by (A.1) cannot be ferromagnetic. This theorem also holds for Hamiltonian (2.1).

Appendix B
The number of string solutions

Here we prove the number of sets $\{I_\alpha^n\}$ which satisfy (8.13) is $C_M^N - C_{M-1}^N$. It is clear that the number of sets is

$$\sum_{\alpha_1 + 2\alpha_2 + \ldots + M\alpha_M = M} \prod_{i=1}^{M} \binom{N - \sum_{j=1}^{M} t_{ij}\alpha_j}{\alpha_j}, \tag{B.1}$$

from the condition (8.13), where

$$C_M^N = \binom{N}{M} \equiv \frac{N(N-1)(N-2)\ldots(N-M+1)}{M!}.$$

The summation (B.1) is rewritten as follows:

(B.1)
$$= \sum_{\alpha_M = 0}^{\infty} \binom{N - 2M + \alpha_M}{\alpha_M} \sum_{\alpha_{M-1}=0}^{\infty} \binom{N - 2M + 2\alpha_M + \alpha_{M-1}}{\alpha_{M-1}} \ldots$$

$$\times \sum_{\alpha_2=0}^{\infty} \binom{N - 2M + 2(\alpha_3 + 2\alpha_4 + \ldots) + \alpha_2}{\alpha_2}$$

$$\binom{N - M + \alpha_3 + 2\alpha_4 + \ldots}{M - 2\alpha_2 - 3\alpha_3 - \ldots}. \tag{B.2}$$

The last sum on the r.h.s. is the coefficient of $x^{M - 3\alpha_3 - 4\alpha_4 - 5\alpha_5 - \ldots}$ of the Taylor expansion of

$$(1 - x^2)^{-(N+1-2M+2(\alpha_3 + 2\alpha_4 + 3\alpha_5 + \ldots))}(1 + x)^{N - M + \alpha_3 + 2\alpha_4 + 3\alpha_5 + \ldots}.$$

This is the M-th order coefficient of

$$[(1 - x^2)(1 + x)]^{M-1}(1 - x)^{-N+1} \prod_{n=3}^{M} \left(\frac{x^n}{(1 - x)^{2(n-2)}(1 + x)^{(n-2)}}\right)^{\alpha_n}.$$

Using the relation

$$\sum_{\alpha=0}^{\infty} \binom{B+\alpha}{\alpha} X^\alpha = (1-X)^{-B-1},$$

we have

$$(\text{B.2}) = \text{the } M\text{-th order coefficient of}$$

$$(1+x)^{N-M} \left\{ \prod_{j=2}^{M} (1 - u_j^{-1}(x)) \right\}^{-N+2M-1}, \tag{B.3}$$

where $u_j(x)$ are functions determined by:

$$(u_{j+1} - 1)^2 = u_j u_{j+2}, \tag{B.4}$$

and

$$u_2 = \frac{1}{x^2}, \quad u_3 = \frac{(1-x)^2(1+x)}{x^3}. \tag{B.5}$$

The general solution of the difference equation (B.4) is

$$u_j = f_j^2, \quad f_j \equiv \frac{ba^j - b^{-1}a^{-j}}{a - a^{-1}}. \tag{B.6}$$

Parameters a and b are determined from the initial conditions (B.5) and we have

$$a = b = \frac{1}{2} \left(\sqrt{\frac{1-3x}{x}} + \sqrt{\frac{1+x}{x}} \right). \tag{B.7}$$

As $u_j^{-1}(x) = O(x^j)$, we have

$$\prod_{j=M+1}^{\infty} (1 - u_j^{-1}(x)) = 1 + O(x^{M+1}),$$

and therefore

$$(\text{B.3}) = \text{the } M\text{-th order coefficient of}$$

$$(1+x)^{N-M} \left\{ \prod_{j=2}^{\infty} (1 - u_j^{-1}(x)) \right\}^{-N+2M-1}. \tag{B.8}$$

Substituting $1 - u_j^{-1} = f_{j-1}f_{j+1}/f_j^2$ we have:

$$\prod_{j=2}^{\infty} (1 - u_j^{-1}(x)) = (1 - u_2^{-1}) \prod_{j=3}^{\infty} \left(\frac{f_{j-1}f_{j+1}}{f_j^2} \right) = (1 - x^2) \frac{f_2}{f_3} \lim_{n\to\infty} \frac{f_{n+1}}{f_n}$$

$$= (1 - x^2) a \sqrt{\frac{u_2(x)}{u_3(x)}} = \frac{2x}{1 - \sqrt{1 - 4x/(1+x)}}.$$

Thus we have

$$(B.8) = N - M + 1\text{-th order coefficient of}$$

$$(1 + x)^{N-M}\left[\frac{1}{2}\left(1 - \sqrt{1 - \frac{4x}{1+x}}\right)\right]^{N-2M+1}. \tag{B.9}$$

Using the binomial theorem, the function on the r.h.s. is

$$2^{-N+2M-1}\sum_{s=0}^{\infty}\sum_{r=0}^{\infty}(-1)^{s+r}\binom{N-2M+1}{s}\binom{s/2}{r}(4x)^r(1+x)^{N-M-r}.$$

As the coefficient of x^{N-M+1} for $(4x)^r(1+x)^{N-M-r}$ is zero for $r \neq N-M+1$, we have

$$(B.9) = 2^{N+1}\sum_{s=0}^{\infty}(-1)^{N-M+1+s}\binom{N-2M+1}{s}\binom{s/2}{N-M+1}$$

$$= \frac{1}{2^{N-2M}}\frac{N!(N-2M+1)!}{(N-M+1)!(N-M)!}$$

$$\times \sum_{r=0}^{\infty}(-1)^r\binom{N-M}{r}\binom{2N-2M-2r}{N}. \tag{B.10}$$

The last sum becomes

$$\sum_{r=0}^{\infty}(-1)^r\binom{N-M}{r}\binom{2N-2M-2r}{N}$$

$$= N - 2M\text{-th order coefficient of }(1-x^2)^{N-M}(1-x)^{-N-1}$$

$$= N - 2M\text{-th order coefficient of }(1+x)^{N-M}(1-x)^{-M-1}$$

$$= \sum_{r=0}^{N-2M}\frac{(N-M)!}{r!(N-2M-r)!M!} = (1+1)^{N-2M}\frac{(N-M)!}{M!(N-2M)!}.$$

Then we have finally

$$(B.1) = (B.10) = \frac{N!(N-2M+1)}{(N-M+1)!M!} = \binom{N}{M} - \binom{N}{M-1}. \tag{B.11}$$

Appendix C

The commuting transfer matrix and spectral parameter

For the transfer matrices defined by (5.3) and (5.24), we consider the condition that two matrices $\mathbf{T}(a,b,c,d)$ and $\mathbf{T}(a',b',c',d')$ commute. The transfer matrix is defined by the trace of a $2 \otimes 2$ matrix,

$$\mathbf{T} = \mathrm{Tr}_\tau \mathbf{R}_1 \cdot \mathbf{R}_2 \cdot ... \mathbf{R}_N,$$
$$\mathbf{R}_l = w_1 \sigma_l^1 \otimes \tau^1 + w_2 \sigma_l^2 \otimes \tau^2 + w_3 \sigma_l^3 \otimes \tau^3 + w_4 \sigma_l^4 \otimes \tau^4,$$
$$w_1 = \frac{c+d}{2}, \quad w_2 = \frac{c-d}{2}, \quad w_3 = \frac{a-b}{2}, \quad w_4 = \frac{a+b}{2}. \tag{C.1}$$

Here $N+1$ spin spaces are defined. The matrix $\mathbf{T}(a,b,c,d)\mathbf{T}(a',b',c',d')$ is given by $\mathrm{Tr}\prod_l U_l$, where U_l is a 4×4 matrix,

$$
\begin{aligned}
U_l \equiv \ & \sigma_l^1 \{ w_1 w_4' \Phi^{14} + w_4 w_1' \Phi^{41} + i w_2 w_3' \Phi^{23} - i w_3 w_2' \Phi^{32} \} \\
& + \sigma_l^2 \{ w_2 w_4' \Phi^{24} + w_4 w_2' \Phi^{42} + i w_3 w_1' \Phi^{31} - i w_1 w_3' \Phi^{13} \} \\
& + \sigma_l^3 \{ w_3 w_4' \Phi^{34} + w_4 w_3' \Phi^{43} + i w_1 w_2' \Phi^{12} - i w_2 w_1' \Phi^{21} \} \\
& + \sigma_l^4 \{ w_1 w_1' \Phi^{11} + w_2 w_2' \Phi^{22} + w_3 w_3' \Phi^{33} + w_4 w_4' \Phi^{44} \}.
\end{aligned}
\tag{C.2}
$$

Here we put $\Phi^{\alpha\beta} \equiv \tau^\alpha \otimes \tau'^\beta$. The matrix $\mathbf{T}(a',b',c',d')\mathbf{T}(a,b,c,d)$ is given by $\mathrm{Tr}\prod_l V_l$ with

$$
\begin{aligned}
V_l \equiv \ & \sigma_l^1 \{ w_1' w_4 \Phi^{14} + w_4' w_1 \Phi^{41} + i w_2' w_3 \Phi^{23} - i w_3' w_2 \Phi^{32} \} \\
& + \sigma_l^2 \{ w_2' w_4 \Phi^{24} + w_4' w_2 \Phi^{42} + i w_3' w_1 \Phi^{31} - i w_1' w_3 \Phi^{13} \} \\
& + \sigma_l^3 \{ w_3' w_4 \Phi^{34} + w_4' w_3 \Phi^{43} + i w_1' w_2 \Phi^{12} - i w_2' w_1 \Phi^{21} \} \\
& + \sigma_l^4 \{ w_1' w_1 \Phi^{11} + w_2' w_2 \Phi^{22} + w_3' w_3 \Phi^{33} + w_4' w_4 \Phi^{44} \}.
\end{aligned}
\tag{C.3}
$$

If we can find a non-singular 4×4 matrix S which satisfies

$$SU_l = V_l S,$$

we can show

$$\mathbf{T}(a,b,c,d)\mathbf{T}(a',b',c',d') = \mathbf{T}(a',b',c',d')\mathbf{T}(a,b,c,d).$$

If we put

$$S = x_1\tau^1 \otimes \tau^{1'} + x_2\tau^2 \otimes \tau^{2'} + x_3\tau^3 \otimes \tau^{3'} + x_4\tau^4 \otimes \tau^{4'},$$

we have

$$
\begin{aligned}
SU_l - V_l S =& \\
2i\sigma_l^1 \{ \tau^2 &\otimes \tau^{3'}(-w_3w_2'x_1 - w_4w_1'x_2 + w_1w_4'x_3 + w_2w_3'x_4) \\
-\tau^3 &\otimes \tau^{2'}(-w_2w_3'x_1 + w_4w_1'x_2 - w_1w_4'x_3 + w_3w_2'x_4)\} \\
+2i\sigma_l^2 \{ \tau^3 &\otimes \tau^{1'}(w_2w_4'x_1 - w_1w_3'x_2 - w_4w_2'x_3 + w_3w_1'x_4) \\
-\tau^1 &\otimes \tau^{3'}(-w_2w_4'x_1 - w_3w_1'x_2 + w_4w_2'x_3 + w_1w_3'x_4)\} \\
+2i\sigma_l^3 \{ \tau^1 &\otimes \tau^{2'}(-w_4w_3'x_1 + w_3w_4'x_2 - w_2w_1'x_3 + w_1w_2'x_4) \\
-\tau^2 &\otimes \tau^{1'}(w_4w_3'x_1 - w_3w_4'x_2 - w_1w_2'x_3 + w_2w_1'x_4)\}. \quad\text{(C.4)}
\end{aligned}
$$

Then the following six equations should be satisfied:

$$
\begin{aligned}
-w_3w_2'x_1 - w_4w_1'x_2 + w_1w_4'x_3 + w_2w_3'x_4 &= 0, \\
-w_2w_3'x_1 + w_4w_1'x_2 - w_1w_4'x_3 + w_3w_2'x_4 &= 0, \\
w_2w_4'x_1 - w_1w_3'x_2 - w_4w_2'x_3 + w_3w_1'x_4 &= 0, \\
-w_2w_4'x_1 - w_3w_1'x_2 + w_4w_2'x_3 + w_1w_3'x_4 &= 0, \\
-w_4w_3'x_1 + w_3w_4'x_2 - w_2w_1'x_3 + w_1w_2'x_4 &= 0, \\
w_4w_3'x_1 - w_3w_4'x_2 - w_1w_2'x_3 + w_2w_1'x_4 &= 0. \quad\text{(C.5)}
\end{aligned}
$$

From the first three equations we can determine the ratio of $x_1 : x_2 : x_3 : x_4$. Substituting this into the last three equations, following three determinants must be zero:

$$
\det \begin{bmatrix}
-w_3w_2' & -w_4w_1' & w_1w_4' & w_2w_3' \\
-w_2w_3' & w_1w_4' & -w_4w_1' & w_3w_2' \\
w_2w_4' & -w_1w_3' & -w_4w_2' & w_3w_1' \\
-w_2w_4' & -w_3w_1' & w_4w_2' & w_1w_3'
\end{bmatrix},
$$

$$
\det \begin{bmatrix}
-w_3w_2' & -w_4w_1' & w_1w_4' & w_2w_3' \\
-w_2w_3' & w_1w_4' & -w_4w_1' & w_3w_2' \\
w_2w_4' & -w_1w_3' & -w_4w_2' & w_3w_1' \\
-w_4w_3' & +w_3w_4' & -w_2w_1' & w_1w_2'
\end{bmatrix},
$$

$$\det \begin{bmatrix} -w_3 w_2' & -w_4 w_1' & w_1 w_4' & w_2 w_3' \\ -w_2 w_3' & w_1 w_4' & -w_4 w_1' & w_3 w_2' \\ w_2 w_4' & -w_1 w_3' & -w_4 w_2' & w_3 w_1' \\ w_4 w_3' & -w_3 w_4' & -w_1 w_2' & w_2 w_1' \end{bmatrix}.$$

The first determinant is factorized as follows:

$$-(w_3 w_4 w_1' w_2' + w_1 w_2 w_3' w_4')(w_3^2 w_1'^2 - w_4^2 w_1'^2 - w_3^2 w_2'^2 + w_4^2 w_2'^2$$
$$-w_1^2 w_3'^2 + w_2^2 w_3'^2 + w_1^2 w_4'^2 - w_2^2 w_4'^2). \tag{C.6}$$

Assume that the second term is zero. $w_4'^2$ is given by

$$w_4'^2 = w_3'^2 + (w_4^2 - w_3^2)(w_1'^2 - w_2'^2)/(w_1^2 - w_2^2). \tag{C.7}$$

The third determinant is factorized as follows:

$$(w_2' w_3' + w_1' w_4')(w_2^2 w_4^2 w_1'^2 - w_3^2 w_4^2 w_1'^2 - w_1^2 w_3^2 w_1'^2 + w_3^2 w_4^2 w_2'^2$$
$$+ x_1^2 w_2^2 w_3'^2 - w_2^2 w_4^2 w_3'^2 - w_1^2 w_2^2 w_4'^2 + w_1^2 w_3^2 w_4'^2). \tag{C.8}$$

If one substitutes (C.7) into the second term, we have

$$\left(\frac{w_1^2 w_3^2 - w_2^2 w_4^2}{w_1^2 - w_2^2}\right)(w_1'^2(x_2^2 - w_3^2) + w_2'^2(x_3^2 - w_1^2) + w_3'^2(x_1^2 - w_2^2)). \tag{C.9}$$

Then we assume that the last parenthesis is zero and therefore $w_3'^2$ is given by

$$w_3'^2 = \frac{w_1'^2(w_3^2 - w_2^2) + w_2'^2(w_1^2 - w_3^2)}{w_1^2 - w_2^2}. \tag{C.10}$$

Substituting this into (C.7) we have

$$w_4'^2 = \frac{w_1'^2(w_4^2 - w_2^2) + w_2'^2(w_1^2 - w_4^2)}{w_1^2 - w_2^2}. \tag{C.11}$$

The second determinant is

$$w_3 w_4 w_1' w_3' [w_1'^2(w_4^2 - w_2^2) + w_2'^2(w_1^2 - w_4^2) + w_4'^2(w_2^2 - w_1^2)]$$
$$+ w_1 w_2 w_1' w_3' [w_2'^2(w_4^2 - w_3^2) + w_3'^2(w_2^2 - w_4^2) + w_4'^2(w_3^2 - w_1^2)]$$
$$+ w_1 w_2 w_2' w_4' [w_1'^2(w_4^2 - w_3^2) + w_3'^2(w_1^2 - w_4^2) + w_4'^2(w_3^2 - w_1^2)]$$
$$+ w_3 w_4 w_2' w_4' [w_1'^2(w_3^2 - w_2^2) + w_2'^2(w_1^2 - w_3^2) + w_3'^2(w_2^2 - w_1^2)]. \tag{C.12}$$

Substituting (C.10) and (C.11) into the above formula we find four brackets vanish. Then three equations are satisfied by only two relations (C.10) and (C.11). This means that we can generate a transfer matrix with one arbitrary parameter except for the trivial scale factor. Conditions (C.10) and (C.11) are equivalent to

$$\frac{w_1^2 - w_2^2}{w_3^2 - w_4^2} = \frac{w_1'^2 - w_2'^2}{w_3'^2 - w_4'^2}, \quad \frac{w_1^2 - w_3^2}{w_2^2 - w_4^2} = \frac{w_1'^2 - w_3'^2}{w_2'^2 - w_4'^2}. \tag{C.13}$$

Using (C.1) we obtain (5.25) and (5.26).

Appendix D
The matrix $\mathbf{Q}(v)$

Assume that $2L\eta = 2m_1K + im_2K'$. Here L, m_1 and m_2 are arbitrary integers. We define $2^N \times 2^N$ matrices $\mathbf{Q}_R(v)$ and $\mathbf{Q}_L(v)$ as follows:

$$\mathbf{Q}_R(v)\Big|_{\alpha|\beta} = \mathrm{Tr}\Big\{\prod_{j=1}^{N} \mathbf{S}(\alpha_j, \beta_j|v)\Big\},$$

$$\mathbf{Q}_L(v)\Big|_{\alpha|\beta} = \mathrm{Tr}\Big\{\prod_{j=1}^{N} \mathbf{S}'(\alpha_j, \beta_j|v)\Big\}. \tag{D.1}$$

Here $\mathbf{S}(\alpha_j, \beta_j|v)$ and $\mathbf{S}'(\alpha_j, \beta_j|v)$ are $L \times L$ matrices with $2L$ non zero elements as follows:

$$\begin{pmatrix}
z_0 & z_{-1} & 0 & 0 & . & & 0 \\
z_1 & 0 & z_{-2} & 0 & . & & . \\
0 & z_2 & 0 & . & . & & . \\
0 & 0 & . & . & . & & 0 \\
. & . & . & . & 0 & & z_{1-L} \\
0 & . & . & 0 & z_{L-1} & & z_L
\end{pmatrix}, \tag{D.2}$$

$$\mathbf{S}(+, \beta|v) : z_m = H(v + K + 2m\eta)\tau_{\beta m},$$
$$\mathbf{S}(-, \beta|v) : z_m = \Theta(v + K + 2m\eta)\tau_{\beta m},$$
$$\mathbf{S}'(\alpha, +|v) : z_m = \Theta(v - K - 2m\eta)\tau'_{\alpha m},$$
$$\mathbf{S}'(\alpha, -|v) : z_m = H(v - K - 2m\eta)\tau'_{\alpha m}. \tag{D.3}$$

$\tau_{\beta m}$ and $\tau'_{\alpha m}$ are arbitrary parameters. One can show that

$$\mathbf{T}(v)\mathbf{Q}_R(v) = \phi(v - \eta)\mathbf{Q}_R(v + 2\eta) + \phi(v + \eta)\mathbf{Q}_R(v - 2\eta), \tag{D.4}$$

$$\mathbf{Q}_L(v)\mathbf{T}(v) = \phi(v - \eta)\mathbf{Q}_L(v + 2\eta) + \phi(v + \eta)\mathbf{Q}_L(v - 2\eta), \tag{D.5}$$

$$\phi(v) = \{\rho\Theta(0)H(v)\Theta(v)\}^N, \tag{D.6}$$

$$\mathbf{Q}_L(u)\mathbf{Q}_R(v) = \mathbf{Q}_L(v)\mathbf{Q}_R(u). \tag{D.7}$$

Assuming $\mathbf{Q}_R(v)$ and $\mathbf{Q}_L(v)$ are non singular we put

$$\mathbf{Q}(v) = \mathbf{Q}_R(v)\mathbf{Q}_R^{-1}(v_0) = \mathbf{Q}_L^{-1}(v_0)\mathbf{Q}_L(v).$$

Using (D.7) we obtain (5.31). The transfer matrix is given by

$$[\mathbf{T}(v)]_{\alpha|\beta} = \mathrm{Tr}\Big\{\prod_{j=1}^{N} \mathbf{R}(\alpha_j, \beta_j|v)\Big\}. \tag{D.8}$$

$\mathbf{T}(v)\mathbf{Q}_R(v)$ is given by

$$[\mathbf{TQ}_R]_{\alpha|\beta} = \mathrm{Tr}\Big\{\prod_{j=1}^{N} \mathbf{U}(\alpha_j, \beta_j)\Big\},$$

where $\mathbf{U}$ are $2L$ by $2L$ matrices,

$$\mathbf{U}(+, \beta) = \begin{pmatrix} a\mathbf{S}(+, \beta), & d\mathbf{S}(-, \beta) \\ c\mathbf{S}(-, \beta), & b\mathbf{S}(+, \beta) \end{pmatrix},$$
$$\mathbf{U}(-, \beta) = \begin{pmatrix} b\mathbf{S}(-, \beta), & c\mathbf{S}(+, \beta) \\ d\mathbf{S}(+, \beta), & a\mathbf{S}(-, \beta) \end{pmatrix}, \tag{D.9}$$

We should look for a $2L \times 2L$ matrix M such that

$$M^{-1}\mathbf{U}(\alpha, \beta)M = \begin{pmatrix} \mathbf{A}(\alpha, \beta), & 0 \\ \mathbf{C}(\alpha, \beta), & \mathbf{B}(\alpha, \beta) \end{pmatrix},$$

where $\mathbf{A}, \mathbf{B}, \mathbf{C}$ are $L \times L$ matrices. If this is satisfied we can decompose $\mathbf{TQ}_R$ as follows:

$$\mathbf{T}(v)\mathbf{Q}_R(v) = \mathbf{H}_1 + \mathbf{H}_2, \tag{D.10}$$

where

$$[\mathbf{H}_1]_{\alpha|\beta} = \mathrm{Tr}\Big\{\prod_{j=1}^{N} \mathbf{A}(\alpha_j, \beta_j)\Big\}, \tag{D.11}$$

$$[\mathbf{H}_2]_{\alpha|\beta} = \mathrm{Tr}\Big\{\prod_{j=1}^{N} \mathbf{B}(\alpha_j, \beta_j)\Big\}. \tag{D.12}$$

If we set M to be the block triangular matrix,

$$M = \begin{pmatrix} E & P \\ 0 & E \end{pmatrix},$$

M^{-1} should be

$$M^{-1} = \begin{pmatrix} E & -P \\ 0 & E \end{pmatrix},$$

where E is the $L \times L$ identity matrix and P is the diagonal matrix,

$$
P = \begin{pmatrix} p_1 & 0 & 0 & . & & 0 \\ 0 & p_2 & 0 & . & & . \\ 0 & 0 & p_3 & . & & . \\ 0 & 0 & & . & & . \\ . & . & . & p_{L-1} & 0 & \\ 0 & . & . & 0 & p_L \end{pmatrix}, \quad p_j = k^{1/2}\mathrm{sn}[K + (2j-1)\eta]. \quad \text{(D.13)}
$$

We find $\mathbf{A}(\alpha, \beta)$ and $\mathbf{B}(\alpha, \beta)$ are matrices of the type (D.2),

$$
\mathbf{A}(\alpha, \beta) : z_m = \rho\Theta(0)H(v-\eta)\Theta(v-\eta)\left(\frac{x_m}{x_{m+1}}\right)q(\alpha, \beta, m|v + 2\eta),
$$

$$
\mathbf{B}(\alpha, \beta) : z_m = \rho\Theta(0)H(v+\eta)\Theta(v+\eta)\left(\frac{x_{m+1}}{x_m}\right)q(\alpha, \beta, m|v - 2\eta),
$$

$$\text{(D.14)}$$

where

$$
x_m = \Theta(K + (2m-1)\eta),
$$
$$
q(+, \beta, m|v) = H(v + \eta + K + 2mn)\tau_{\beta m},
$$
$$
q(-, \beta, m|v) = \Theta(v + \eta + K + 2mn)\tau_{\beta m}.
$$

Then we can write

$$
\mathbf{A}(\alpha, \beta) = \rho\Theta(0)H(v-\eta)\Theta(v-\eta)X^{-1}\mathbf{S}(\alpha, \beta|v + 2\eta)X,
$$
$$
\mathbf{B}(\alpha, \beta) = \rho\Theta(0)H(v-\eta)\Theta(v-\eta)X\mathbf{S}(\alpha, \beta|v - 2\eta)X^{-1}, \quad \text{(D.15)}
$$

where X is the $L \times L$ diagonal matrix with elements $x_m\delta_{m,n}$. We should note that X and P are independent of v. So we have equation (D.4).

In the same way we can show equation (D.5).

Finally we should prove (D.7). The matrix elements of $Q_L(u)Q_R(v)$ are given by

$$
[Q_L(u)Q_R(v)]_{\alpha|\beta} = \mathrm{Tr}\left\{\prod_{j=1}^{N} \mathbf{W}(\alpha_j, \beta_j|u, v)\right\}, \quad \text{(D.16)}
$$

where the $\mathbf{W}$ are $L^2 \times L^2$ matrices,

$$
[\mathbf{W}(\alpha, \beta|u, v)]_{mm'|nn'} = \sum_{\gamma=\pm} [\mathbf{S}'(\alpha, \gamma|u)]_{mn}[\mathbf{S}(\gamma, \beta|v)]_{mm'}. \quad \text{(D.17)}
$$

The number of non-zero matrix elements is $4L^2$ and has the form as follows:

$$
(H(u - K - 2j'\eta)\Theta(v + K + 2j\eta) + \Theta(u - K - 2j'\eta)H(v + K + 2j\eta))\tau'_{\alpha j'}\tau_{\beta j}.
$$

The function $H(A)\Theta(B) + \Theta(A)H(B)$ is decomposed as $f(A+B)g(A-B)$. The function $g(x)$ satisfies

$$g(x) = g(-x) = -g(x + 4K).$$

Then the ratio of the non-zero elements of $\mathbf{W}(u,v)$ and $\mathbf{W}(v,u)$ is

$$\frac{g(u - v - 2K - 2\eta(j + j'))}{g(v - u - 2K - 2\eta(j + j'))}.$$

Thus we find that

$$[\mathbf{W}(\alpha, \beta | u, v)]_{mm'|nn'} = y_{m,m'} [\mathbf{W}(\alpha, \beta | v, u)]_{mm'|nn'} / y_{n,n'}, \tag{D.18}$$

where

$$y_{m,m'} = t_{m+m'} t_{m-m'+1}, \tag{D.19}$$

and the t_m are defined by the recurrence relation

$$t_m/t_{m+2} = g(u - v + 2K + 2m\eta)/g(v - u + 2K + 2m\eta). \tag{D.20}$$

Equation (D.18) is written in matrix form as

$$\mathbf{W}(\alpha, \beta | u, v) = Y \mathbf{W}(\alpha, \beta | v, u) Y^{-1}. \tag{D.21}$$

Thus we can show (D.7).

Appendix E
Special functions

AE.1 Elliptic integrals

The complete elliptic integrals $K(k)$ and $E(k)$ are defined by

$$K(k) = \int_0^1 \frac{\mathrm{d}x}{\sqrt{(1-x^2)(1-k^2x^2)}} = \int_0^{\pi/2} \frac{\mathrm{d}\phi}{\sqrt{1-k^2\sin^2\phi}}, \tag{E.1}$$

$$E(k) = \int_0^1 \sqrt{\frac{1-k^2x^2}{1-x^2}}\,\mathrm{d}x = \int_0^{\pi/2} \sqrt{1-k^2\sin^2\phi}\,\mathrm{d}\phi. \tag{E.2}$$

These are represented by the hypergeometric function,

$$K(k) = \frac{\pi}{2}F\left(-\frac{1}{2},\frac{1}{2},1;k^2\right), \quad E(k) = \frac{\pi}{2}F\left(\frac{1}{2},\frac{1}{2},1;k^2\right). \tag{E.3}$$

AE.2 Elliptic functions

AE.2.1 Elliptic theta functions

To define elliptic functions it is convenient to use the following elliptic theta functions through nome q with $|q| < 1$,

$$\vartheta_1(x,q) = -i\sum_{n=-\infty}^{\infty} (-1)^n q^{(n+1/2)^2} e^{(2n+1)xi}, \tag{E.4}$$

$$\vartheta_2(x,q) = \sum_{n=-\infty}^{\infty} q^{(n+1/2)^2} e^{(2n+1)xi}, \tag{E.5}$$

$$\vartheta_3(x,q) = \sum_{n=-\infty}^{\infty} q^{n^2} e^{2nxi}, \tag{E.6}$$

$$\vartheta_4(x,q) = \sum_{n=-\infty}^{\infty} (-1)^n q^{n^2} e^{2nxi}. \tag{E.7}$$

239

These functions are doubly quasi-periodic with respect to the shifts $x \to x+\pi$ and $x \to x + i\ln(1/q)$. Then nome q characterizes the periodicity of elliptic functions. These satisfy

$$\vartheta_1(x + \pi, q) = -\vartheta_1(x, q), \tag{E.8}$$

$$\vartheta_2(x + \pi, q) = \vartheta_2(x, q), \tag{E.9}$$

$$\vartheta_3(x + \pi, q) = \vartheta_3(x, q), \tag{E.10}$$

$$\vartheta_4(x + \pi, q) = \vartheta_4(x, q), \tag{E.11}$$

$$\vartheta_1(x - i\ln q, q) = -q^{-1}e^{-2ix}\vartheta_1(x, q), \tag{E.12}$$

$$\vartheta_2(x - i\ln q, q) = q^{-1}e^{-2ix}\vartheta_2(x, q), \tag{E.13}$$

$$\vartheta_3(x - i\ln q, q) = q^{-1}e^{-2ix}\vartheta_3(x, q), \tag{E.14}$$

$$\vartheta_4(x - i\ln q, q) = -q^{-1}e^{-2ix}\vartheta_4(x, q). \tag{E.15}$$

These functions are represented by infinite products,

$$\vartheta_1(x, q) = -iq^{1/4}(z - z^{-1}) \prod_{n=1}^{\infty}(1 - q^{2n}z)(1 - q^{2n}z^{-1})(1 - q^{2n}), \tag{E.16}$$

$$\vartheta_2(x, q) = q^{1/4}(z + z^{-1}) \prod_{n=1}^{\infty}(1 + q^{2n}z)(1 + q^{2n}z^{-1})(1 - q^{2n}), \tag{E.17}$$

$$\vartheta_3(x, q) = \prod_{n=1}^{\infty}(1 + q^{2n-1}z)(1 + q^{2n-1}z^{-1})(1 - q^{2n}), \tag{E.18}$$

$$\vartheta_4(x, q) = \prod_{n=1}^{\infty}(1 - q^{2n-1}z)(1 - q^{2n-1}z^{-1})(1 - q^{2n}), \tag{E.19}$$

where $z = e^{ix}$. For nome q, the corresponding modulo k and k' are determined by

$$k = \left(\frac{\vartheta_2(0, q)}{\vartheta_3(0, q)}\right)^2, \quad k' = \left(\frac{\vartheta_4(0, q)}{\vartheta_3(0, q)}\right)^2. \tag{E.20}$$

Elliptic integrals $K(k)$, $K(k')$ are

$$K(k) = \frac{\pi}{2}(\vartheta_3(0, q))^2, \quad K(k') = \frac{\ln(1/q)}{2}(\vartheta_3(0, q))^2. \tag{E.21}$$

Jacobian elliptic theta functions $H(x, k), \Theta(x, k), H_1(x, k), \Theta_1(x, k)$ are defined by

$$H(x, k) = \vartheta_1\left(\frac{\pi x}{2K}, q\right), \quad \Theta(x, k) = \vartheta_4\left(\frac{\pi x}{2K}, q\right), \tag{E.22}$$

$$H_1(x, k) = \vartheta_2\left(\frac{\pi x}{2K}, q\right), \quad \Theta_1(x, k) = \vartheta_3\left(\frac{\pi x}{2K}, q\right). \tag{E.23}$$

By the imaginary transformation we have

$$H(x,k) = -i(K/K')^{1/2} \exp\left(-\frac{\pi x^2}{4KK'}\right) H(ix,k'), \qquad \text{(E.24)}$$

$$\Theta_1(x,k) = (K/K')^{1/2} \exp\left(-\frac{\pi x^2}{4KK'}\right) \Theta_1(ix,k'), \qquad \text{(E.25)}$$

$$H_1(x,k) = (K/K')^{1/2} \exp\left(-\frac{\pi x^2}{4KK'}\right) \Theta(ix,k'), \qquad \text{(E.26)}$$

$$\Theta(x,k) = (K/K')^{1/2} \exp\left(-\frac{\pi x^2}{4KK'}\right) H_1(ix,k'). \qquad \text{(E.27)}$$

AE.2.2 Liouville's theorems

We call a complex function with two periods a doubly periodic function,

$$f(x) = f(x+\omega) = f(x+\omega'), \quad \Im\omega/\omega' \neq 0.$$

We call a doubly periodic function which is rational an elliptic function. The parallelogram $a, a+\omega, a+\omega', a+\omega+\omega'$ is called the fundamental period-parallelogram. The sum of order of poles in the fundamental period-parallelogram is called the order of the elliptic function.

(i) (Liouville's first theorem)
An elliptic function which has no pole in the period-parallelogram is a constant.
(ii) (The second theorem)
In the parallelogram the sum of the residues of poles is zero.
(iii) (The third theorem)
An elliptic function with order n takes arbitrary value n times in the parallelogram.
(iv) (The forth theorem)
The difference of the sum of poles and the sum of zeros is equal to one period.

AE.2.3 Jacobian elliptic functions

The Jacobian elliptic functions $\mathrm{sn}(x,k), \mathrm{cn}(x,k), \mathrm{dn}(x,k)$ are

$$\mathrm{sn}(x,k) = \frac{1}{\sqrt{k}} \frac{H(x,k)}{\Theta(x,k)}, \qquad \text{(E.28)}$$

$$\mathrm{cn}(x,k) = \sqrt{\frac{k'}{k}} \frac{H_1(x,k)}{\Theta(x,k)}, \qquad \text{(E.29)}$$

$$dn(x,k) = \sqrt{k'}\frac{\Theta_1(x,k)}{\Theta(x,k)}. \tag{E.30}$$

We sometimes write these as $sn_k(x), cn_k(x), dn_k(x)$ or omit modulus k. These are doubly periodic functions,

$$sn(x+2K) = -sn(x), \quad sn(x+2iK') = sn(x), \tag{E.31}$$

$$cn(x+2K) = -cn(x), \quad cn(x+2iK') = -cn(x), \tag{E.32}$$

$$dn(x+2K) = dn(x), \quad dn(x+2iK') = -dn(x). \tag{E.33}$$

These satisfy

$$sn^2(x) + cn^2(x) = 1, \quad k^2sn^2(x) + dn^2(x) = 1, \tag{E.34}$$

and

$$sn(x)' = cn(x)dn(x), cn(x)' = -sn(x)dn(x), dn(x)' = -k^2sn(x)cn(x). \tag{E.35}$$

Addition theorems for these functions are

$$sn(x+y) = \frac{sn(x)cn(y)dn(y) + cn(x)dn(x)sn(y)}{1 - k^2sn^2(x)sn^2(y)}, \tag{E.36}$$

$$cn(x+y) = \frac{cn(x)cn(y) - sn(x)dn(x)sn(y)dn(y)}{1 - k^2sn^2(x)sn^2(y)}, \tag{E.37}$$

$$dn(x+y) = \frac{dn(x)dn(y) - k^2sn(x)dn(x)sn(y)dn(y)}{1 - k^2sn^2(x)sn^2(y)}. \tag{E.38}$$

The elliptic amplitude function is a multi-valued function,

$$am(x,k) \equiv \sin^{-1} sn(x,k). \tag{E.39}$$

This function satisfies

$$am'(x,k) = dn(x,k), \quad am(x+2K,k) = am(x,k) + \pi,$$
$$am(x+2iK',k) = am(x,k). \tag{E.40}$$

AE.3 The gamma function and Riemann's zeta function

$\Gamma(x)$ is defined by

$$\Gamma(x) = \int_0^\infty e^{-t}t^{x-1}dt, \quad \text{at } \Re x > 1. \tag{E.41}$$

The following relations stand for this function:

$$\Gamma(x+1) = x\Gamma(x), \quad \Gamma(1) = 1, \quad \Gamma(1/2) = \sqrt{\pi}. \tag{E.42}$$

The following formula is useful for Wiener–Hopf factorization:

$$\Gamma\left(\frac{1}{2}+x\right)\Gamma\left(\frac{1}{2}-x\right) = \frac{\pi}{\cos \pi x}, \tag{E.43}$$

$$\cos(x) = \frac{\pi}{\Gamma\left(\frac{1}{2}+\frac{x}{\pi}\right)\Gamma\left(\frac{1}{2}-\frac{x}{\pi}\right)}, \tag{E.44}$$

$$\sin(x) = \frac{x}{\Gamma\left(1+\frac{x}{\pi}\right)\Gamma\left(1-\frac{x}{\pi}\right)}. \tag{E.45}$$

The asymptotic behaviour at big $|x|$ and $|\arg x| < \pi$ is

$$\Gamma(x) = x^{x-\frac{1}{2}}e^{-x}\sqrt{2\pi}\left[1 + \frac{1}{12x} + O(x^{-2})\right]. \tag{E.46}$$

Riemann's zeta function is defined by

$$\zeta(x) = \sum_{n=1}^{\infty}\frac{1}{n^x}, \quad \Re x > 1. \tag{E.47}$$

This function has the integral representation

$$\zeta(x) = \frac{1}{\Gamma(x)}\int_0^{\infty}\frac{t^{x-1}}{e^t - 1}dt, \quad \Re x > 1, \tag{E.48}$$

$$\zeta(x) = \frac{1}{(1 - 2^{1-x})\Gamma(x)}\int_0^{\infty}\frac{t^{x-1}}{e^t + 1}dt, \quad \Re x > 0. \tag{E.49}$$

There is a functional relation

$$\zeta(1 - x) = 2^{1-x}\pi^{-x}\Gamma(x)\cos(\pi x/2)\zeta(x). \tag{E.50}$$

Values at special points are

$$\zeta(0) = -\frac{1}{2}, \quad \zeta(2) = \frac{\pi^2}{6}, \quad \zeta(4) = \frac{\pi^4}{90} \tag{E.51}$$

$$\zeta(2n) = 2^{2n-1}\pi^{2n}\frac{B_n}{(2n)!}, \quad \zeta(1 - 2n) = (-1)^n\frac{B_n}{2n}, \tag{E.52}$$

$$\zeta(-2n) = 0 \quad \text{for} \quad n = 1, 2, ... \tag{E.53}$$

Here B_n is the Bernoulli number defined by

$$\frac{x}{e^x - 1} + \frac{x}{2} = 1 + \sum_{n=1}^{\infty}\frac{(-1)^{n-1}}{(2n)!}B_n x^{2n},$$

$$B_1 = \frac{1}{6}, \quad B_2 = \frac{1}{30}, \quad B_3 = \frac{1}{42}, \quad B_4 = \frac{1}{40}, \quad B_5 = \frac{5}{66}, ... \tag{E.54}$$

$\zeta(x)$ has a pole at $x = 1$ and

$$\lim_{x\to1}\left[\zeta(x) - \frac{1}{x - 1}\right] = \gamma = 0.57721 \text{ (Euler constant)}. \tag{E.55}$$

The Bose–Einstein integral function $F(x, v)$ is defined by

$$F(x, v) = \frac{1}{\Gamma(x)} \int_0^\infty \frac{t^{x-1}}{e^{t+v} - 1} dt = \frac{e^{-v}}{1^x} + \frac{e^{-2v}}{2^x} + \dots \qquad (E.56)$$

If x is not a positive integer, we have

$$F(x, v) = \Gamma(1 - x)v^{x-1} + \sum_{n=0}^\infty \frac{\zeta(x - n)(-v)^n}{n!}. \qquad (E.57)$$

When x is a positive integer, we have

$$F(x, v) = \frac{(-v)^{x-1}}{(x - 1)!} \left[\sum_{r=1}^{x-1} \frac{1}{r} - \ln v \right] + \sum_{n \neq x-1} \frac{\zeta(x - n)(-v)^n}{n!}. \qquad (E.58)$$

AE.4 The Bessel function and modified Bessel function

The Bessel function is defined by

$$J_v(x) = \left(\frac{x}{2}\right)^v \sum_{j=0}^\infty \frac{(-1)^j (x/2)^{2j}}{j! \Gamma(v + j + 1)}. \qquad (E.59)$$

For integer v this function has the integral representation,

$$J_n(x) = \frac{1}{2\pi} \int_0^{2\pi} e^{i(n\theta - z \sin \theta)} d\theta \qquad (E.60)$$

The spherical Bessel function $j_n(x)$ is defined by

$$j_n(x) = \sqrt{\frac{\pi}{2x}} J_{n+\frac{1}{2}}(x) = (-1)^l x^l \left(\frac{1}{x} \frac{d}{dx}\right)^l \frac{\sin x}{x}. \qquad (E.61)$$

The modified Bessel function $I_v(x)$ is defined by

$$I_v(x) = e^{-iv\pi/2} J_v(ix) = \left(\frac{x}{2}\right)^v \sum_{j=0}^\infty \frac{(x/2)^{2j}}{j! \Gamma(v + j + 1)}. \qquad (E.62)$$

The hypergeometric function $F(\alpha, \beta, \gamma; x)$ is defined by

$$F(\alpha, \beta, \gamma; x) = \frac{\Gamma(\gamma)}{\Gamma(\alpha)\Gamma(\beta)} \sum_{n=0}^\infty \frac{\Gamma(\alpha + n)\Gamma(\beta + n)}{n! \Gamma(\gamma + n)} x^n$$

$$= 1 + \frac{\alpha\beta}{1!\gamma} x + \frac{\alpha(\alpha + 1)\beta(\beta + 1)}{2!\gamma(\gamma + 1)} x^2 + \dots \qquad (E.63)$$

The Legendre polynomials $P_n(x)$ are defined by

$$P_n(x) = \frac{1}{2^n n!} \frac{d^n}{dx^n} (x^2 - 1)^n. \qquad (E.64)$$

The associated Legendre polynomials are defined by

$$P_n^m(x) = \frac{1}{2^n n!}(1 - x^2)^{|m|/2}\frac{\mathrm{d}^{n+|m|}}{\mathrm{d}x^{n+|m|}}(x^2 - 1)^n. \tag{E.65}$$

The spherical harmonic function $Y_l^m(\theta, \phi)$ is defined by

$$Y_{lm}(\theta, \phi) = (-1)^{\frac{m+|m|}{2}}\sqrt{\left(l + \frac{1}{2}\right)\frac{(l - |m|)!}{(l + |m|)!}}P_l^{|m|}(\cos\theta)e^{im\phi}. \tag{E.66}$$

Bibliography

[1]   Affleck, I., Kennedy, T., Lieb, E.H. and Tasaki, H. (1987) *Phys. Rev. Lett.* **59**, 799; (1988) *Commun. Math. Phys.* **115**, 477.

[2]   Anderson, P.W. (1952) *Phys. Rev.* **86**, 694.

[3]   Anderson, P.W. (1959) *Phys. Rev.* **115**, 2.

[4]   Baker, G.A., Rushbrooke, G.S. and Gilbert, H.E. (1964) *Phys. Rev.* **135**, A 1272.

[5]   Bariev, R.Z. (1982) *Theor. Math. Phys.* **49**, 1021.

[6]   Barma, M. and Shastry, B.S. (1978) *Phys. Rev.* **B18**, 3351.

[7]   Baxter, R. (1971) *Stud. Appl. Math.* (Mass. Inst. of Technology) **50**, 51.

[8]   Baxter, R. (1972) *Ann. Phys.(N.Y.)* **70**, 193.

[9]   Baxter, R. (1972) *Ann. Phys.(N.Y.)* **70**, 323.

[10] Baxter, R. (1973) *J. Stat. Phys.* **9**, 145.

[11] Baxter, R. (1976) *J. Stat. Phys.* **15**, 485.

[12] Baxter, R. and Kelland, S. (1976) *J. Phys. C* **7**, L403.

[13] Bethe, H.A. (1931) *Z. Physik* **71**, 205.

[14] Betsuyaku, M. (1984) *Phys. Rev. Lett.* **53**, 629.

[15] Betsuyaku, M. (1985) *Prog. Theor. Phys.* **73**, 319.

[16] Bonner J.C. and Fisher,M.E. (1964) *Phys. Rev.* **135**, A640.

[17] Botet, R. and Jullien, R. (1983) *Phys. Rev.* **B27**, 613; Kolb, M., Botet, R. and Jullien, R. (1983) *J. Phys. A* **16**, L673; Botet, R., Jullien, R. and Kolb, M. (1983) *Phys. Rev.* **B28**, 3914; Glaus, U. and Schneider, T. (1984) *Phys. Rev.* **B30**, 215; Parkinson, J.B. and Bonner, J.C. (1985) *Phys. Rev.* **B32**, 4703.

[18] Buyers, W.J.L., Morra, R.M., Armstrong, R.L., Gerlach, P. and Hirakawa, K. (1986) *Phys. Rev. Lett.* **56**, 371; Morra, R.M., Buyers, W.J.L., Armstrong, R.L. and Hirakawa, K. (1988) *Phys. Rev.* **B38**, 543; Steiner, M., Kakurai, K., Kjems, J.K., Petitgrand, D. and Pynn, R. (1987) *J. Appl. Phys.* **61**, 3953; Tun, Z., Buyers, W.J.L., Armstrong, R.L., Hirakawa, K. and Briat, B. (1990) *Phys. Rev.* **B42**, 4677; Tun, Z., Buyers, W.J.L., Harrison, A. and Rayne, J.A. (1991) *Phys. Rev.* **B43**, 13331.

[19] des Cloizeaux, J. and Pearson, J. (1962) *Phys. Rev.* **128**, 2131.

[20] des Cloizeaux, J. and Gaudin, M. (1966) *J. Math. Phys.* **7**, 1384.

[21] Cullen, J.J. and Landau, D.P. (1983) *Phys. Rev.* **B27**, 297.

[22] Economou, E.N. and Poulopoulos, P.N. (1979) *Phys. Rev.* **B20**, 4756.

[23] Eggert, S., Affleck, I. and Takahashi, M. (1994) *Phys. Rev. Lett.* **73**, 332.

[24] Essler, F., Korepin, V. and Schoutens, K. (1992) *Nuc. Phys.* **B384**, 431.

[25] Fisher, M.E. (1964) *Am. J. Phys.* **32**, 343.

[26] Flicker, M. and Lieb, E.H. (1967) *Phys. Rev.* **161**, 179.

[27] Fowler, M. and Zotos, X. (1981) *Phys. Rev.* **B24**, 2634.

[28] Frischmuth, B., Ammon, B. and Troyer, M. (1996) *Phys. Rev.* **B54**, 3714.

[29] Gaudin, M. (1967) *Phys. Lett.* **24A**, 55.

[30] Gaudin, M. (1971) *Phys. Rev. Lett.* **26**, 1301.

[31] Girardeau, M. (1960) *J. Math. Phys.* **1**, 516.

[32] Griffiths, R.B. (1964) *Phys. Rev.* **133A**, 768.

[33] Haldane, F.D.M. (1983) *Phys. Lett.* **93A** 464.

[34] Haldane, F.D.M. (1983) *Phys. Rev. Lett.* **50**, 1153.

[35] Haldane, F.D.M. (1988) *Phys. Rev. Lett.* **60**, 635; Shastry, B.S. (1988) *Phys. Rev. Lett.* **60**, 639.

[36] Haldane, F.D.M. (1991) *Phys. Rev. Lett.* **66**, 1529.

[37] Hida, K. (1981) *Phys. Lett.* **84A**, 338.

[38] Hirsch, J.E., Sugar, R.L., Scalapino, D.J. and Blankenbecler, R. (1982) *Phys. Rev.* **B26**, 5033.

[39] Hulthen, L. (1938) *Arkiv Math. Astron. Fys.* **26A**, No11.

[40] Ishikawa, M. and Takayama, H. (1980) *J. Phys. Soc. Jpn.* **49**, 1242.

[41] Ishimura, N. and Shiba, H. (1980) *Prog. Theor. Phys.* **63**, 743.

[42] Jimbo, M. and Miwa, T. (1993) 'Algebraic Analysis of Solvable Lattice Models', CBMS No. 85.

[43] Jimbo, M., Miwa, T. and Nakayashiki A. (1993) *J. Phys. A* **26**, 2199.

[44] Johnson, J.D., Krinsky, S. and McCoy, B. (1973) *Phys. Rev. A* **8**, 2526.

[45] Johnson, J.D. and McCoy, B. (1972) *Phys. Rev. A* **6**, 1613.

[46] Joyce, G.S. (1967) *Phys. Rev.* **155**, 478.

[47] Katsura, S. (1962) *Phys. Rev.* **127**, 1508.

[48] Katsura, S. (1965) *Ann. Phys. (N.Y.)* **31**, 325.

[49] Kawakami, N., Usuki, T. and Okiji, A. (1989) *Phys. Lett.* **137A**, 287.

[50] Kawano, K. and Takahashi, M. (1995) *J. Phys. Soc. Jpn.* **64**, 4331.

[51] Koma, T. (1987) *Prog. Theor. Phys.* **78**, 1213.

[52] Koma, T. (1989) *Prog. Theor. Phys.* **81**, 783.

[53] Kubo K. and Takada, S. (1986) *J. Phys. Soc. Jpn.* **55**, 438; Moreo, A. (1987) *Phys. Rev.* **B35**, 8562; Betsuyaku, M. (1987) *Phys. Rev.* **B36**, 799; Delica, T., Kopinga, K., Leshke, H. and Mon, K.K. (1991) *Europhys. Lett.* **15**, 55; Golinelli, O., Jolicœur Th. and Lacaze, R. (1992) *Phys. Rev.* **B45**, 9798.

[54] Kubo, R. (1952) *Phys. Rev.* **87**, 568.

[55] Lai, C.K. (1971) *Phys. Rev. Lett.* **26**, 1472.

[56] Lai, C.K. (1973) *Phys. Rev.* **A8**, 2567.

[57] Lai, C.K. and Yang, C.N. (1971) *Phys. Rev.* **A3**, 393.

[58] Lenard, A. (1964) *J. Math. Phys.* **5**, 960.

[59] Lieb, E.H. (1963) *Phys. Rev.* **130**, 1616.

[60] Lieb, E.H. and Liniger, W. (1963) *Phys. Rev.* **130**, 1605.

[61] Lieb, E.H., Schultz, T. and Mattis, D. (1961) *Ann. Phys. (N.Y.)* **16**, 417.

[62] Lieb, E.H. and Wu, F.Y. (1968) *Phys. Rev. Lett.* **20**, 1443.

[63] Lyklema, J.W. (1983) *Phys. Rev.* **27**, 3108.

[64] Ma, S., Broholm, C., Reich, D.H., Sternlieb, B.J. and Erwin, R.W. (1992) *Phys. Rev. Lett.* **69**, 3571.

[65] McCoy, B. (1968) *Phys. Rev.* **173**, 531.

[66] McGuire, J.B. (1965) *J. Math. Phys.* **6**, 432.

[67] McGuire, J.B. (1966) *J. Math. Phys.* **7**, 123.

[68]   Metzner, W. and Vollhardt, D. (1989) *Phys.Rev.* **B39**, 4462.
[69]   Motoyama, N., Eisaki, H. and Uchida, S. (1996) *Phys. Rev. Lett.* **76**, 3212.
[70]   Musurkin, I.N. and Ovchinnikov, A.A. (1970) *Fiz. Tver. Tela* **12**, 2524;
       (1971) *Solid State Phys. (Soviet Physics)* **12**, 2031.
[71]   Nakamura, H., Hatano, N. and Takahashi, M. (1995) *J. Phys. Soc. Jpn.* **64**,
       1955, 4142.
[72]   Nakamura, H. and Takahashi, M. (1994) *J. Phys. Soc. Jpn.* **63**, 2563.
[73]   Nakamura, T. (1952) *J. Phys. Soc. Jpn.* **7**, 264.
[74]   Nakano, H. and Takahashi, M. (1994) *J. Phys. Soc. Jpn.* **63**, 926, 4256;
       (1994) *Phys. Rev.* **B50**, 10331.
[75]   Nightingale, M.P. and Blöte, H.W.J. (1986) *Phys. Rev.* **B33**, 659.
[76]   Nomura, K. (1989) *Phys. Rev.* **B40**, 2421; Liang, S. (1990) *Phys. Rev. Lett.*
       64, 1597.
[77]   Orbach, R. (1958) *Phys. Rev.* **112**, 309.
[78]   Ovchinnikov, A.A. (1969) *Zjur. Eksp. Theor. Fiz.* **57**, 2137; (1970) *JETP
       (Soviet Physics)* **30**, 1160.
[79]   Renard, J.P., Verdaguer, M., Regnault, L.P., Erkelens, W.A.C.,
       Rossa-Mignod, J. and Stirling, W.G. (1987) *Europhys. Lett.* **3**, 945; Renard,
       J.P., Verdaguer, M., Regnault, L.P., Erkelens, W.A.C., Rossa-Mignod, J.,
       Ribas, J., Stirling, W.G. and Vettier, C. (1988) *J. Appl. Phys.* **63**, 3538;
       Regnault, L.P., Rossa-Mignod, J., Renard, J.P., Verdaguer, M. and Vettier,
       C. (1989) *Physica* **B156 & 157**, 247.
[80]   Sakai T. and Takahashi, M. (1989) *J. Phys. Soc. Jpn.* **58**, 3131.
[81]   Schlottmann, P. (1985) *Phys. Rev. Lett.* **54**, 2131.
[82]   Schlottmann, P. (1986) *Phys. Rev.* **B33**, 4880 .
[83]   Shiba, H. (1972) *Phys. Rev.* **B6**, 930.
[84]   Sutherland, B. (1968) *Phys. Rev. Lett.* **20**, 98.
[85]   Sutherland, B. (1975) *Phys. Rev.* **B12**, 3795.
[86]   Suzuki, J., Akutsu, Y. and Wadati, M. (1990) *J. Phys. Soc. Jpn.* **59**, 2667.
[87]   Suzuki, M. (1969) *Prog. Theor. Phys.* **42**, 1076.
[88]   Takada, S. and Kubo, K. (1986) *J. Phys. Soc. Jpn.* **55** 1671.
[89]   Takagi, S., Deguchi, H., Takeda, K., Mito, M. and Takahashi, M. (1996)
       J. Phys. Soc. Jpn. **65**, 1934.
[90]   Takahashi, M. (1969) *Prog. Theor. Phys.* **42**, 1098; **43**, 860(E).
[91]   Takahashi, M. (1970) *Prog. Theor. Phys.* **43**, 1619.
[92]   Takahashi, M. (1970) *Prog. Theor. Phys.* **44**, 899.
[93]   Takahashi, M. (1971) *Prog. Theor. Phys.* **46**, 401.
[94]   Takahashi, M. (1971) *Prog. Theor. Phys.* **46**, 1388.
[95]   Takahashi, M. (1972) *Prog. Theor. Phys.* **47**, 69.
[96]   Takahashi, M. (1973) *Prog. Theor. Phys.* **50**, 1519.
[97]   Takahashi, M. (1974) *Prog. Theor. Phys.* **51**, 1348.
[98]   Takahashi, M. (1974) *Prog. Theor. Phys.* **52**, 103.
[99]   Takahashi, M. (1975) *Prog. Theor. Phys.* **53**, 386.
[100] Takahashi, M. (1976) *Prog. Theor. Phys.* **55**, 33.
[101] Takahashi, M. (1977) *J. Phys. C* **10**, 1289.
[102] Takahashi, M. (1986) *Prog. Theor. Phys. (Suppl.)* **87**, 233.
[103] Takahashi, M. (1987) *Phys. Rev. Lett.* **58**, 168.
[104] Takahashi, M. (1988) *Phys. Rev.* **B38**, 5188.
[105] Takahashi, M. (1989) *Phys. Rev. Lett.* **62**, 2313.
[106] Takahashi, M. (1993) *Phys. Rev.* **B48**, 311.

[107] Takahashi, M. and Sakai, T. (1991) *J. Phys. Soc. Jpn.* **60**, 760; Sakai, T. and Takahashi, M. (1991) *Phys. Rev.* **B43**, 13383; (1991) *J. Phys. Soc. Jpn.* **60**, 3615.

[108] Takahashi, M. and Suzuki, M. (1972) *Prog. Theor. Phys.* **46**, 2187.

[109] Takahashi, M. and Yamada, M. (1985) *J. Phys. Soc. Jpn.* **54**, 2808.

[110] Takahashi, M., Turek, P., Nakazawa, Y., Tamura, M., Nozawa, K., Shiomi, D., Ishikawa, M. and Kinoshita, M. (1991) *Phys. Rev. Lett.* **67**, 746; **69**, 1290(E).

[111] Takeuchi, T., Hori, H., Yosida, T., Yamagishi, A., Katsumata, K., Renard,J.P., Gadet, V., Verdaguer, M. and Date, M. (1992) *J. Phys. Soc. Jpn.* **61**, 3262.

[112] Takhtajan, L. (1982) *Phys. Lett.* **87A**, 479; Babujian, J. (1982) *Phys. Lett.* **90A**, 479; (1983) *Nucl. Phys.* **B215**, 317.

[113] Truong, T.T. and Schotte, K.D. (1983) *Nucl. Phys.* **B220**, 77.

[114] Walker, L.R. (1959) *Phys. Rev.* **116**, 1089.

[115] Yamada, M. (1990) *J. Phys. Soc. Jpn.* **59**, 848 ;

[116] Yamada, M and Takahashi, M. (1986) *J. Phys. Soc. Jpn.* **55**, 2024.

[117] Yang, C.N. (1967) *Phys. Rev. Lett.* **19**, 1312.

[118] Yang, C.N. (1989) *Phys. Rev. Lett.* **63**, 2144.

[119] Yang, C.N. and Yang, C.P. (1966) *Phys. Rev.* **147**, 303.

[120] Yang, C.N. and Yang, C.P. (1966) *Phys. Rev.* **150**, 321.

[121] Yang, C.N. and Yang, C.P. (1966) *Phys. Rev.* **150**, 327.

[122] Yang, C.N. and Yang, C.P. (1966) *Phys. Rev.* **151**, 258.

[123] Yang, C.N. and Yang, C.P. (1969) *J. Math. Phys.* **10**, 1115.

[124] Yang, C.P. (1970) *Phys. Rev.* **A2**, 154.

Index